STATISTICS: AN INTUITIVE APPROACH

Third Edition

STATISTICS: AN INTUITIVE APPROACH

Third Edition

GEORGE H. WEINBERG
Consulting Psychologist, New York City

JOHN A. SCHUMAKER
Professor of Mathematics, Rockford College

BROOKS/COLE PUBLISHING COMPANY
Monterey, California

A Division of Wadsworth Publishing Company, Inc.

ISBN: 0-8185-0113-8
L.C. Catalog Card No.: 73-85595
Printed in the United States of America

6 7 8 9 10—78

What is now proved was once only imagined.

William Blake

This book is designed primarily as a text for use in an elementary statistics course given on either the undergraduate or the graduate level. It is aimed especially at the growing number of students in the fields of psychology, education, and the social sciences who have little knowledge of mathematics. Our intent is to impart to such students a clear understanding of the various topics usually covered in a first course without forcing them to grapple with elaborate mathematical symbols or notations.

Though gratified by the success of the first two editions of this book, we have felt the need to revise it, especially by supplying many new end-of-chapter problems and by incorporating suggestions made by a number of college teachers of statistics in the United States and Canada. In addition, this new edition is accompanied by a useful study guide developed by Freeman F. Elzey. As a supplementary aid to working through this text, the study guide highlights the key terms, symbols, concepts, and procedures covered in each chapter and gives readers an opportunity to become actively involved in their own learning by responding to a series of statements, questions, and problems.

The primary emphasis in this text is on proper insight into concepts; only after each new concept is developed are problems posed. Our hope is that students who use this book and study guide will be able not only to make use of the basic concepts but to think in their terms. We believe students will learn the subject more firmly through this approach than if they had merely gone the mathematical route. The book is self-contained and should also be useful to the individual who wishes to develop statistical literacy but who is not actually taking a formal statistics course. The study guide expands the reader's awareness and understanding of the material. The reader who is willing to study the book and the study guide carefully will, we believe, understand and actually enjoy this introduction to statistics.

The text is suitable for courses of various lengths and different emphases. The instructor can tailor his own course to suit the backgrounds and needs of the students in his class. For example, the emphasis placed on the various parts of Chapter 6 may well be determined by the amount of such computation that the students are likely to be faced with in their line of endeavor. In fact, in our age of computers some instructors may find it preferable to omit Chapter 6 entirely. In a one-semester course there will probably be time to consider at least the first eleven chapters along with selections from Chapters 12 to 20. Chapter 11 should be included if at all possible, for the introduction to decision making and risk is one of the unique features of the text. With the exception of Section 18.6, Chapters 16 to 18 do not depend on Chapters 11 to 15.

There are two sets of problems appearing at the end of each chapter beginning with Chapter 2. These problem sets give the reader an opportunity to apply the statistical techniques introduced in the chapter. Each set gives full coverage of the topics in the chapter. The problem sets are parallel, providing comparable problems in both sets. Answers to the problems in Set A are given at the end of the text; answers for Set B are available to the instructor. Thus the reader can use Set A as a self-test of the material in each chapter. Once Set A is mastered, the reader should have no trouble working the problems in Set B. The latter set can be used by the instructor as a formal test of the reader's ability to perform the statistical techniques covered in the chapter.

For contributing many new problems for this edition we are deeply grateful to Freeman F. Elzey of Far West Laboratory for Educational Research and Development, San Francisco. We believe that his work adds a freshness to the book and makes it more useful to teachers and students.

For their careful readings of the previous edition for revision purposes, we are grateful to several people whose insights were extremely helpful: William Frederickson of Central State University, Oklahoma, William Ballock of the College of Lake County, J. Barnard Gilmore of the University of Toronto, David Gustafson of Rock Valley College, John E. Klingensmith of Arizona State University, E. I. Sawin of San Francisco State University, William L. Sawrey of California State College, Hayward, and Robert A. Smith of the University of Southern California. We scrutinized every comment made by these men and incorporated many of their suggestions, with the result, we believe, that the book is clearer and more accurate than we alone would have made it. And again we wish to express our gratitude to those many instructors and students who supplied comments on the manuscripts and published versions of the first and second editions.

Finally we are thankful to Sir Ronald A. Fisher, F.R.S., Cambridge, to Dr. Frank Yates, F.R.S., Rothamsted, and to Messrs. Oliver & Boyd, Edinburgh, for permission to reprint, in abridged form, Tables III and IV from their book *Statistical Tables for Biological, Agricultural, and Medical Research*; to the Institute of Mathematical Statistics for permission to reprint, in abridged form, a table from page 66 of volume 14 of the *Annals of Mathematical Statistics*; and to Professors George W. Snedecor and William G. Cochran and the Iowa State University Press for permission to use the *F* table from their book *Statistical Methods, Sixth Edition*.

George H. Weinberg
John A. Schumaker

contents

STATISTICS:
AN INTUITIVE
APPROACH

orientation
and introduction

1.1 "Statistics" and You. You, the reader, are probably beginning your study of statistics without any previous knowledge of the subject. You are likely to think of a mass of data whenever you hear the word "statistics." You have heard references by news analysts, politicians, and others to "the latest statistics on the cost of living," "statistics on industrial production," "statistics on population growth," "statistics on baseball," etc. You may imagine these as long lists of figures which would be rather discouraging to view. You may realize, however, that some of the original figures or data have already been condensed into more meaningful and useful figures, such as "indexes" of the cost of living and baseball "averages." These numbers are also appropriately called "statistics." In fact, the singular, "statistic," may properly be used to refer to a specific value of one of these measures. This is a usage that we shall develop more fully later in the text.

The word "statistics" is also frequently used as a short way of saying "statistical methods." *Statistical methods* may be described as methods used in the collection, presentation, analysis, and interpretation of data. An example would be the methods used in going from the data on each separate time "at bat" for the baseball player to his final "batting average."

We shall be concerned primarily with developing an understanding of the basic statistical concepts and methods. Most of these methods may be understood in the light of common-sense reasoning, and it is our major purpose to show how this reasoning can be done. In fact, *statistical methods*

1

may be described as an application of common-sense reasoning to the analysis of data.

Statistical methods are essential whenever useful information is to be distilled from large masses of data. The examples mentioned previously, as well as others that you can think of, will help you to appreciate the importance of this aspect of statistics. You may be embarking on a field of endeavor in the natural or social sciences in which you will need to derive useful results from data that you or others have collected. On the other hand, you may not foresee that you will ever have to do this and, under these circumstances, you may well ask the value of a knowledge of statistical concepts. We would call your attention to two major reasons why you should become statistically literate.

The literature of most professional fields contains the results of experiments conducted by researchers in the field. These published reports often use the language of statistics in such a way that they require statistical literacy on the part of the reader if he is to make profitable use of them. Such literacy may be essential to a decision as to whether a newly tested method should be used by the reader in his own practice of his professional field.

One need not turn to professional literature to find widespread use of statistics. The daily newspapers and the popular magazines often use statistical terminology, especially in presenting "averages" and graphs. In many instances statistics are not only used but also abused, as we shall see in a later section. The alert and conscientious citizen needs to be an *intelligent consumer* of statistics when reading such material.

This need for statistical literacy has been kept constantly in mind in the writing of this book. Our goal has been to present the major concepts in statistical method in a manner that will create the understanding needed for such literacy.

It is, of course, necessary in the study of elementary statistics to use a certain amount of arithmetic as well as a little symbolic notation. The latter is kept to a minimum in this text and is carefully explained when introduced. This book is in no sense a mathematical treatise. We simply want to help you become statistically literate.

1.2 The History of Statistics. The development of statistics has been like that of language itself. Its origins are ancient. For instance, the use of the mean or average was well known at the time of Pythagoras, and mention of statistical surveys was made in biblical times. Like language, statistics developed gradually where it was needed. As society became more complex, there developed more demand for accurate summary statements and inferences made in numerical form. The discipline of statistics has in recent years gained momentum both in its mathematical development and through its many applications in new fields.

The modern development of statistics began in the sixteenth century, when governments of various western European countries became interested in collecting information about their citizens.   By the seventeenth century, surveys that closely resemble our modern census were already being conducted. By that time, insurance companies were beginning to thrive and were already compiling mortality tables to determine their life insurance rates.   Different vested interests were slowly turning toward enterprises that necessitated the treatment of data and that demanded the use of statistics.

The mathematical basis of statistics is certainly not the subject matter of this book, but its existence, like the engine of an automobile, is vital.   Thus Isaac Newton (1642–1727), whose contribution to the invention of calculus was an outstanding event in mathematics, was perhaps the most necessary figure for the development of modern statistics, though Newton could scarcely have heard of the subject.   Other mathematicians whose contributions have been primarily in the field of pure mathematics have done more for the development of statistics indirectly than many of those whose names are associated specifically with the field of statistics itself.   Perhaps the two most prominent ones are Abraham DeMoivre (1667–1754) and Carl Gauss (1777–1855).

As for the statisticians themselves, Adolph Quetelet (1796–1874) a Belgian, was the first to apply modern methods to collected data.   Quetelet is sometimes referred to as the father of modern statistics, less because of his contributions than because of his continued emphasis on the importance of using statistical methods.   After studying with the best-known mathematicians of his day, Quetelet established a Central Commission for Statistics which became a model for similar organizations in other countries.

Unexpected as it sounds, Florence Nightingale (1820–1910) was an ardent proponent of the use of statistics all of her life.   She argued that the administrator could be successful only if he were guided by statistical knowledge and that both the legislator and the politician often failed because their statistical knowledge was insufficient.

Two other important contributors to statistics were the Englishmen Sir Francis Galton (1822–1911) and Karl Pearson (1857–1936).   Galton, a cousin of Charles Darwin, became deeply interested in the problem of heredity, to which he soon applied statistical tools.   Among other things, he developed the use of percentiles.   Pearson made many statistical discoveries, and both Galton and Pearson contributed greatly to the development of correlation theory, which we shall consider in detail later.

The most prominent contributor to the field of statistics in the twentieth century has been Sir Ronald Fisher (1890–1962).   Fisher made continuous contributions from 1912 to 1962 and many of them have had great impact on contemporary statistical procedures.*

The twentieth century has seen the birth and growth of formal statistics instruction in the United States.   At the turn of the century only a handful of

---

* The reader who wants an interesting account of the history of statistics up to 1929 should consult Walker, Helen M., *Studies in the History of Statistical Method* (Baltimore: Williams & Wilkins Co., 1929).

statistics courses were given in all the colleges in America. In fact, most of these were given in economics departments and by instructors whose focus made them unique among their colleagues. During the first thirty years of this century the emphasis on applications of statistics to problems of psychology slowly increased; but it should be noted that during this period psychology itself had a shallow status and was frequently considered only as a branch of philosophy.

Shortly before World War II the number of applications of statistical methods in the social sciences began to increase. The number of surveys of all kinds increased, and the need to interpret data in psychology and education made it necessary for workers to have at least a basic understanding of statistics. One by one, statistics courses were added as requirements in psychology departments and in schools of education. Today the worker in any of the human behavior fields, such as psychology, sociology, or anthropology, is expected at least to have what is called statistical literacy. In fact, it is virtually impossible to major in any of these fields without having to take at least one statistics course.

One reason for the rapid growth of statistics in recent years is the increasing ease of processing large masses of data. Modern electronic computers make it possible to analyze in a short time huge collections of data that would have been entirely unmanageable just a few years ago. Such data can now be analyzed and the results made available for public and professional consumption while they are still pertinent to the current situation. If you should ever have to make an extensive statistical survey, you will undoubtedly enlist the aid of modern methods of computation. Familiarity with a desk calculator is useful for less extensive computations.

One implication of the existence of rapid computational devices is that there is much less need for an emphasis on this aspect in the teaching and learning of elementary statistical concepts. The time is much more profitably spent in furthering a real understanding of the ideas involved. On the other hand, a certain amount of practice in working with the various statistical measures is actually helpful in developing this understanding. The problems at the ends of chapters are designed for this purpose.

1.3 Descriptive and Sampling Statistics. A given set of values or measures of the same thing is called a *distribution*. For example, the scores on a particular examination constitute a distribution. Most of the discussion in this book is concerned with various types of distributions and their properties.

There are two main approaches that comprise the subject matter of statistics. One of them is the approach of *descriptive statistics* and the other is that of *sampling statistics*. The methods of descriptive statistics entail specifying a population of interest and then collecting the measurements of all the members of that population. These original measurements or scores are called *raw data*. The raw data themselves are descriptive, but the science of descriptive statistics deals with methods of deriving from raw data measurements that are more tersely descriptive of the original population. In fact, it is the type of measure once removed from the raw data that is of prime importance to the statistician and research worker. For instance, the *average* IQ of members of an army battalion is obviously much more comprehensible and meaningful than the lengthy list of IQ scores as they were originally obtained. But it almost goes without saying that an understanding of the exact meaning of an average is necessary to interpret an average in any particular case.

Let us emphasize what has been said. The descriptive statistical approach makes use of all the data concerning a population, and it entails deriving descriptive statistical measurements from data. The word "population" does not imply vastness; for instance, the marks of 22 students in a section of an elementary statistics course might constitute a population of interest to us. Where our interest is wholly in this population of 22 cases, the average thus computed is a descriptive statement about a population. We are content that our interpretation of the average is a meaningful statement about the class itself. For instance, we might be able to say, "The average mark was 87, indicating that the class grasped the material." Note that we are making use of all the data comprising the population of interest, and, because our concern does not extend beyond the group we have measured, we must be said to be using descriptive statistics.

The approach of sampling statistics becomes relevant when we have access to only a sample drawn from the population of interest, and when we do not have in our possession the raw scores comprising the total population. In the case described, we might have the same data for the 22 students comprising the elementary statistics course, but we might wish to generalize about the level of all students of elementary statistics in the college. We again have 22 raw scores in our possession, but now they comprise a sample drawn from a much larger population. Under this condition one might say, "The average score in this section was 87. Judging from this sample, I conclude that the students in the college are tending to grasp the material." A statement of this kind is quite different from the one cited earlier, for it is a generalization from a sample to a population that has not been fully surveyed.

The science of sampling statistics, or *statistical inference* as it is alternatively called, consists of procedures that tell us how much we may generalize and what kinds of statements we may make. The situation is virtually always of the sampling kind when we are doing problems of research, since our investigation is limited to a small group of subjects; but our desire is to make statements about a vast population of similar subjects, including even those as yet nonexistent.

In sum, the distinction is that the methods of descriptive statistics are those of describing a population given all the relevant data. The methods of sampling statistics are those of making inferences from a sample about a larger population that remains unseen but that yielded the sample. The next seven chapters of this book deal exclusively with problems of descriptive statistics; the focus of the last twelve chapters is primarily on sampling problems.

1.4 The Random Sample and the Stratified Random Sample. The ability to do accurate research is dependent upon at least some insight into statistical methods. It is becoming increasingly hard to separate the domain of the research worker from that of the statistician. In particular, the issue of how to gather a sample is, strictly speaking, one for the research worker; but real insight into the problem of sampling depends upon at least some appreciation of how the sample is to be treated and used.

A basic concept is that of *randomness*, and the best way to introduce this concept is by illustration. Suppose we wish to get a fair sample of the opinions of all the students in a college. Obviously, it would not be satisfactory merely to question the first 20 students who come out of a room after a freshman English class, or the first 20 students whom we might meet at the entrance to the mathematics building. If we are to generalize from a sample it must be representative, and such samples would not be—since, among other things, they would not represent the total college population.

We can see, merely by applying common-sense reasoning, that our procedure should be such that no one student's opinion is given a better opportunity than any other's of being included in the sample. Such a sample, namely one that is gathered so that each member of the population is an equally likely candidate to be included in it, is called a *random sample*.

There are various techniques for gathering random samples. For instance, in the situation described one might put the name of each of the students on a separate card. Then one might shuffle all the cards in a large container and draw at random the number of cards desired for the sample. The final step would be to consider only those students selected and ask them for their opinions, which would constitute the sample.

A variation of the technique described is that of collecting a *stratified random sample*. This technique ensures in advance that a sample will contain the same proportions of members of different groups as are in the larger population. For instance, in the situation described one might wish to ensure that the same fraction of juniors appears in the sample as is enrolled in the college. That is, if one-fifth of the students in the college are juniors, then one would randomly select exactly enough juniors to comprise one-fifth of the sample. The same procedure would be followed for each of the other classes and perhaps even for the different subject matter areas, if they were considered relevant to the opinions being surveyed.

The word "stratified" is applied to the method described because it entails actually going to different strata or levels in order to complete the sample. The procedure of stratifying a sample is necessary only where subgroups within the population have meaning for the data being gathered, in which case nonproportionality would make a sample nonrepresentative and misleading. The method of stratifying samples is used for such surveys as election polls, where a stratified sample is crucial since different subpopulations often manifest markedly different voting preferences.

1.5 The Abuses of Statistics. The abuses of statistics are many; insight into basic statistical concepts is the best defense against them.   Abuses are frequently found in popular publications as well as in technical journals. Often they are the result of ignorance rather than design.   For instance, as we shall see, there is quite a difference among the mean, the median, and the mode, though all three are apt to be described as averages.   It often occurs, especially during disputes, that different vested interests each use the word "average" but have in mind different measures that were actually computed. For instance, where labor and management have clashed, leaders on each side have cited averages that were computed correctly but that were markedly different.   Naturally, each of the vested interests cites the value that is most advantageous to its own case.

One of the most famous abuses of research method was the result of a fallacious sampling procedure.   This time the flaw was in selecting cases rather than in statistically treating the data incorrectly, and it was undoubtedly an actual error rather than a purposeful distortion.   The error was made in 1936 by the *Literary Digest* magazine during its pre-election presidential poll. The *Digest* predicted that Alfred Landon would win easily over Franklin Roosevelt, but in the end Roosevelt carried 46 of the 48 states and many of them by a landslide.   The *Digest* had made its error in choosing a sample from ten million persons originally selected from telephone listings and from the list of its own subscribers.   In the depression year of 1936, the persons who could afford magazine subscriptions or telephones did not constitute a random sample of the voters in the United States.   The voters in the sample selected were predominantly for Landon, whereas the majority in the larger population was for Roosevelt. ˙ Incidentally, the *Literary Digest* rapidly lost status after the election, and, undoubtedly as a partial result of its error, it soon ceased to exist.   In subsequent years election pollsters took careful note of how the *Digest* had blundered and developed stratified sampling techniques to insure representativeness.

Obviously we are limited in our discussion of the abuses of statistics, since we have not as yet presented the subject matter of statistics.   It is enough to say that nearly any statistical procedure may be done incorrectly either through ignorance or by design.   The reader who wishes to look further into the possible abuses of ordinary statistical methods should consult the little book entitled *How to Lie with Statistics.**

---

* Huff, Darrell, *How to Lie with Statistics* (New York: W. W. Norton & Co., Inc., 1954).

1.6   Skepticism and the Counter-Resistance to Statistics.   The use of the statistical argument is far from academic, and in fact the man in the street can scarcely get through a day without having statistics cited to him.   Surveys of all kinds are done or at least reported, and one reads findings in the newspaper and hears them quoted on the radio and on television.   Exhortations from sponsors are often statistical and are framed so as to leave little doubt of the implications of the studies conducted.

Quite naturally, the enormity of the push of statistics has mobilized what is almost a necessary self-protective attitude on the part of the individual.   A new orientation is becoming quite common: "Keep statistics away from me." This attitude is unfortunate.   Although distortions are in fact often carried out with the aid of statistics, the fault is that of individuals using inappropriate methods and it makes no sense to blame the statistical orientation itself.

There seem to be several main objections to the use of statistics, and it is worthwhile to verbalize them now and to deal with them briefly.   These objections are (1) that statistics are untrustworthy since by using statistics one can prove anything, (2) that statistical thinking is unfair because it is anti-individual, and (3) that the use of statistics in connection with human problems is cold and unfeeling.

The distrust of statistics as a vehicle that can "prove" anything is undoubtedly a reaction to many current abuses.   To go back, it is apparent that to conduct accurate research is a painstaking task.   Designing an experiment and actually gathering data require careful planning.   Improper procedures may alter results and make an experiment invalid.   Obviously, at least some statistical knowledge is needed to interpret data.   Therefore, when an organization that obviously has a vested interest omits the details of an experiment because of "time shortage" and merely reports a finding that favors its own product, one must certainly be curious if not suspicious of the report.

It does appear quite remarkable that each of many organizations with competing products can conduct research and find from obtained data that its own product is superior.   Obviously it is possible to "slant" research by varying data-gathering techniques or misinterpreting findings.   The man in the street hears conflicting statistics reflecting obvious vested interests nearly every day.   Quite understandably he comes to distrust statistical reports altogether.   Thus one finds great distrust of statistical methods expressed by college students and by others, and no wonder.

The fact is that the science of statistics is an enormously powerful tool to facilitate learning.   But, like any powerful instrument, it may be used to increase man's effectiveness and happiness or it may be abused through conscious misapplication.   No statistical method is in itself right or wrong, but each has its time and place.   Large organizations, like advertisers, sometimes deliberately misapply statistical methods.   The best defense against being hoodwinked is education.   Typically, the brief statements of findings issued by vested interests cannot be properly interpreted without considerably more information. Where sufficient details are given, there is great advantage in being able to evaluate them. In any event, where there are abuses, the situation is not that figures lie but that liars are apt to figure. The solution is for the intelligent reader to be able to figure too.

The next objection—that statistical thinking is unfair because there is no acknowledgment of the individual—is a subtle one that is often felt but seldom verbalized. As defined, statistics deals with properties in populations; but the great pride of modern man, particularly in American culture, is in his emphasis on the individual. The impulse is to say, "Don't tell me that the majority in my subculture is voting for the candidate whom I dislike." It is to say, "Don't tell me what the average man does or that my expectancy based on my education is to make ninety dollars a week." Statistics, with its emphasis on the population as opposed to the individual, seems to lead to statements that are subtly abrasive to individual freedom. The feeling behind the old maxim "Never generalize" is that to do so is unfair.

Once again the answer is that a general statement, if true, cannot itself be at fault. The statements that the majority in a community is voting for a particular candidate and that the average person in a subgroup makes ninety dollars a week do not imply that any individual must also vote for that candidate or make ninety dollars a week. A general statement that is true is apt to be quite informative. Difficulty arises only when the listener mistakenly interprets such a general statement as if it had to be true for each individual in the group. In each of the cases cited, the listener's objection would undoubtedly be based on such a misinterpretation. It is his own error, or desire to conform, that leads him to conclude that he must operate as the majority or be the expectancy.

The objection to general statements when they are true typically reflects either a misinterpretation of these statements or a fear of them. One has absolute proof of this where the listener replies by pointing out an "exception." Since the general statement is never an invariable one, there is no such thing as an exception. Where the listener cites a contradictory instance he is evidencing that he has mistakenly translated the general statement into an invariable one. The solution is for the individual to learn to acknowledge general statements where he believes they are true, and at the same time to appreciate that a general statement is not an attempt to deprive him of his freedom. "Yes, I understand that the majority in my community are voting for that candidate, but I am not."

Another objection to statistics is that data themselves are numbers; they are "cold." The feeling is that "pure numbers" can never do justice to the material when human beings are concerned. Can numbers really describe human problems or suffering? The answer is that they cannot and are not intended to do so. But enumeration has led to social and medical discovery that has already been relevant in alleviating suffering. Numbers themselves obviously do not tell the whole story of man but only the enumerable part of it.

9

As for the complaint that statistics are "cold," one might consider what his own position would be in the following situation. Suppose it was shown that a two-minute lie detector test led to a higher per cent of correct convictions and acquittals in murder trials than when live juries were used. In other words, an innocent man would have a better chance of escaping sentence by taking the test than by pleading his case before a jury. One may ask himself whether as an innocently accused person he would prefer to be tried by the lie detector test or to plead his case, in which case his chance of proving his innocence would be less. The question is obviously posed to contrast cold statistical ·findings with human judgment, and obviously the answer is not easy.

Fortunately, however, there would always be other alternatives. Were the situation to occur, a careful statistical analysis would show where the lie detector test and the jury disagreed in their verdicts. This analysis might prompt further investigations into where juries make their mistakes so that they might be better educated in those areas. The point is that the "coldness" of statistics should not imply that they be disregarded, because the gathering and interpreting of data are procedures aimed at specific ends. Where one has both integrity and knowledge of correct procedures, the benefits are most apt to be great.

PROBLEMS

1.1 Listen for mention of "statistics" on radio and TV newscasts. In what context is the word used?

1.2 Look for references to "statistics" in newspapers and magazines. In what context is it used?

1.3 Let any baseball fan explain how "batting averages" are computed. With respect to a player's batting average, discuss how many hits a player may be expected to make in his next 10 times at bat.

1.4 Find newspaper or magazine statements about "average" values. Is any mention made of the type of average being used?

1.5 Comment on a sales manager's remark that no salesman should have a monthly sales amount below the average for that month.

1.6 Bring in any other examples that you think illustrate abuses of statistical method and discuss them in class.

1.7 An advertising claim states, "Product X is preferred by 9 out of 10 men." State why you would or would not accept this as a valid claim to product superiority.

1.8 A television commercial proclaims that Product X is "now 30 per cent better, faster acting." What criticism do you have of this commercial as "statistical proof of product superiority"?

1.9 Bring in the latest survey or poll result from the Sunday newspaper. Is there any mention of the sampling technique used?

1.10 Discuss the success or failure of the predictions made about the outcome of the last presidential election.

1.11 Report on some abuse of statistical method discussed in *How to Lie with Statistics*.

1.12 Find out anything you can about the technique known as "area sampling."

1.13 Describe a method not mentioned in the text that might be useful in obtaining a fair random sample of the students in a college.

1.14 Find references to statistics and statistical data in a text in your major field.

1.15 Obtain more information about Quetelet, Galton, Pearson, Fisher, and Nightingale and their work or interest in statistics.

1.16 Consult the indexes in the library and ascertain the names of five governmental agencies that publish statistics.

1.17 Examine a daily report of New York Stock Exchange transactions. The usual data given for a particular stock are total shares traded, high price, low price, and closing price for the day. Is the closing price an average of the high and low prices? Can you calculate from the given data the dollar value of the shares traded for a particular stock? Why?

1.18 How would you proceed to obtain a sample of one-syllable words used in this textbook? How would you apply the sample results to cover the whole textbook?

1.19 Your college administration undertakes a study of the total hours a student devotes to homework per week. How would you select a suitable sample of students for this study?

1.20 A new type electric light bulb is developed and produced in small quantities for experimental study. Suggest a suitable sampling procedure for determining the life of these new bulbs.

two

the mode,
the median,
and the mean

2.1 Definitions. The most common statistical concepts to indicate the middle of a distribution are the *mean*, the *median*, and the *mode*. The best way to convey the meaning of these words is to illustrate them simply. Consider the following set of five terms or observations:

$$4$$
$$4$$
$$5$$
$$7$$
$$10$$

Definition 2.1 *The mode is the value of the term that appears most frequently.* In the example, the mode is 4 since the term 4 appears twice and each of the other terms appears once.

Mode = 4

Definition 2.2 *The median is the value of the term that is larger than or equal to half of the other terms and equal to or smaller than half of them.** In the example, the third term from the top, 5, is the median. The reason is that 5 is larger than two terms (4 and 4) and smaller than two terms (7 and 10).

Median = 5

_____

* See page 14 for computation of the median in the case where the number of terms is even.

Definition 2.3 *The mean is the value that is obtained by adding the terms and then dividing their sum by the number of terms.* In the example there are five terms and their sum is 30. The mean is 30 divided by 5, or 6.

$$\text{Mean} = \frac{30}{5} = 6$$

The mean is by far the most important of the three terms defined. To see why, consider the terms again:

4	
4	Mean = 6
5	Median = 5
7	Mode = 4
10	

Suppose that the term inside the square was increased from 7 to 17. The terms would then be:

4

4

5

17

10

The median is still 5 and the mode is still 4. But the mean of the new set of terms is now 8, not 6. The change has affected the mean but not the median or the mode. Some changes would have affected the other two measures as well as the mean. For instance, adding 1 to the middle term would have increased the median to 6. However, the mean responds to *every* change whereas the median and the mode respond only to *some* changes. For this reason, the mean is often described as "sensitive" and as "reflecting the entire distribution." Its sensitivity seen in this way gives a hint of its importance. We shall see that the sensitivity of the mean may be a disadvantage to its representativeness when the distribution contains a few outlying terms at one extreme.

It is noteworthy that the mean of the original distribution was not actually a term in the distribution. That is, none of the five terms was a 6. The mean should be thought of as a statement about the distribution itself. The mean may or may not also be the same number as a term in the distribution.

For convenience, the terms in the distribution were presented in ascending order. That is, each term was equal to or larger than every term above it. Remember that any order might have been used, and in practice, data are not ordered when collected. Here are some other orders:

10	7	10	5
4	5	5	4
5	4	7	10
7	10	4	4
4	4	4	7

Each of these columns represents the same distribution and in each case the median (in the square) is 5 by the definition given earlier. It is easy to make the mistake of thinking that the median in each case is the third number from the top. Remember that the definition does not say anything about placement on the page but specifies the median simply by its relative size.

When there is an even number of terms, there exists no actual median— that is, no term larger than exactly half of the others and smaller than half. No matter which term one picks there are an odd number of others. But now it is possible to find two middle terms, and the median is defined as the mean of these two middle terms. Note that under this condition the median is not the same number as a term in the distribution (unless the two terms are identical). A more detailed discussion of the median for grouped data appears in Section 6.3.

There is little to say about identifying the mode. However, note that in some distributions there is no mode, because no one term occurs more than once.

3	2	9
5	5	3
2	7	2
0	3	1
8		0

A collection of terms may have more than one mode. Here is an example of a collection with two modes. It is said to have a "bimodal distribution."

9
3
3
8
7
9
2

The two modes are 3 and 9, since each appears twice and no other number appears that often.

2.2 Two Properties of the Mean. We will now give our full attention to the mean. An important property of the mean is clear in a physical analogy. Imagine a long wooden plank with numbers on it at equal intervals from 4 to 10. Suppose five one-pound weights are placed on this plank in positions corresponding to the distribution of numbers already discussed: 4, 4, 5, 7, 10.

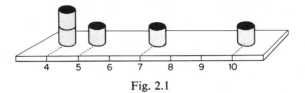

Fig. 2.1

This distribution is illustrated by placing two weights on 4, another on 5, another on 7, and another on 10 (see Fig. 2.1).

Now suppose we are given a fulcrum and told to balance the plank with its weights on this fulcrum, assuming the plank itself has no weight. After several attempts that fail (Fig. 2.2), we manage to find the balance point (Fig. 2.3).

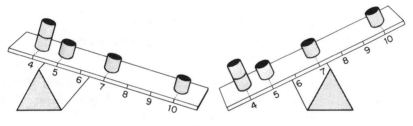

Fig. 2.2

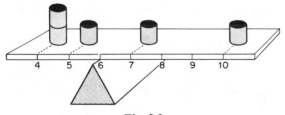

Fig. 2.3

It turns out that the fulcrum has been placed under the mean of the terms. In Fig. 2.3 it is under the point 6. The mean of any collection of terms is their balance point and, furthermore, the mean is the only balance point.

Consider the weights on the left side of the mean. Let us see how far from the fulcrum (mean) each of them is. The two weights on the extreme left are each 2 units away from the mean and the third is 1 unit away. The sum of these three distances is, therefore, 5. The weights on the right side also total 5 units away from the mean. It is this equivalence that balances the plank.

15

The sum of the distances of the weights from the mean is always the same on the left side as it is on the right side. The fact that the distances on either side of the mean "balance out" should be kept in mind. The way to obtain these distances is to subtract the mean from each of the terms. In Table 2.1 the five terms in the distribution are listed in the first column, and the mean, 6, has been written into the second column. The third column, called "Differences," has been found by subtracting the mean from each of the original terms. Instead of "differences," the word "distances" might have been used. Notice that some of the differences are negative numbers.

TABLE 2.1   DIFFERENCES FROM MEAN

TERMS IN DISTRIBUTION	MEAN	DIFFERENCES
4	6	-2
4	6	-2
5	6	-1
7	6	$+1$
10	6	$+4$
		0 total of differences

When the differences are added, with attention paid to the plus and minus signs, the sum turns out to be zero. This makes sense if we think of each of the differences as a distance from the balance point, the mean. Differences with minus signs are results of weights on the left side of the mean and those with plus signs are results of weights on the right side.

The important facts to remember are that the sum of the differences from the mean equals zero and that the property of the differences totaling zero is unique to the mean. That the differences total zero means that no other number would have served as a balance point. For instance, in Table 2.2 the number 5 is tested as the balance point.

TABLE 2.2   DIFFERENCES FROM TEST BALANCE POINT

TERMS IN DISTRIBUTION	TEST BALANCE POINT	DIFFERENCES
4	5	-1
4	5	-1
5	5	0
7	5	2
10	5	5
		5 total of differences

The sum of the differences is not zero. Using 5, the imaginary plank has fallen to one side.

The fact that the mean is the only number with the property described can help to find it. Just as one finds a balance point, in theory one may find the mean by trying different numbers until he discovers the number that makes all the differences total zero.

For purposes of later reference, what has been said may now be summed up as a theorem.

Theorem 2.1 *In any set of terms, the sum of the differences from the mean equals zero. Conversely, if the sum of differences from some number in a distribution equals zero, that number is the mean.*

It should be noted that we have not proved Theorem 2.1 but have only illustrated it and given intuitive justification for it. We shall often proceed in this way.

Now let us turn our attention to the set of differences that appeared in the third column of Table 2.1. These differences were:

$$-2$$
$$-2$$
$$-1$$
$$1$$
$$4$$

For our next theorem we square each of these differences, as shown in Table 2.3.

TABLE 2.3 SQUARES OF DIFFERENCES FROM MEAN

DIFFERENCES	SQUARES OF DIFFERENCES
−2	4
−2	4
−1	1
1	1
4	16
	26

The sum of the squares of the differences is 26. Note that these squares that we have just added were determined by the choice of the mean as a reference point. If we had chosen any reference point other than the mean, the sum of the squares of the differences from that point would have been larger than 26. In other words, to keep the sum of the squares of the differences from some point as small as possible we should choose the mean as the reference point.

What would have happened if we had chosen 5 as the reference point is shown in Table 2.4.

TABLE 2.4   SQUARES OF DIFFERENCES FROM
REFERENCE POINT

ORIGINAL TERMS	5 AS A REFERENCE POINT	DIFFERENCES	SQUARES OF DIFFERENCES
4	5	-1	1
4	5	-1	1
5	5	0	0
7	5	2	4
10	5	5	25
			31

The sum of the squares of the differences has turned out to be larger than 26.

We are now ready for a second theorem about the mean.

Theorem 2.2   *In any collection of terms, the sum of the squared differences from the mean is less than the sum of squared differences from any other point.   Conversely, the point that minimizes the sum of the squared differences is the mean.*

Obviously one makes no use of Theorem 2.2 in finding the mean.   However, the "least squares" property of the mean has importance, as will be seen later.

2.3   How Changing the Terms Affects the Mean.   Suppose now that the five terms in our distribution are the ages of the five Smith children, the mean age of a Smith child being 6 years.   One might ask, "What will be the mean age of the Smith children 9 years from now?"   The answer seems almost too obvious: In 9 years, the mean age of the Smith children will be 15 years.   Obvious or not, this answer is correct.

Now, one more question.   How did we arrive at this answer?   The reader who did arrive at it will probably agree that his method was easier to use than to explain.   To get the answer, we made use of a valuable property of the mean.   It is essential to focus our attention on this property, for it will be helpful in less obvious problems.

The question was, what will be the average age of the Smith children 9 years from now.   One might have added 9 years to each of the terms in the distribution of ages of Smith children.   In this way a new age would have been found for each child.   The mean of these new ages would still have turned out to be 15.   However, it is not necessary to use this method.   We may correctly assume that when each child is 9 years older, the mean age of the children also increases by 9.

Our physical analogy suggests why this happens.   Once more consider the five numbers 4, 4, 5, 7, 10 as weights on an imaginary plank.   The fulcrum is the mean, 6.   The plank, as usual, is considered to have no weight.

Our problem tells us to consider each child as being 9 years older. To depict the ages, we move each weight 9 units to the right. The plank has been extended in Fig. 2.4 to show how the weights look in their new positions, with X's to denote their original positions.

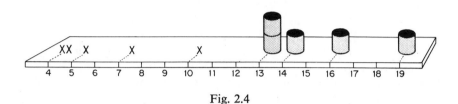

Fig. 2.4

We may direct our attention to the portion of the plank that extends under the weights in their new positions. The new balance point is the new mean. This balance point is under the point 15 in Fig. 2.5.

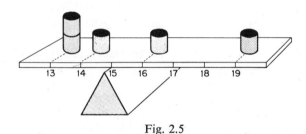

Fig. 2.5

The mean has shifted along with the scores. It is exactly 9 units to the right of where it was. Thus the new mean is 9 units larger than the old mean. Notice that the scores have the same positions with respect to each other, even though each has been increased, and that the mean has the same relative position with respect to the scores as it formerly had.

Of course, there is nothing special about the number 9. Any one number that had been added to each of the scores would have preserved their relative positions while shifting them all. Whatever number had been added would have told exactly how much the mean had increased.

If a constant number had been subtracted from each of the scores, the weights would have been shifted to the left and the mean would have been decreased by that number. In general, when we decrease each term in a distribution by a constant amount, we are in effect decreasing the mean by that amount. Accordingly, we may say that the average age of a Smith child a year ago was 5.

We will describe the addition of a given number to each term in a distribution as "adding a constant" and we will also have occasion to speak of 19 "subtracting a constant." What has been said may now be summarized as a theorem.

Theorem 2.3 *If each term in a collection is increased by a constant, the mean is increased by the same constant. If each term is decreased by a constant, the mean is decreased by the same constant.*

Theorem 2.3 is not only intrinsically interesting but, more than that, we can "cash in" on it. Suppose we are after the mean of a distribution made up of awkwardly large numbers. It is often convenient to deliberately change these numbers by subtracting a given number or constant from each of them; thus arriving at a new distribution that yields an obvious mean. Our rule tells that this new mean is too small by exactly the amount of the constant we have subtracted from the original distribution. So we add back the constant to get the mean of the original distribution.

Suppose we are to find the mean of 92, 94, and 99. Subtract 90 from each term and find the mean of 2, 4, and 9. This mean is 5. Therefore, the mean of the original terms equals $5 + 90 = 95$.

The choice of the constant to be subtracted has no effect on the final result, since that constant is added back. Thus we might have subtracted 91 from each of the original terms 92, 94, and 99. The result would be a new distribution, 1, 3, and 8, which yields a mean of 4. Adding 91 to 4 again results in 95, the mean of the original terms.

Suppose that a statistician subtracted a constant exactly *equal to the mean.* That is, suppose he subtracted 95 from each of the three terms. He would end with Table 2.5.

TABLE 2.5 SUBTRACTION OF MEAN FROM
EACH TERM

ORIGINAL TERMS	CONSTANT SUBTRACTED	DIFFERENCES
92	95	-3
94	95	-1
99	95	4

This time the remaining differences have a sum of zero (Theorem 2.1) and the mean of the original terms is $0 + 95 = 95$. Notice how simple the arithmetic was in this last case.

When using the method of subtracting a constant to find the mean, it is usually preferable to subtract a constant that promises to be as close to the mean as possible. The nearer the constant is to the mean itself, the simpler the computations, because the differences are then small and easy to compute. It is a good idea to estimate what the mean will be before beginning to work, and then to use this estimate as the constant to be subtracted. The better the estimate, the more work saved, although the final answer is the same whatever the estimate.

The short cut just recommended may seem roundabout, for admittedly the device of subtracting a constant to find the mean is seldom worthwhile when a distribution has only three terms. However, the method proves invaluable when the distribution is large and computation must be done without a calculating machine.

Another technique sometimes used is to make the lowest term in the distribution the estimate, knowing in advance that it will not be the mean. The advantage of this technique is that it insures that none of the differences will be negative, and thus does away with one possible source of error, the inadvertent mishandling of minus signs. For instance, if we used the value 92 as our constant, we would obtain differences of 0, 2, and 7. Their mean being 3, we add 92 and 3, and once again get 95 as our computed mean of the distribution.

Next we ask what happens to the mean when each term in a distribution is multiplied by a constant. To begin with, observe what happens when each term in our distribution is multiplied by 2 (Table 2.6).

The effect has been to double the sum of the terms. That is, 60 is the new sum instead of 30. The multiplication by 2 in the new case led to the determination of a mean *twice* as large as formerly. By analogous reasoning, if we had multiplied each term by 3, the mean would have trebled its size. In general, when each term in a distribution is multiplied by a constant, the effect is to multiply the mean by the same constant.

TABLE 2.6   MULTIPLICATION OF TERMS BY TWO

TERMS	NEW DISTRIBUTION WHEN TERMS ARE DOUBLED
4	8
4	8
5	10
7	14
10	20
Sum 30	60
Mean 30/5 = 6	60/5 = 12

Finally, we can show that when each term in a distribution is divided by the same number, the mean becomes divided by that number. The process of dividing by a number is equivalent to that of multiplying by 1 over that number. For example, dividing each term by 3 is the same as multiplying each term by one-third. By our multiplication rule we know that when each term is multiplied by one-third, the mean becomes multiplied by one-third also. Thus dividing each term by 3 results in shrinking the mean to one-third of its original size. In general, when each of the terms in a distribution is divided by a constant, the mean is also divided by that constant.

To summarize, we can state the following theorem.

21

Theorem 2.4 *If each of the terms in a collection is multiplied by a constant, the mean becomes multiplied by the same constant. If the terms are divided by a constant, the mean becomes divided by that constant.*

2.4 Some Applications of the Mean, the Median, and the Mode. The sensitivity of the mean frequently makes it a good measure of the size of all the terms in a group. Loosely speaking, the mean is often used as a measure of how high the terms in a group "are running." A frequent way to compare the sizes of the terms in two groups is to compare their means. Thus when we say that the mean number of home runs hit by players on the home team is larger than the mean number hit by players on the visiting team, we are saying that the home-team players are better home-run hitters as a group than the visiting-team players. Consider the statement that the average number of children per family in one country is 2.3, whereas in a second country it is 1.5. Very likely the mean has been used as the basis of comparison. The implication is that families tend to be larger in the first country mentioned.

The fact that the median (or middle term) is not affected by the size of the extreme terms makes it useful for certain purposes. The median provides important information, especially about a distribution in which a relatively small number of terms are extreme in one direction or the other. For instance, consider the distribution of the incomes of 6,000 citizens in a mining community. Virtually all of these citizens earn less than $5,000 per year. However, a few of them, the mine owners, earn as much as nearly all the others combined.

The median of this distribution of incomes gives a better picture of the economic level of the community than the mean does. The fact that the median is not influenced by the vastness of the incomes of a few makes it a representative measure. For instance, the median yearly income might be $4,400 per year. (That is, half of the citizens make less than $4,400 per year and half of them make more than this amount.) The mean, because it is swollen by the incomes of the few, may be $8,000. Thus it might turn out that only a tiny percentage of the people in the community earn as much as the mean income. The median is often a more revealing measure than the mean, especially when both are obtained from a distribution that contains a relatively small number of terms with values at one extreme.

The value of the modal term provides important information to producers, designers, and storekeepers who aim their commodities at specific markets. For instance, a producer might wish to know the most frequent price being paid for men's watches so that he can devote his attention to making watches in the particular price range. A clothing salesman might wish to know the most popular sizes of men's suits so that he can purchase his stock accordingly. Naturally, many other factors enter into both of these illustrations. The point is merely that the mode indicates maximum frequency, which is often synonymous with maximum popularity.

22

2.5 Notation. The number of terms comprising a distribution is symbolized by the letter N. Thus for the following three distributions:

DISTRIBUTION A	DISTRIBUTION B	DISTRIBUTION C
5	5	5
3 $N=3$	3 $N=4$	3 $N=5$
2	2	2
	4	4
		5

The terms in a particular set may be given a letter value to distinguish them (usually the letter X or Y). For instance, we may use the letter X to stand for any term in Distribution A. When we specify a particular order for the terms, we may attach a subscript to indicate a particular term. For instance, suppose we are using the letter X to refer to any term in Distribution A and we think of them in the order indicated above. Then X_1 stands for the first term, 5. $X_2=3$ and $X_3=2$. Similarly, we may use the letter Y to refer to any term in Distribution B. Then $Y_1=5$, $Y_2=3$, $Y_3=2$, and $Y_4=4$.

The symbol $\sum$ (Greek letter capital *sigma*) is used to indicate "the sum of." It tells us to add certain expressions. Remember that we used the letter X to denote a term in Distribution A. Thus $\sum X=10$ and $\sum Y=14$.

The symbol for the mean of a population is μ (Greek letter *mu*). Thus for Distribution A, $\mu=\dfrac{\sum X}{N}=\dfrac{10}{3}$. For Distribution B, $\mu=\dfrac{\sum Y}{N}=\dfrac{14}{4}$. In general,

$$(2.1) \qquad \mu = \frac{\sum X}{N} \text{ for any distribution having terms denoted by } X.$$

When we are considering, several distributions simultaneously we use subscripts to avoid ambiguity. For instance, we can refer to the mean of Distribution A as μ_X and to the number of terms in Distribution A as N_X. Similarly μ_Y refers to the mean of Distribution B and N_Y refers to the number of terms in Distribution B. Thus:

$$N_X = 3; \quad \mu_X = \frac{\sum X}{N_X} = \frac{10}{3}$$

$$N_Y = 4; \quad \mu_Y = \frac{\sum Y}{N_Y} = \frac{14}{4}$$

As stated in the Preface, for this and subsequent chapters there are two sets of problems. Each set gives full coverage of the topics in the chapter. Answers to the problems in Set A are included in the Appendix and thus serve as a self-test of material in the chapter. If you are able to master the problems in Set A you should have no trouble solving those in Set B.

PROBLEM SET A

2.1 Find the mean, median, and mode of the following set of IQ scores: 100, 83, 88, 81, 83, 96, 105, 108, 78, 102, 97, 113, 126, 94, 85, 119, 67, 91, 88, 99, 88, 72, 77, 88, 114.

2.2 A researcher collected the following achievement-test scores from 20 students in an experimental program: 72, 74, 75, 71, 73, 73, 77, 75, 76, 73, 72, 74, 76, 74, 75, 76, 75, 73, 71, 75. Determine the mean, median, and mode of these scores.

2.3 Seventeen third-grade children were given a social-awareness test and received the following scores: 28, 32, 34, 36, 39, 40, 36, 41, 41, 42, 43, 44, 46, 46, 49, 50, 50. Compute the mean, median, and modal scores of the children.

2.4 The numbers of dental cavities found in a group of individuals are 3, 11, 5, 12, 9, 8, 16, 13, 12, 11, 6, 19, 16, 11, 15, 12, and 0. Find the mean, median, and mode for this distribution.

2.5 Thirty newspapers are purchased at 8¢ each, 20 at 10¢ each, and 10 at 15¢ each. Find the mean price per newspaper.

2.6 There are 25 executives in a computer center. Fifteen of them earn $12,000 per year, five earn $20,000, one earns $22,000, one earns $25,000, and three earn $30,000. Find the mean, median, and mode of the salaries.

2.7 For the scores 11, 13, 13, 14, 17, 17, 17, and 18, compute the sum of the squared differences from the mean *and* from each of the numbers 13, 14, 17, and 18. Thus verify in part that Theorem 2.2 is satisfied.

2.8 Find the mean of the following numbers and check that the sum of their deviations from the mean is zero: 13, 39, 17, 14, 100, 58, 4, 43, 113, 92.

2.9 Calculate the mean of the numbers given in Problem 2.8 by applying Theorem 2.3 and first subtracting 15 from each of the numbers, later adding 15 to the mean of the new numbers.

2.10 Suppose that you are buying a thermometer and there are many on display. You note that not all of these indicate the same temperature. How will you select one?

PROBLEM SET B

2.11 The ages (to the nearest year) of the members of the graduating class of a certain high school are 17, 19, 17, 18, 17, 16, 17, 17, 16, 17, 18, 15, 18, 18, 17, 19, 17, 17, 15, 18, 17, 17, 17, 17, 17, 18, 19, 16, 18, 16, 16, 16, 18, 19, 17, 16, 17, 18, and 17. Find the mean, median, and modal ages for the group.

2.12 Find the mean, median, and mode for the following data: 4, −5, 3, −7, −8, 2, −6, 5, −4, −8, −4, −4, 5, 0, 3, 0, −4, −2, −5.

2.13 Find the mean, median, and mode of the following numbers: 2, 4, 8, 16, 32, 64, 128, 256, 512, 1024.

2.14 Find the mean, median, and mode of the following distribution of ages: 20, 17, 16, 15, 19, 19, 18, 12, 13, 13, 17, 16, 14, 14, 16, 15, 18.

2.15 For the data of Problem 2.14, compute the sum of the squared differences from the mean *and* from each of the numbers 14, 15, 17, and 18. Thus verify in part that Theorem 2.2 is satisfied.

2.16 The mean yearly income of workers in a geophysical research laboratory is $9,000. If each worker obtains a $400 yearly raise, what is the new mean income? Suppose, instead, that each worker is granted a 10 per cent pay increase; what is the resulting mean yearly income?

2.17 There are six faculty representatives in a faculty senate. Two of these have served for 2 years and the others have served for 8, 11, 13, and 18 years. Find the mean, median, and modal years of service for those representatives. What will the mean, median, and modal years of service of the same group be 10 years from now?

2.18 The mean of a set of 35 numbers is 55. Find the sum of the 35 numbers.

2.19 (a) Find the mean, median, and mode of the following numbers: 15, 18, 18, 21, 45, 63, 69, 78, 45, 45, 27, 36, 60. (b) Calculate the mean of the numbers by applying Theorem 2.4 and first dividing each number by 3.

2.20 (a) Find the mean and median of the following numbers: 6.2, 5.1, 8.7, 6.2, 4.1, 3.3, 5.4, 6.2, 6.7, 9.6. (b) Calculate the mean of the numbers by applying Theorem 2.4 and first multiplying each number by 10, later dividing the mean of the new numbers by 10.

three

variability and two measures of variability

3.1 Variability. The terms in a distribution may vary widely or they may be very close to each other. To appreciate the difference, consider a distribution in which the terms are 8, 9, 10, 10, 13. We shall call it Distribution A. The representation of these terms as weights on a plank is shown in Fig. 3.1.

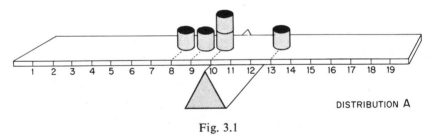

DISTRIBUTION A

Fig. 3.1

Now consider Distribution B made up of the terms 1, 5, 10, 16, 18. This distribution is illustrated in Fig. 3.2.

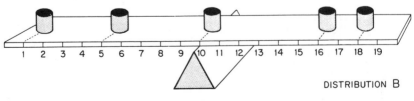

DISTRIBUTION B

Fig. 3.2

In Distribution A the weights appear closer together or more clustered, whereas in Distribution B they are relatively scattered or spread apart. Note that the means of the two distributions are the same. The statistician's way of describing the difference is to say that the terms in Distribution B are *more variable* than those in Distribution A.

Variability is an extremely important idea to grasp. The property of variability refers to the relationship among all the terms in a distribution considered together. Thus a change in any term changes the variability of the distribution. Suppose, for example, we were to change the 18 to 20 in Distribution B. In the illustration this would mean moving the weight on number 18 two units to the right. The shift would spread out the weights even more than they were, with the result that their variability would be increased. On the other hand if the same weight had been shifted two units to the left—to 16, for instance—the variability would have been reduced.

Now suppose we were to change the number 10 to 11 in Distribution B. Would shifting the weight on 10 one unit to the right increase or decrease the variability of the distribution? The answer is not apparent, since we cannot see whether the shift increases or decreases the variability or "scatter," as we sometimes call it.

Frequently our eye cannot tell us which of two distributions is more variable. Compare Distribution C with Distribution D in Fig. 3.3.

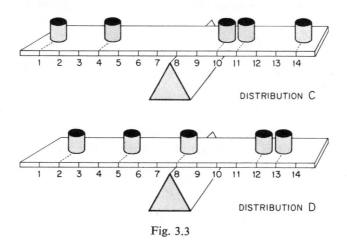

Fig. 3.3

How will we decide which distribution is more variable? Thus far we have not defined a measure of variability, so no decision is possible. Our next step is to define an appropriate measure.

3.2 The Variance. Our measure of variability must be some number that can be computed from the terms in any distribution. When computed, this number will tell us about the spread of the terms in the distribution. To

27

satisfy our demands, the measure we are after must have certain properties. Our way of finding the measure will be to specify these properties in advance. Once we have specified them, they will guide us toward the measure we are seeking.

One property is that our measure must have a small value when it is computed from a distribution in which the terms are close together numerically. Its value must be large when the terms are widely scattered.

A second property is that the value of the measure should not be related to the number of terms in the distribution. Specifically, we would not want a measure that became large just because there were *many* terms in the distribution. Our measure should reflect only the similarity or the dissimilarity of the numbers themselves, and this has nothing to do with how many there are. We might have a thousand terms but, if they were all very similar, our measure should have a small value.

The third property is that our measure should be independent of the mean. We are interested only in how *scattered* the terms are. Distributions A and B in the previous section have the same mean even though Distribution B is more variable. Even if we were told that the mean was one million, we still could not say whether or not the weights representing the terms were far apart. The mean tells us nothing about the variability and its size should not influence our measure.

We now decide instead that when we say Distribution B is more variable than Distribution A, we mean that the weights in Distribution B *tend to be greater distances from the mean*. *Scatter* suggests wide distances from the fulcrum and *cluster* suggests that the weights are huddled around the fulcrum. Thus the measure of variability which we are seeking should be some measure of how far the weights in our distribution are from the fulcrum or mean. Notice that we are not considering the numerical value of the mean itself, but merely the set of distances from the mean.

Now let us examine the distribution of five terms introduced in Chapter 2. This time the distances from each weight to the mean are indicated in Fig. 3.4. The distances to the left of the mean receive minus signs.

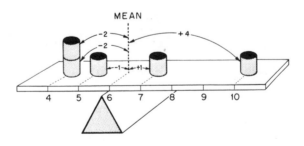

Fig. 3.4

Our impulse might be to add up all the distances and to use their sum as an indication of variability. But the minus signs balance out the plus signs.

We recall from Theorem 2.1 that no matter how long the distances are, their sum is always equal to zero.

To avoid the result of the differences balancing out to zero, we treat the differences as though they were all positive. By summing the distances in this way and then taking their mean, an old-fashioned measure of variability, called the *average deviation*, was developed. However, the measure we are after has so many applications that the average deviation is seldom used.

Our way of escaping the minus signs is to square each of the differences— that is, to multiply each difference by its own value. We will profit by the fact that the square of any nonzero number is positive. Then, having squared each difference, we add up the squares. The computations performed on the distribution just illustrated are shown in Table 3.1. (See Appendix I for a discussion of squares and square roots.)

TABLE 3.1   COMPUTATION OF SQUARES OF
DIFFERENCES FROM MEAN

(1) ORIGINAL TERMS	(2) MEAN	(3) (ORIGINAL TERM) MINUS (MEAN) DIFFERENCES	(4) SQUARES OF DIFFERENCES
4	6	-2	4
4	6	-2	4
5	6	-1	1
7	6	1	1
10	6	4	16
			26

The column containing the squares of the differences includes no negative terms. May we now consider the sum of these squares, 26, as our measure of variability? Not quite. For suppose the number of terms in the distribution was increased from five to eleven. That is, suppose six more weights were placed on the plank. The addition of these weights, unless we placed them all at the mean, would introduce new distances or differences. When these distances were squared and included, the sum of the squares would increase. Thus the *sum* of squares of distances reflects the *number* of weights or terms in the distribution as well as their variability. Since we want our measure not to be sensitive to the number of terms in the distribution, this sum is unsatisfactory as a measure of variability.

It is easy to "purge" our measure of this unwanted influence. We simply divide the sum of squares of the differences by the number of terms in the distribution. The quotient becomes our measure of variability. This measure is called the *variance*. Hence, the variance for the distribution in Table 3.1 is $26/5 = 5.2$.

We have traversed a long road to find the variance, but our steps have not been difficult. In general, given a set of terms, our first step is to subtract the mean from each of them to obtain a set of differences. Then each difference is squared and the total of the squares is found. This total is then divided by the number of terms. The quotient, called the variance, becomes the measure of variability, for it satisfies all the demands we originally made.

Definition 3.1 *The variance is the mean of the squared differences from the mean of the distribution.*

We may now decide whether Distribution C or Distribution D is more variable. We find the variance of each in Tables 3.2 and 3.3.

TABLE 3.2 FINDING THE VARIANCE OF
DISTRIBUTION C

TERMS	MEAN	DIFFERENCES	SQUARES OF DIFFERENCES
1	8	−7	49
4	8	−4	16
10	8	2	4
11	8	3	9
14	8	6	36
			‾‾‾
			114

Variance $= 114/5 = 22.8$

TABLE 3.3 FINDING THE VARIANCE OF
DISTRIBUTION D

TERMS	MEAN	DIFFERENCES	SQUARES OF DIFFERENCES
2	8	−6	36
5	8	−3	9
8	8	0	0
12	8	4	16
13	8	5	25
			‾‾
			86

Variance $= 86/5 = 17.2$

Distribution C has a larger variance, so we say it is more variable.

The variance may be found by short-cut methods, which are indicated at the end of this chapter. However, the method we have given here is the only one that literally follows the definition. The method of computation that we have discussed must be grasped before we can understand the meaning of the measure.

3.3 The Standard Deviation. The variance enables us to tell which of a set of distributions is the most variable. But there is still another measure of variability called the *standard deviation*.

Definition 3.2 *The standard deviation is the positive square root of the variance.*

Admittedly we have as yet seen no purpose for this second measure of variability, but the standard deviation has various uses which will be discussed later. The standard deviation has all of the properties thus far ascribed to the variance. And the requirements that led to the development of the variance are met equally well by the standard deviation.

We may, for example, compare standard deviations to determine which of two distributions is more variable. Our conclusion will always be the same as if we had compared variances.* The reason is that if one distribution has a larger variance than another, it will also have a larger standard deviation. For instance, recall the distribution of Table 3.1 with its variance of 5.2. Suppose that Distribution B has a variance of 16. The former distribution has a standard deviation of 2.3 which is smaller than Distribution B's standard deviation of 4. The reader should verify that the standard deviation of Distribution C is 4.8 and the standard deviation of Distribution D is 4.1. Appendix I contains a table of squares and square roots.

By now it should be clear that computing the standard deviation entails going one step beyond computing the variance. Yet the standard deviation is a much more familiar measure to most students of statistics, for it has more generally known applications. The standard deviation may be computed by the following four steps:

1. Subtract the mean from each term in the distribution to obtain a set of differences.
2. Square each difference and add the squares.
3. Divide the sum of the squared differences by the number of terms in the distribution.
4. Find the square root of the quotient just obtained; the positive square root is the standard deviation.

At this point we should think of the standard deviation as a line segment, or perhaps a ruler, of a given length. When the terms in the distribution are clustered, the ruler is relatively short (Fig. 3.5).

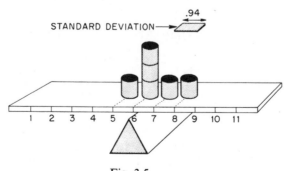

Fig. 3.5

---

* Note that neither the standard deviation nor the variance allows us to say that one distribution is a *certain number of times* more or less variable than another.

When the terms in the distribution are scattered, the line segment or ruler that applies to the distribution is relatively long (Fig. 3.6).

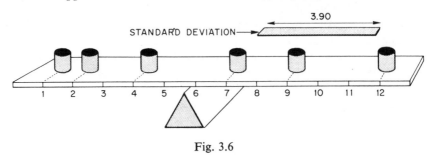

Fig. 3.6

3.4 How Changing the Terms in a Distribution Affects the Standard Deviation and the Variance. We have already discussed the effect of adding a constant to each term in a distribution. In our picture analogy, the corresponding operation is to shift each weight some specified distance to the right. The weights in their new positions remain the same distance from each other and each weight preserves its original distance from the mean. Remember that the mean shifts along with the weights. Thus, adding a constant to each term leaves the variability of a distribution unchanged. Since the standard deviation and the variance measure this variability, it can be seen that adding a constant to every term does not change either the standard deviation or the variance.

For example, we may try adding 9 to each of the terms in the distribution 4, 4, 5, 7, 10. The new distribution has the terms 13, 13, 14, 16, 19. These two distributions were presented in Figs. 2.3 and 2.4. To verify that each has the same standard deviation and variance, it suffices to show that the set of distances from the mean in each distribution is identical. The variance and the standard deviation are each derived entirely from these distances. The computations are in Table 3.4.

TABLE 3.4   COMPARISON OF DISTRIBUTIONS

ORIGINAL DISTRIBUTION			DISTRIBUTION OBTAINED BY ADDING 9 TO EACH TERM		
(1)	(2)	(3)	(1)	(2)	(3)
TERMS	MEAN	DISTANCES	TERMS	MEAN	DISTANCES
4	6	−2	13	15	−2
4	6	−2	13	15	−2
5	6	−1	14	15	−1
7	6	1	16	15	1
10	6	4	19	15	4

The sets of distances (Column 3) are the same in each distribution. We may now write as a theorem:

Theorem 3.1 *Increase or decrease of each term in a distribution by a constant amount does not alter the variability of the distribution, and therefore does not affect the variance or the standard deviation of the distribution.*

On the other hand, multiplication of each term by a constant does affect the variability of a distribution. For instance, let us compare the illustration of our original distribution with that of the distribution obtained by doubling each of the terms (Fig. 3.7).

The new distribution is clearly more variable. Note that the distances between terms have doubled. The standard deviation of the new distribution turns out to be twice as large as that of the old distribution, a fact that the reader should verify. In general, when each term is multiplied by a constant, the standard deviation of the distribution becomes multiplied by the absolute value of that constant; for example, if each term is multiplied by 3, the standard deviation becomes three times as large as it was. If each term is multiplied by -2, the standard deviation becomes twice as large as it was.

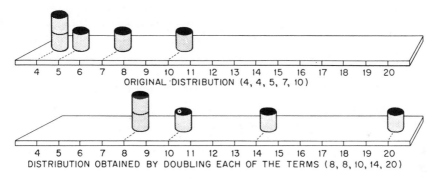

Fig. 3.7

Note that if the constant is a fraction, the effect is to diminish the standard deviation. Multiplication by a fraction pulls the weights closer together, whereas multiplication by a number greater than one spreads them out. Division by a constant effects a division of the standard deviation by the same constant.

Since the variance is the square of the standard deviation, any change in the standard deviation becomes squared in the variance. Thus multiplication of each term by a constant results in multiplication of the variance by the square of that constant; for instance, multiplication of each term by 3 results in multiplication of the variance by 9. Division by a constant results in division of the variance by the square of that constant. Division of each term by 3 affects a cutting down of the variance to 1/9 of its original size.

In sum, we may now write:

Theorem 3.2 *Multiplication of each term in a distribution by a constant results in multiplication of the standard deviation by the absolute value*   33 *of that constant, and in multiplication of the variance by the square of that constant. The same holds when multiplication is replaced by division throughout.*

3.5 Applications of the Variance. The variance is important as a measure of heterogeneity. For instance, the scatter of intelligence test scores in an elementary school class indicates the heterogeneity of the children with respect to intelligence. The variance of a single individual's different scores on a performance test given each day for ten days indicates the unevenness of his performances. Think of two cities in each of which the average yearly income is $5,000. The city in which the distribution of incomes has the larger variance is the one in which the individuals differ from each other more in earning power. Similarly, the variance of the heights of trees is a measure of how different from each other these heights are. Note that the standard deviation might also have been used in any of the above contexts.

3.6 Notation.* Suppose we are using the letter X to refer to the terms in a particular group. For instance, say that the group consists of three terms and that $X_1 = 1$, $X_2 = 3$, and $X_3 = 5$. Now suppose we square the value of each term, however many there are. For our particular group, $X_1^2 = 1$, $X_2^2 = 9$, and $X_3^2 = 25$. Now we add up the squares of the terms (*e.g.*, $1 + 9 + 25 = 35$). The sum of the squares of the terms in the group is symbolized $\sum X^2$. (In our particular group, $\sum X^2 = 35$.)

We have seen that the symbol denoting the sum of the terms themselves is $\sum X$. In our particular group:

$$\sum X = X_1 + X_2 + X_3 = 1 + 3 + 5 = 9$$

The square of the sum of the terms is symbolized by $(\sum X)^2$. In our particular group $(\sum X)^2 = (9)^2 = 81$.

Note the distinction between the *sum of the squares* of the terms denoted by $\sum X^2$, and the *square of the sum* of the terms denoted by $(\sum X)^2$. In the former case, the terms are squared individually and the individual squares are added up. In the latter case, the terms are added first and their sum is squared.

The order of operations is crucial. A graphic demonstration of the difference in a particular case will help us to remember the distinction. Watch what happens when all the terms are positive. We shall perform both operations on the group of three terms, 1, 3, and 5, this time using illustrations. We shall depict each term as a line segment designating its size. For instance, the first term would be a line segment one unit long, the second term a line segment 3 units long, and so on. We add the terms to find the sum of X by laying off the line segments end to end (Fig. 3.8).

*The formulas given in this book are listed in Appendix VIII.

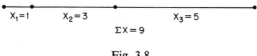

$$X_1 = 1 \qquad X_2 = 3 \qquad X_3 = 5$$
$$\Sigma X = 9$$

Fig. 3.8

Our next step is to square the value of each term.   The square of a number is the (area of the) square built on the line segment representing that number. The squares of the three terms are indicated in Fig. 3.9.

To find ΣX^2 we now add up the areas of the three squares. $\Sigma X^2 = 1 + 9 + 25 = 35.$

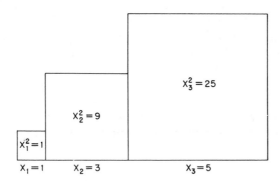

Fig. 3.9

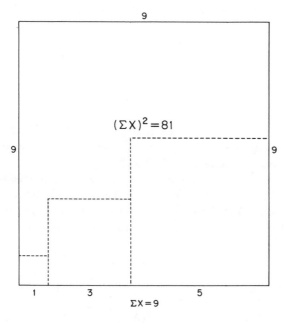

Fig. 3.10

35

Remember that we have represented the sum of all the terms, $\sum X$, by the line segment 9 units long. The *square of the sum* of all the terms, $(\sum X)^2$, is the area of the square built on this line segment (Fig. 3.10). $(\sum X)^2 = 81$.

The larger area is obtained by building up the baseline first and then building a square on it. Thus $(\sum X)^2$ is greater than $\sum X^2$ (when the terms are each positive). The purpose of this demonstration was merely to emphasize that the order of operations is important when there is squaring and adding to be done.

It is customary to use letters toward the end of the alphabet to refer to values of a variable and to use letters near the beginning of the alphabet to stand for constants. For instance, suppose the Greek letter α (read "alpha") represents a particular constant. Consider the operation of subtracting α from each term in the group. For instance, if the terms are X_1, X_2, and X_3 and we subtract α from each term, we get the three differences: $(X_1 - \alpha)$, $(X_2 - \alpha)$, and $(X_3 - \alpha)$. We may write the operation in three columns, as in Table 3.5.

We designate the sum of the differences obtained by subtracting α from each of the terms by $\sum (X - \alpha)$. Now suppose we square all the differences—that is, in the particular group of three terms we obtain $(X_1 - \alpha)^2$, $(X_2 - \alpha)^2$

TABLE 3.5 SUBTRACTION OF A CONSTANT

TERM	CONSTANT	TERM MINUS CONSTANT
X	α	$X - \alpha$
X_1	α	$(X_1 - \alpha)$
X_2	α	$(X_2 - \alpha)$
X_3	α	$(X_3 - \alpha)$
		$\sum(X - \alpha)$

and $(X_3 - \alpha)^2$. We have obtained the squared differences of each of the terms from the fixed reference point with the value α. The sum of these squared differences is designated $\sum (X - \alpha)^2$.

The mean of a group is a constant, which is often subtracted from each of the terms. The mean is denoted by μ. Theorem 2.1 states that, for any group, $\sum (X - \mu) = 0$. The sum of the squared differences from the mean is symbolized $\sum (X - \mu)^2$. Thus the formula for the variance of a set of terms is

$$\frac{\sum (X - \mu)^2}{N}$$

The symbol for the variance is σ^2 (Greek small *sigma* squared), so that we can write

$$\sigma^2 = \frac{\sum (X - \mu)^2}{N}$$

The symbol for the standard deviation is σ (*sigma*), so that

$$\sigma = \sqrt{\frac{\Sigma (X-\mu)^2}{N}}$$

3.7 Computational Formula for the Variance. We have stated that:

(3.1)
$$\sigma^2 = \frac{\Sigma (X-\mu)^2}{N}$$

Two equivalent statements are:

(3.2)
$$\sigma^2 = \frac{\Sigma X^2}{N} - \left(\frac{\Sigma X}{N}\right)^2$$

$$\sigma^2 = \frac{\Sigma X^2}{N} - \frac{(\Sigma X)^2}{N^2}$$

This is sometimes written as

$$\sigma^2 = \frac{\Sigma X^2}{N} - \mu^2$$

The mean is likely to be a fraction even when the terms are each whole numbers, and then the differences obtained by using Formula (3.1) may be quite unwieldy. A tremendous advantage of computing the variance by Formula (3.2) is that there is no need to subtract the mean from each term in the group. When the original terms are whole numbers, Formula (3.2) enables the worker to compute the variance without using fractions.

PROBLEM SET A

3.1 A class of 25 students receives spelling scores of 102, 85, 109, 83, 112, 105, 98, 115, 91, 117, 88, 95, 116, 105, 130, 80, 115, 84, 95, 121, 85, 89, 94, 111, and 106. Find the variance and the standard deviation of these scores, first by using Formula (3.1) and then by using Formula (3.2).

3.2 Compute the variance and standard deviation of the following distribution: 58, 70, 36, 56, 72, 60, 30, 44, 50, 72, 64, 50.

3.3 Compute the variance and standard deviation of the following data: 46, 12, 18, 52, 38, 24, 48, 16, 18, 30, 26.

3.4 Compute the variance and standard deviation of the following distribution of ages: 17, 19, 17, 18, 17, 16, 17, 17, 16, 17, 18, 15, 18, 18, 17, 19, 17, 17, 15, 18, 17, 17, 17, 17, 18, 19, 16, 18, 16, 16, 16, 18, 19, 17, 16, 17, 18, 17.

3.5 On a test of physical fitness the following scores were obtained: 13, 39, 17, 14, 100, 58, 4, 43, 113, 92. Compute the variance and standard deviation of these scores.

3.6 Compute the variance and standard deviation of the following data: -5, 3, -7, -8, 4, 2, -6, 5, -4, -8, -4, -4, 5, 0, 3, 0, -4, -2, -5.

3.7 Compute the variance and standard deviation of the following spelling scores: 11, 13, 13, 14, 17, 17, 17, 18.

3.8 Add a constant of 5 points to each of the scores in Problem 3.7 and compute the variance of these new scores. How does the new variance differ from the variance of the original scores?

3.9 Divide each of the scores in Problem 3.7 by 2 and compute the variance of these new scores. How does the new variance differ from the variance of the original scores?

3.10 Is it possible for the variance to be equal to the standard deviation? If so, when?

PROBLEM SET B

3.11 The weights (in pounds) of 10 people are 156, 162, 170, 177, 180, 181, 183, 196, 205, 209. Find the variance and standard deviation of these weights.

3.12 Compute the variance and standard deviation of the following data: 23, 14, 20, 19, 28, 10, 12, 26, 22, 19, 17.

3.13 Compute the variance and standard deviation of the following set of IQ scores: 100, 83, 88, 81, 83, 96, 105, 108, 78, 102, 97, 113, 126, 94, 85, 119, 67, 91, 88, 99, 88, 72, 77, 88, 114.

3.14 Compute the variance and standard deviation of the following distribution of ages: 20, 17, 16, 15, 19, 19, 18, 12, 13, 13, 17, 16, 14, 14, 16, 15, 18.

3.15 Compute the variance and standard deviation of the following set of ratings: 6.2, 5.1, 8.7, 6.2, 4.1, 3.3, 5.4, 6.2, 6.7, 9.6.

3.16 The numbers of typing errors made by a group of students on a final typing exam are: 3, 11, 5, 12, 9, 8, 16, 13, 12, 11, 6, 19, 16, 11, 15, 12, and 0. Compute the variance and standard deviation of these scores.

3.17 Subtract a constant of 10 points from each of the scores in Problem 3.16 and compute the variance of these new scores. How does the new variance differ from the variance of the original scores?

3.18 Multiply each of the scores in Problem 3.16 by 3 and compute the variance of these new scores. How does the new variance differ from the variance of the original scores?

3.19 A group of 10 students took a spelling test and the distribution of their marks turned out to have a mean of 8 and a standard deviation of 0. What were the 10 spelling marks obtained?

3.20 Is it possible for the variance to be less than the standard deviation? If so, when?

3.21 See whether you can derive Formula (3.2) from Formula (3.1).

3.22 Combine the data from Problems 3.14 and 3.16 and compute the variance and standard deviation for the combined distribution. If σ_1^2 and σ_2^2 are the variances in 3.14 and 3.16 and m_1, m_2 are the respective means, verify that the combined variance is

$$\sigma^2 = \frac{n_1\sigma_1^2 + n_2\sigma_2^2 + n_1(m_1 - m)^2 + n_2(m_2 - m)^2}{n_1 + n_2}$$

where m is the mean of the combined distribution and n_1 and n_2 are the respective numbers of observations in 3.14 and 3.16.

four

ways to describe
the position of a term
in a distribution

4.1 Percentile Ranks and Percentiles. To describe the position of a term in a distribution is often important. For instance, the terms may be the numerical grades of students in a class and we may wish to communicate how a particular student is doing. We may say something like "John got the third highest mark in the class," or "Arthur got next to the lowest mark." These two statements each locate a term by telling how many others there are on one side of it.

However, the information given does not enable us to determine how successful John or Arthur was. There may have been three students in the class or a hundred. We must know this total to determine how well the particular students fared. To indicate the status of a term, it does not suffice to announce the number of terms above or below it. The number of terms in the entire distribution would also have to be made known.

We are going to take a different approach, which enables us to disregard the actual number of terms in the distribution.

We shall now consider how to locate the position of a term in a distribution by assigning to it a value called a percentile rank. To begin with a specific example, suppose that Bill is one of ten students in a class and got a mark higher than exactly five of his fellow students. So far we have accounted for six students, Bill and the five others whom he surpassed. The remaining four students got marks higher than Bill's.

Here is the percentage breakdown. Fifty per cent of the marks (5 out of 10) were lower than Bill's. Bill's mark itself (1 out of 10) comprises 10 per cent of the marks in the distribution. Forty per cent of the marks (4 out of 10) are higher than Bill's mark. These percentages are illustrated in Fig. 4.1.

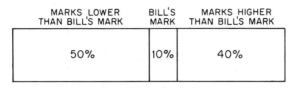

Fig. 4.1

Now we are going to represent Bill's place in the percentage picture by a single line. We choose the midpoint of the segment representing his mark and draw a vertical line through it, as shown in Fig. 4.2.

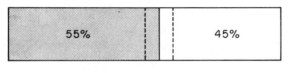

Fig. 4.2

Fifty-five per cent of the terms in the distribution are below this point and 45 per cent of them are above it.

We say that the percentile rank of Bill's mark is the per cent of terms in the distribution below the line we have just found to represent it. The percentile rank of Bill's mark is 55.

Definition 4.1 *The percentile rank of a term in a distribution is the sum of the per cent of terms below it and one-half of the per cent that the particular term comprises.*

As another illustration, suppose that Jim, one of 50 students in his class, received a higher mark than 17 of his classmates. We wish to find the percentile rank of Jim's mark. We begin by noting that 34 per cent of the marks in the distribution are lower than Jim's. Jim's mark comprises 2 per cent of the distribution of marks.

Thus, the percentile rank of Jim's mark is 34 plus one-half of 2, or 35.

Note that Bill's mark has a percentile rank of 55 whereas Jim's mark has a percentile rank of 35. We may now say that Bill did better in his class than Jim did in his. Percentile ranks have enabled us to compare the two students even though factors such as the severity of the two teachers and their methods of marking may have been different in the two classes. In general, percentile ranks enable some comparisons to be made between terms in different distributions. Judging by percentile ranks, Bill's accomplishment in his class is superior even though Jim happened to surpass more students in his particular class. Bill's performance was superior since he surpassed a larger proportion of his competitors than did Jim. However, we are assuming that the levels of competition are equivalent in the two classes. In practice this assumption is often not justified. Note that we have paid no heed to the numerical values of the terms themselves. That is, we have not considered the actual values of the marks that Bill and Jim received.

The percentile rank of a term is often more meaningful than its actual value. For instance, suppose you are told that Bill Jones received a score of 113 on the Kent-Smith Aptitude Test for carpenters and then asked whether you think Bill Jones will be a good carpenter. You cannot possibly answer without more information. But if you are told instead that Bill Jones' score is at the ninety-fourth percentile in a distribution of scores of carpentry students, you may say, judging from the test, that he is a good prospect for the trade of carpentry.

The word *percentile* refers directly to a term in the distribution or to a value intermediate to two terms. The term with a percentile rank of 35 is called the thirty-fifth percentile. The term with a percentile rank of 74 is called the seventy-fourth percentile, and so on. In Chapter 6 we shall see how to determine the value of any specified percentile in a distribution.

4.2 z Scores. Percentile ranks leave some questions unanswered. For instance, suppose each of the two students to be compared is the most outstanding in his respective class. One may dwarf his competitors, whereas the other's superiority in his class may be slight. The percentile ranks of the two students might be the same, giving us no clue to this difference. In this section we shall see a new way to describe the position of a term in its distribution. This new method, unlike the one just described, is sensitive to differences. We will pay no attention to percentile ranks. Instead, our way of locating a term will be to indicate how far it is from the balance point (the mean) and on which side of the mean it is.

Consider the distribution of six terms, 4, 6, 7, 10, 15, 18 with its mean of 10. By subtracting the mean from any term, we can determine the distance of that term from the mean. The sign of the distance indicates whether the term is above or below the mean. The term 15, for instance, is five units above the mean, so we may give it a "distance score" of $+5$. The term 7 is three units below the mean, so its "distance score" becomes -3. In the same way, we can convey the location of each of the terms in any distribution.

Suppose that the terms in the distribution given are the test marks obtained by the six students in a class. We now write the students' names, their marks, and their "distance scores" (Table 4.1).

Each student's distance score indicates the position of his mark in the distribution. However, we cannot use distance scores to compare students in two different classes. One reason is that, in another class, the marks may run differently and a few points above or below the mean may have a very different meaning. For instance, consider the marks in Class B (Table 4.2).

Here the marks are much more scattered than in Class A. The distance scores of the various students tend to be larger. For instance, Marty has

TABLE 4.1 DISTANCE SCORES FROM MEAN

Class A

STUDENT	MARK	MEAN	DISTANCE SCORE
Robert	4	10	−6
Arthur	6	10	−4
Bert	7	10	−3
Edward	10	10	0
Richard	15	10	5
Tom	18	10	8

(Mean = 10.   Standard Deviation = 5)

TABLE 4.2 DISTANCE SCORES FROM MEAN

Class B

STUDENT	MARK	MEAN	DISTANCE SCORE
Frank	20	50	−30
James	30	50	−20
Tony	35	50	−15
Bill	50	50	0
Marty	75	50	25
David	90	50	40

(Mean = 50.   Standard Deviation = 25)

obtained a distance score of +25, which is considerably larger than any of the distance scores obtained by students in Class A. Yet Marty was not even the best student in his class. Furthermore, the distance score of +8, which is very outstanding in Class A, would be run-of-the-mill in Class B.

In order to compare a student in one class with a student in the other we cannot simply compare their distance scores. We must do something to make up for the fact that distance score points may be "cheap" in one distribution and "expensive" in another. Suppose we wish to compare Marty's performance (in Class B) with Tom's (in Class A). Marty's distance score is +25 and Tom's is only +8, but we cannot conclude that Marty did better since the two students are in different classes. If we ask who did better, we are in effect asking: Was Marty's distance score more outstanding

43

in Class B than Tom's distance score was in Class A? Therefore, to proceed we must find a way to determine how outstanding a distance score is in its own distribution.

Note that when the distance scores in a distribution are large as a group, the standard deviation of the distribution is large. When the distance scores are small, the standard deviation is small. The standard deviation, after all, is a measure of how far from the mean the terms are (or how large the distance scores are as a group). It follows that a student whose mark has a distance score of, say, +10 may not be outstanding if the standard deviation of marks in his class is large, for this means that some other distance scores in the distribution are also large. But a student whose mark has a distance score of +10 might be quite outstanding. If the standard deviation of marks in his class is small, this indicates that some other distance scores in the class are small. Since the standard deviation tells us about the size of the other distance scores, to appraise a distance score we should consider the size of the standard deviation of the distribution from which it came. If we were appraising the height of a six-foot aborigine we would think of him one way if his compatriots were all giants and another way if they were all dwarfs.

The extent to which a distance score is large compared to the standard deviation of the distribution indicates the extent to which that distance score is outstanding. For instance, if a distance score is twice as large as the standard deviation, then the distance score is outstanding; if the distance score is less than the standard deviation, then it is commonplace. Remember that the sign of the distance score simply tells whether the original term was above or below the mean.

We may now ask how outstanding Tom was in his class. What we actually ask is: How does Tom's distance score compare with the standard deviation of marks in his class? Tom's distance score is +8 and the standard deviation is 5. So we answer that Tom's distance score is 8/5 as large as the standard deviation of his class.

Definition 4.2 *The ratio—distance score divided by standard deviation—is called a z score.*

We say that Tom's z score is $8/5 = 1.6$. To find Marty's z score, we divide his distance score of +25 by the standard deviation of marks in his class (Class B). This standard deviation is 25, so Marty's z score is 1.0. Since Tom's z score is higher, we may now say that Tom was more outstanding in his class than Marty was in his.

To repeat, the z score of a term tells how many standard deviations that term is above or below the mean of its distribution. For instance, if a term has a z score of 1.2, the term is 1.2 standard deviations above its mean. A term with a z score of $-.3$ is 3/10 of a standard deviation below the mean of its distribution. One important use of z scores is to compare terms in different distributions. Unlike distance scores, z scores are not affected by the "cheapness" or "expensiveness" of points. The reason is that a z score

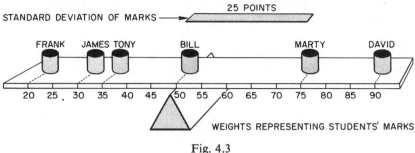

Fig. 4.3

locates a term while taking into consideration the variability of the entire distribution.

We pointed out that the standard deviation is like a ruler that reflects the scatter of the terms in the distribution (Fig. 4.3).

The z score of a term tells how many ruler lengths the particular term is from the mean. For instance, the z score of James' mark is − .80 (Fig. 4.4). The z score of Marty's mark is 1.0.

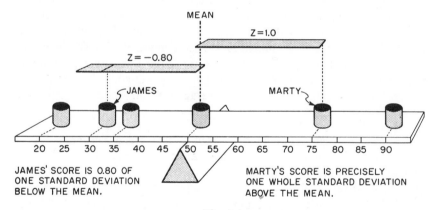

JAMES' SCORE IS 0.80 OF ONE STANDARD DEVIATION BELOW THE MEAN.

MARTY'S SCORE IS PRECISELY ONE WHOLE STANDARD DEVIATION ABOVE THE MEAN.

Fig. 4.4

4.3 z Score Averages. z scores are sometimes useful for finding averages when differing units of measurement are in use. Suppose a teacher gives two tests, A and B, to a class. The highest possible mark on Test A is 100 and the highest possible mark on Test B is 10. One student, Tom, gets a 76 on Test A and an 8 on Test B. Another student, Bob, gets an 83 on Test A but completely fails Test B with a mark of 1. The teacher gives Tom the higher grade, reasoning that he has done adequately on both tests whereas Bob has passed only one of them.

Our problem begins the next day when Bob returns with his grade card, complaining bitterly: "My two test marks, 83 and 1, give me an average of 42.

45

Tom's two marks were 76 and 8, so his average is also 42. But Tom got the higher mark. It isn't fair."

Bob's complaint seems somewhat reasonable. His average is as high as Tom's. Yet Tom did adequately on both tests whereas Bob completely failed the second one. Our feeling that Tom deserved a better mark is not substantiated by a simple comparison of their means. As a matter of fact, it was inappropriate to average the two tests in the first place because score points on one test were worth more than score points on the other. To compute the mean in the usual way is possible only when the score points being averaged have approximately the same worth. Bob surpassed Tom by seven points on Test A but Tom came out seven points higher on Test B, where each test point counted considerably more. In a sense, Bob picked up his seven-point advantage when points were *cheap*, whereas Tom's seven-point superiority indicated a markedly superior performance. By comparing means, Bob gave himself an unfair advantage, since he was, in effect, counting each point he had earned on Test A as heavily as each point he had lost on Test B. Bob was like a man with $2.08 claiming to be as rich as a man with $8.02.

By converting the marks of each student into z scores, it is possible to obtain equal units of measurement despite the fact that the original units of measurement may have been different. The reason is that a given z score has the same meaning in any distribution. Thus, instead of comparing direct means, Bob's mean z score should have been compared with Tom's mean z score. In general, when different terms are based upon different units of measurement, the ordinary average loses its meaning and it is advisable to convert the terms into z scores before averaging them.

4.4 No Inherent Relationship between z Scores and Percentile Ranks. z scores and percentile ranks do not have an inherent equivalence; thus it is impossible to determine either one from the other unless other information (to be discussed later) is given. In theory, a term with a particular percentile rank may have nearly any z score. Furthermore, two terms in different distributions with identical z scores may have markedly different percentile ranks. Let us also consider the distribution of the five terms 2, 3, 6, 8, 11, with its mean of 6 (Fig. 4.5).

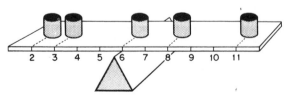

46   Fig. 4.5

Two terms or weights (2 and 3) are to the left of the mean, so they have negative z scores. Two others (8 and 11) are to the right of the mean, so they have positive z scores. One term (6) is equal to the mean, so its z score is zero.

The five terms are listed in Table 4.3 with their z scores and percentile ranks.

The term 8 has a percentile rank of 70. The z score of this term is .61. The fact that this z score is positive tells us that the term is above the mean. However, it is possible for a term to have a percentile rank of 70 and still be below the mean (*i.e.*, to have a negative z score). For example, consider the distribution of the five terms 0, 1, 2, 3, 14, with its mean of 4 (Fig. 4.6).

Table 4.4 shows the terms, their z scores, and their percentile ranks.

TABLE 4.3   COMPARISON OF z SCORES AND
PERCENTILE RANKS

TERMS	z SCORES	PERCENTILE RANKS
2	−1.21	10
3	−0.91	30
6	0.00	50
8	.61	70
11	1.52	90

TABLE 4.4   COMPARISON OF z SCORES AND
PERCENTILE RANKS

TERMS	z SCORES	PERCENTILE RANKS
0	−.78	10
1	−.59	30
2	−.39	50
3	−.20	70
14	1.96	90

The term 3 has a percentile rank of 70 because it is higher than 70 per cent of the terms of this distribution. But it has a negative z score because it is below the mean. In even more extremely "unbalanced" distributions, a term may have a percentile rank as high as 90 and still be below the mean.

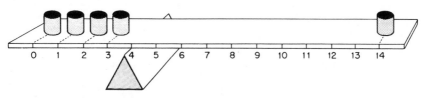

Fig. 4.6

To summarize: *There are no hard and fast rules for a relationship between z scores and percentile rank. Each locates a term in a different way. The percentile rank of a term tells what per cent of terms in the distribution are below it, but says nothing about its relationship to the mean. The z score of a term indicates how it stands in relationship to the mean of the distribution but does not indicate what per cent of terms are above or below it.*

Admittedly, in nearly every distribution that the reader is likely to meet, a term with a percentile rank as high as 70 or 80 does have a positive z score. Also, when the z score of a term is as high as 1.0, that term is generally larger than at least half of the remaining terms. But the exact relationship between percentile ranks and z scores depends upon the distribution.

4.5 The Mean and Standard Deviation of a Set of z Scores. Recall that the z score of a term is found in two steps. The first is to subtract the mean from the term to find its distance score. The second is to divide the distance score by the standard deviation of the distribution.

To demonstrate several properties of z scores, let us find the z scores of all the terms in a distribution. Suppose we are given the distribution 4, 7, 8, 10, 11, 20, which has a mean of 10 and a standard deviation of 5. In Table 4.5

TABLE 4.5   FINDING z SCORES

(1) ORIGINAL TERMS	(2) DISTANCE SCORES	(3) z SCORES
4	− 6	− 1.2
7	− 3	− .6
8	− 2	− .4
10	0	0.0
11	1	.2
20	10	2.0

the original terms have been listed in Column 1. The mean, 10, has been subtracted from each of them, leaving the distance scores listed in column 2. These distance scores have each been divided by the standard deviation, 5, and the quotients or z scores are listed in Column 3.

The distribution of distance scores (in Column 2) has been obtained by subtracting the mean from each of the terms in Column 1. In other words, the constant, 10, was taken from each term in Column 1 to yield the set of numbers in Column 2. We have seen that subtracting a constant from each term in a distribution reduces the mean by that constant (Theorem 2.3). Therefore, the mean of Column 2 equals the mean of Column 1 minus 10. Thus the mean of Column 2 is $10 - 10 = 0$. Since subtracting a constant does not change the standard deviation of a distribution (Theorem 3.1), the standard deviation of Column 2 is the same as that of Column 1. Thus the standard deviation of Column 2 equals 5.

Now that we know the mean and the standard deviation of Column 2, we may temporarily forget Column 1.

Now we are after the mean and the standard deviation of Column 3, the column of z scores. Remember that we divide each term in Column 2 by 5 to get the terms in Column 3. We have seen that the process of dividing the terms in a distribution by a constant has the effect of dividing both their mean and their standard deviation by the same constant (Theorems 2.4 and 3.2). Thus the mean of the terms in Column 3 equals the mean of the terms in Column 2 divided by 5, and the standard deviation of the terms in Column 3 equals the standard deviation of the terms in Column 2 divided by 5. The mean of the terms in Column 3 is zero and the standard deviation of the terms in Column 3 is one. By the same reasoning, it can be shown that in general:

Theorem 4.1   *The mean of a distribution of z scores is zero and the standard deviation is one.*

Note that the steps we took did not depend upon any coincidental factors in the distribution we began with. We might have begun with any set of terms listed in Column 1. Since the set of distance scores was obtained by subtracting the mean from each of them, these distance scores inevitably had a mean of zero and a standard deviation equal to that of the original terms. The process of dividing these distance scores by the standard deviation produced a mean that still was equal to zero, since zero divided by any nonzero number is zero. Thus the mean of the z scores had to be zero. Since we divided the terms in Column 2 by their own standard deviation, the effect was to divide the standard deviation of the terms in Column 2 by itself. This inevitably yielded a new standard deviation of one. Thus we might have started with any set of terms, and still we would have been able to show that their z scores have a mean of zero and a standard deviation of one.

4.6   Standard Scores or T Scores.   Since z scores contain decimal points and since they may be either positive or negative, it is easy to make errors when doing computations with them. For some purposes it is convenient to change a distribution of z scores into a new distribution consisting entirely of positive whole numbers. The first step toward doing this is to multiply each z score by 10 and to round off each resulting term to the nearest whole number. By this process, a z score of 1.23 becomes converted to 12, and a z score of $-.67$ becomes converted to -7. The terms thus obtained are free from decimal points, although some of them are negative.

Now by adding 50 to each of the terms, still another distribution may be created in which all the terms are positive whole numbers. These terms are called *standard scores* or *T scores*. (In theory it is possible for a standard score or T score to be negative, but a negative standard score would corre-   49 spond to a z score of less than -5.00. Later on we shall see that such z scores are virtually nonexistent.)

A standard score or T score is conventionally obtained from a z score by multiplying the z score by 10, rounding off, and then adding 50. For instance, a z score of 1.23 becomes a standard score of 62, and a z score of $-.67$ becomes a standard score of 43. Standard scores or T scores may be used in many of the same ways as z scores. For example, the performances of students in different classes may be compared by comparing their standard scores. The mean of a distribution of standard scores or T scores is 50 and its standard deviation is 10.

4.7 Applications. Percentiles and percentile ranks are useful measures in the fields of education and psychology and in other fields in which scores or marks are obtained. For instance, college entrance committees usually consider the percentile ranks of applicants' marks rather than the numerical mark values themselves. The reason is that a good average in one preparatory school is apt to be a poor one in the next. The percentile rank of a student's average indicates how he did with respect to his classmates, whereas the numerical value of his average reflects the stringency of the marking system in his school as well as the level of his own performance.

Economists sometimes speak of the percentile rank of an individual's income in the community or in the country. They might speak also of the percentile rank of production by a city in its particular state.

z scores play an important part in a multitude of procedures, as we shall see. However, they seldom serve as descriptive measures or ends in themselves. When data are to be presented, z scores are often converted into standard scores or T scores. Note that a single standard score conveys considerable information. Thus, if a student is told that his standard score on a particular test is 60, he may infer that his mark is one standard deviation above the class average. As we shall see, when certain other information is present, one can make rather precise inferences from knowledge of a standard score.

4.8 Notation. The letter P with a subscript is used to denote a particular percentile. Thus P_{10} stands for the tenth percentile. P_{90} is the value of the ninetieth percentile, and so on.

The z score of any particular term in the distribution is obtained by subtracting the mean from the term and dividing the obtained difference by the standard deviation of the distribution. For instance, the eighth term in the distribution, X_8, has a corresponding z score, z_8, which is found by the following formula:

50   (4.1)
$$z_8 = \frac{X_8 - \mu}{\sigma}$$

More generally, the relationship between the kth term, X_k, and its corresponding z score, z_k, is:

(4.2)
$$z_k = \frac{X_k - \mu}{\sigma}$$

The standard score of the kth term, which we shall symbolize T_k, is equal to 50 plus 10 times the z score of the kth term. Thus:

(4.3)
$$T_k = 50 + 10z_k$$

When we substitute the value of z_k given in (4.2) in Formula (4.3), we get a longer expression for the standard score of a term in a distribution.

(4.4)
$$T_k = 50 + 10\left(\frac{X_k - \mu}{\sigma}\right)$$

PROBLEM SET A

4.1 A class of 25 students receives spelling scores of 102, 85, 109, 83, 112, 106, 98, 115, 91, 117, 88, 95, 116, 105, 130, 80, 115, 84, 95, 121, 85, 89, 94, 111, and 105. Compute the percentile ranks of the following scores: (a) 80, (b) 95, (c) 109.

4.2 For the data in Problem 4.1, compute the percentile ranks of (a) 88, (b) 102, (c) 112.

4.3 For the data in Problem 4.1, compute the T scores for (a) 80, (b) 95, (c) 109.

4.4 For the data in Problem 4.1, compute the T scores for (a) 88, (b) 102, (c) 112.

4.5 Trading stamps are collected by 17 third-grade children to purchase a new projection screen. They collect 28, 32, 34, 36, 36, 39, 40, 41, 41, 42, 43, 44, 46, 46, 49, 50, and 50. Compute the percentile ranks of the following numbers of stamps: (a) 36, (b) 44, (c) 49.

4.6 For the data in Problem 4.5, compute the percentile ranks of (a) 41, (b) 46, (c) 50.

4.7 For the data in Problem 4.5, compute the T scores for (a) 36, (b) 44, (c) 49.

4.8 For the data in Problem 4.5, compute the T scores for (a) 41, (b) 46, (c) 50.

4.9 The numbers of typing errors made by a group of students taking a typing test are 3, 11, 5, 12, 9, 8, 16, 13, 12, 11, 6, 19, 16, 11, 15, 12, and 0. Compute the percentile ranks of the following numbers of errors: (a) 5, (b) 8, (c) 11, (d) 15.

4.10 For the data in Problem 4.9, compute the percentile ranks of (a) 6, (b) 9, (c) 12, (d) 16.

4.11 For the data in Problem 4.9, compute the T scores for (a) 5, (b) 8, (c) 11, (d) 15.

4.12 For the data in Problem 4.9, compute the T scores for (a) 6, (b) 9, (c) 12, (d) 16.

PROBLEM SET B

4.13 The weights (in pounds) of 10 people are 156, 162, 170, 177, 180, 181, 183, 196, 205, and 209. Compute the percentile ranks of the following weights: (a) 162, (b) 183, (c) 205.

4.14 For the data in Problem 4.13, compute the percentile ranks of (a) 177, (b) 196, (c) 209.

4.15 For the data in Problem 4.13, compute the T scores for (a) 162, (b) 183, (c) 205.

4.16 For the data in Problem 4.13, compute the T scores for (a) 177, (b) 196, (c) 209.

4.17 A group of fifth-grade students were given a reading achievement test and received the following scores: 100, 83, 88, 81, 83, 96, 105, 108, 78, 102, 97, 113, 126, 94, 85, 119, 67, 91, 88, 99, 88, 72, 77, 88, 114. Compute the percentile ranks of the following scores: (a) 72, (b) 91, (c) 114.

4.18 For the data in Problem 4.17, compute the percentile ranks of (a) 83, (b) 105, (c) 113.

4.19 For the data in Problem 4.17, compute the T scores for (a) 72, (b) 91, (c) 114.

4.20 For the data in Problem 4.17, compute the T scores for (a) 83, (b) 105, (c) 113.

4.21 A researcher determined the speed of 17 rats running a maze to be (in seconds): 20, 17, 16, 15, 19, 19, 18, 12, 13, 13, 17, 16, 14, 14, 16, 15, 18. Compute the percentile ranks of the following times: (a) 12, (b) 14, (c) 16, (d) 18.

4.22 For the data in Problem 4.21, compute the percentile ranks of (a) 13, (b) 15, (c) 17, (d) 19.

4.23 For the data in Problem 4.21, compute the T scores for (a) 12, (b) 14, (c) 16, (d) 18.

4.24 For the data in Problem 4.21, compute the T scores for (a) 13, (b) 15, (c) 17, (d) 19.

grouping data and drawing graphs

5.1 Discrete and Continuous Variables. We recall that the terms in a distribution may be regarded as values of a variable. For instance, when we consider a distribution of people's ages, we may regard the terms as values of the variable, age. Similarly, a distribution of students' marks is a set of values of the variable, mark. A variable is said to *assume* different values; for example, in the case of age, the variable may assume the value 15 for one person and 40 for his father. We wish to define several kinds of variables, and to do so we introduce the concept of an *interval* between two numbers. Our numbers may be illustrated as extending along a line of infinite length, as in Fig. 5.1.

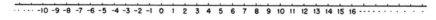

Fig. 5.1

The interval between any two numbers consists of all the numbers that lie between them. An interval is often illustrated as a line segment. For instance, Fig. 5.2 shows the interval 10–20.

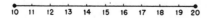

Fig. 5.2

53

Numbers on this line segment are said to be *inside* the interval and all other numbers are said to be *outside* the interval. For instance, the numbers 11 and 16.59 are inside the interval and the number 63.5 is outside of it. The lower endpoint of the interval is 10 and the upper endpoint is 20. We shall include the endpoints of an interval as part of the interval. The midpoint of an interval is the value that is halfway between the endpoints; the midpoint of the interval mentioned is 15.

Definition 5.1 *A variable is called discrete in an interval if in theory it can assume only values that are members of some defined set inside the interval.*

For example, the number of "yes" votes by the members of a ten-man committee is a discrete variable that must assume one of 11 possible values at each observation. These values are the whole numbers 0, 1, 2, and so on up to 10. The variable, number of people married each hour at City Hall, is also a discrete variable. This variable may assume only the even-numbered values, and within any specified interval the number of values that it may assume is limited.

The values of a discrete variable are often categories. For instance, marital status of women is a discrete variable that may assume one of four values: single, married, divorced, and widowed. The variable, army status, is also a discrete variable, for it must assume one of a limited number of categories at each observation.

Definition 5.2 *A variable is continuous in an interval if in theory it can assume any value within that interval.*

In theory a continuous variable may assume any one of an infinitude of values in an interval. The variable, age, is continuous in the interval 10–20 because it may assume any value between 10 and 20. That is, any number between 10 and 20 (like 14.2387 years) is apt to be a person's age. Other examples of continuous variables, or variables that may exist in any quantity, are height and weight. Note that in practice we are able to record only a limited number of values of a continuous variable within any interval. Our measuring instruments do not give us absolute precision, and, even if they did, we would always prefer to round off our values at some point. For instance, the medical researcher might measure age to the nearest year, thereby restricting the number of admissible values. Similarly, the psychologist measures intelligence to the nearest point, even though he views it as a continuous variable. A subject must get one of 158 possible scores when he takes the Wechsler Adult Intelligence Scale. Special abilities are also believed to be continuous, but the number of possible test scores or observations of these abilities is necessarily limited.

A useful distinction is that when a continuous variable is to be measured, successive refinements of the measuring instrument yield increasingly precise values. When a discrete variable is to be measured, refinements of the measuring instrument fail to increase the precision of the observations once a certain measure of accuracy is achieved.

5.2 Tabulating and Graphing the Values of a Discrete Variable. The distribution of values of a discrete variable may be conveniently summarized in what is called a *frequency table*. To begin with, we shall define the word *frequency*.

Definition 5.3 *The frequency of any value which a variable may assume in a distribution is the number of times that value appears in the distribution.*

As an example, suppose that among female employees in a factory 10 are single, 20 are married, 15 are divorced, and 5 are widowed. The frequency of single females is said to be 10, the frequency of married females is 20, and so on.

Table 5.1 is a frequency table of the distribution of marital status in the factory. The possible values of the variable, marital status, are listed in the first column and the frequencies of these values are listed in the second column.

TABLE 5.1   DISTRIBUTION OF MARITAL STATUS

MARITAL STATUS	FREQUENCY
Single	10
Married	20
Divorced	15
Widowed	5

The terms in the distribution are illustrated as weights on a plank in Fig. 5.3.

Note that the four sections on the plank corresponding to the four possible values are separated so that the weights in different piles do not touch each other. The frequency of each value is indicated by the number of weights piled up at that value—that is, the 10 weights piled up over the value "single" indicate that there are 10 single women in the distribution, and so on.

We may simplify the representation by drawing a front-view diagram of the weights. The plank is represented by a horizontal line and the weights are represented by rectangles. A vertical axis serves to indicate the number of weights in each pile (Fig. 5.4).

Now we are ready to simplify the representation into its most usual form. Fig. 5.5 is called a *bar graph*.

The bar graph is the most convenient and popular kind of graph used to depict the distribution of values of a discrete variable. The first step toward creating a bar graph is always to draw up the frequency table of the distribution. Then the bar graph is drawn from this table. Note that there are spaces between the different bars.

5.3 Rounding Off. In practice, the values of a continuous variable must always be rounded off. Measurements are never made with perfect

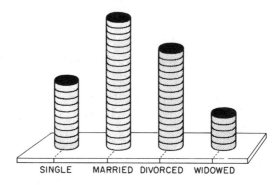

Fig. 5.3

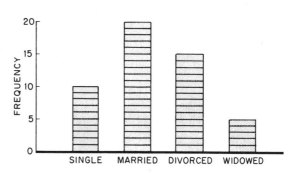

Fig. 5.4

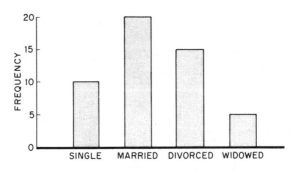

Fig. 5.5

accuracy and, even if they were, the task of working with exact values would be prohibitive. The process of rounding off facilitates many computational procedures and it is often serviceable to round off values of a discrete variable as well as a continuous one.

When we round off values of a variable, we are in effect "forcing" the variable to assume one of a limited (or finite) number of values. For instance, suppose we round off heights to the nearest inch in the interval 4 feet $11\frac{1}{2}$ inches up to but not including 6 feet 1/2 inch. We are forcing the variable, height, to assume one of 13 rounded-off values. These values are 5 feet, 5 feet 1 inch, 5 feet 2 inches, and so on. We picture these rounded-off values as spots along a plank (Fig. 5.6).

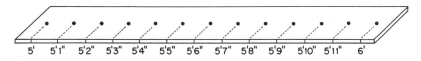

Fig. 5.6

Rounding off a term means giving it the value of the spot or "rounded value" on the plank to which it is closest. Each rounded-off value on the plank is the midpoint of an interval that includes observations extending 1/2 inch on each side of it. Altogether, there are 13 intervals, each an inch wide, stretched across the plank (Fig. 5.7).

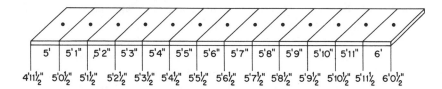

Fig. 5.7

When we round off a term, we are in essence locating it in one of the 13 intervals and then assigning it the value of the midpoint of that interval. Thus the term 5 feet $5\frac{3}{4}$ inches falls in the indicated interval (Fig. 5.8).

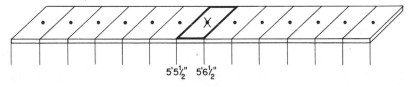

Fig. 5.8

57

We assign it the value of the midpoint of that interval, which is 5 feet 6 inches. In general, the process of rounding off the terms in a distribution involves grouping the terms into intervals and then assigning the terms in each particular interval the value of the midpoint of that interval. Techniques of working with rounded-off values in order to simplify computations are called *grouping methods.*

Note that the width of the intervals used indicates the degree to which the terms are being rounded off. In our example, the intervals are each one inch wide, reflecting the fact that heights are being rounded off to the nearest inch.

5.4 Grouping Data and Drawing the Group Frequency Table. We shall now consider the formal procedure of putting a mass of data into intervals for purposes of drawing graphs and simplifying computing procedures. In effect, we are considering a formal procedure for rounding off terms in a distribution in order to obtain intervals with midpoints that are convenient numbers to work with.

A four-step procedure is generally used to group a mass of observations and to tabulate them.

1. The first step is to compute the *range*, which is the difference between the largest and smallest terms in the distribution. As an illustration, suppose the data are the marks on a one-hundred-word spelling test given to all fourth grade children in a community. The highest mark turns out to be 100 and the lowest mark turns out to be 3. The range of marks is $100 - 3 = 97$.

2. The next step is to decide upon the number of intervals and the size of the intervals to be used. It is a great convenience to make the width of the intervals an odd number so that the midpoint of the intervals will be a single whole number. Thus in this case we might decide to make the intervals 9 units wide. We shall use 11 intervals. The expanse of 11 intervals, laid end to end, is 99 units. It is essential that the expanse of intervals be at least one more than the range. Note that in the spelling test above the possible scores, which are from 3 to 100, include 98 whole numbers, or one more than the value of the range. So far we have specified a convenient width for our intervals and we have decided to use a manageable number of them.

3. The next step is to specify the endpoints of the highest interval. In our illustration, we made the highest interval 92–100. Note that when we state the endpoints of an interval we intend to include these endpoints in the interval. Thus the interval 92–100 is 9 units wide. The next highest interval becomes 83–91. We keep on constructing intervals until we find the lowest interval, which is 2–10.

The 11 intervals and their midpoints are shown in Table 5.2. Remember that each interval is 9 points wide since the lower endpoint is included. We have constructed 11 consecutive equal-sized intervals and have indicated their midpoints.

TABLE 5.2   GROUP DISTRIBUTION PATTERN

INTERVALS	MIDPOINTS
92–100	96
83–91	87
74–82	78
65–73	69
56–64	60
47–55	51
38–46	42
29–37	33
20–28	24
11–19	15
2–10	6

4. The final step is to tabulate the frequencies of the different intervals. In other words, we tabulate the number of terms that fall into each interval and record these numbers to complete the group frequency table. For instance, the group frequency table of the distribution of children's scores on the spelling test might look like Table 5.3.

TABLE 5.3   GROUP FREQUENCY TABLE

INTERVALS	MIDPOINTS	FREQUENCIES
92–100	96	60
83–91	87	140
74–82	78	160
65–73	69	120
56–64	60	140
47–55	51	80
38–46	42	119
29–37	33	81
20–28	24	50
11–19	15	32
2–10	6	18
		1,000

Table 5.3 tells us that 60 children got spelling marks between 92 and 100 (including the mark of 92). It tells us that 140 children got marks between 83 and 91 (including the mark of 83) and so on.

We can now proceed as if the distribution were made up of 60 terms with the value of 96, 140 terms with the value of 87, and so on. That is, if we wish we can assign to all the terms in each interval the value of the midpoint of that interval. Doing this is, in essence, rounding off the terms to the nearest 9-point units just as we rounded off heights to the nearest inch in the previous section. We shall see how to compute various measures by group methods in Chapter 6.

59

5.5   The Histogram and the Frequency Polygon.   Next we shall consider the method of drawing a graph of a continuous variable using the information in a group frequency table.   To be specific, look at the data in Table 5.3. Remember that we conceive of spelling ability as a continuous variable even though our observations, namely the spelling scores, are restricted to whole-number values.

So far as *spelling scores* are concerned, the highest interval in Table 5.3 includes the whole number values from 92 to 100.   The next highest interval includes the whole numbers from 83 to 91.   However, we view *spelling ability* as a continuous variable that may assume values between 91 and 92.   In other words, the fact that we do not obtain values between 91 and 92 is caused by the limitation of our measuring instrument, the spelling test.   In theory, a more sensitive test might reveal that the score of 91 is really 90.71 or 91.3.   It might show that the score of 92 is really 91.6 or 92.4.

In reality, the scores above 91.5 are recorded as being 92 or more.   Thus the lower boundary of the top interval is 91.5.   Similarly we have designated 91.5 as the upper boundary of the interval 83–91.   The lower boundary of the interval 83–91 is 82.5.   Thus the spelling performances, which we classified in the interval 83–91 on the basis of our spelling test, would upon closer inspection be found to range between 82.5 and 91.5.

By an analogous process, we can find the boundaries of all the intervals listed in Table 5.3.   These boundaries are indicated in Table 5.4.

We are now ready to draw a graph of the distribution of children's spelling scores using the data in Table 5.4.   Along the horizontal axis of Fig. 5.9 we note the boundaries of the successive intervals and also their midpoints.   We use the vertical axis to indicate frequencies as we did when we drew the bar graph.   Once again we draw bars over the intervals to indicate the frequencies or number of scores in each interval.   The bar over each interval extends from its lower boundary to its upper boundary so that adjacent bars just touch each other.   The heights of the various bars in Fig. 5.9 indicate the frequencies in the various intervals.

TABLE 5.4   GROUP FREQUENCY TABLE

INTERVALS	BOUNDARIES	MIDPOINTS	FREQUENCIES
92–100	91.5–100.5	96	60
83–91	82.5–91.5	87	140
74–82	73.5–82.5	78	160
65–73	64.5–73.5	69	120
56–64	55.5–64.5	60	140
47–55	46.5–55.5	51	80
38–46	37.5–46.5	42	119
29–37	28.5–37.5	33	81
20–28	19.5–28.5	24	50
11–19	10.5–19.5	15	32
2–10	1.5–10.5	6	18
			1,000

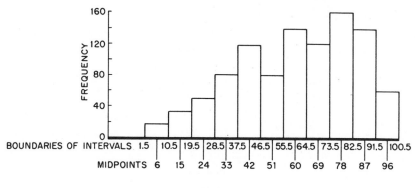

Fig. 5.9

We now remove the vertical lines separating successive bars in order to indicate that we are graphing a continuous variable.   The graph in Fig. 5.10 is called a *histogram*.

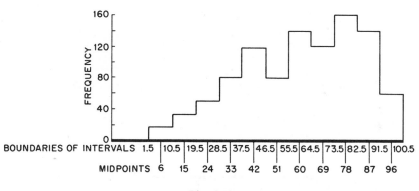

Fig. 5.10

The histogram, which is a graph of a continuous variable, resembles the bar graph in various ways.   Values are recorded along the horizontal axis in each case and frequencies are indicated by the heights of the various columns. However, the bars are quite separate in a bar graph, indicating that the underlying variable is discrete.   Successive bars are "run together" in a histogram to indicate that the underlying variable is continuous.

A second way to graph values of a continuous variable is by means of a *frequency polygon*.   The frequencies of the various intervals are indicated by points drawn over the midpoints of the intervals and these points are connected by straight lines as shown in Fig. 5.11.

61

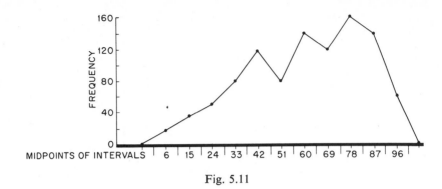

Fig. 5.11

5.6 The Curve as an Approximate Graph. We shall now compare several histograms that depict an identical distribution. This time we shall consider for our illustration the distribution of heights of all American adult males between 5′ 3½″ and 5′ 11½″. The histogram of these heights rounded off to the nearest inch is shown in Fig. 5.12.

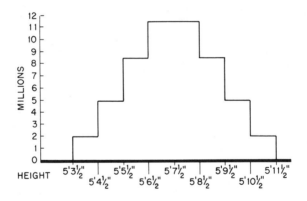

Fig. 5.12

This histogram tells us that two million men are between 5′ 3½″ and 5′ 4½″. It tells us that five million men are between 5′ 4½″ and 5′ 5½″, etc. Twenty per cent of the area is to the left of 5′ 5½″ and 50 per cent is to the left of 5′ 7½″. We may conclude from the graph that one-fifth of the men are shorter than 5′ 5½″ and that one-half of them are shorter than 5′ 7½″.

Now let us look at another graph of the same distribution. The intervals in the histogram in Fig. 5.13 are half inches.

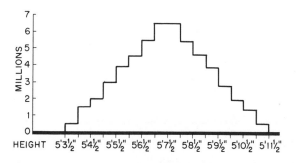

Fig. 5.13

The effect of narrowing down the intervals has been to increase the number of bars and to cause successive bars to change more gradually. As we break down the intervals further, the changes in height from each bar to the next become more and more gradual. The histogram begins to look like a curve. In Fig. 5.14 we see how it looks when the intervals are 1/10 of an inch long.

If we are free to choose intervals as small as we like, we can reduce the difference in area between the histogram and the curve that it is approaching to as small an amount as we like. However, there must be a sufficiently large number of terms in the distribution for us to do this. The reason is that we are creating a multitude of intervals as we cut down their size, and it must

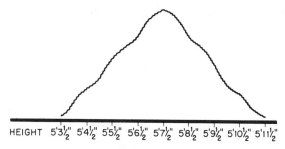

Fig. 5.14

be possible to pour enough terms into each interval for the approximation to hold. Otherwise, many intervals will be empty or nearly empty, leaving breaks in the graph.

There are times when decreasing the intervals does not lead to an approximation of a curve as described. But such times are exceptional. In any event, we shall later discuss conditions under which the approximation is guaranteed to occur.

The purpose of this section was simply to give the reader an intuitive explanation of how a curve can be used to represent a distribution of values of a continuous variable. Thus a highly accurate graph of the distribution of heights that we have been discussing would look like Fig. 5.15.

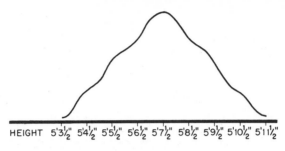

HEIGHT 5'3½" 5'4½" 5'5½" 5'6½" 5'7½" 5'8½" 5'9½" 5'10½" 5'11½"

Fig. 5.15

5.7 The Meaning of Area. The interpretation of the area of a graph is extremely important. Later on we shall see why. The best way to illustrate this meaning is to imagine that we have just come across a bar graph that was published without a vertical axis to indicate the number of terms in the various piles (Fig. 5.16).

A careful examination of this graph reveals that the total area of the columns to the left of the number 8 is exactly equal to the total area of the columns to the right of 8. (The area of the light columns equals the area of the shaded columns.) We now ask what does this tell us about the distribution being described?

Without a vertical axis, we have no idea how many weights are being represented or how large the weights are supposed to be. However, we do

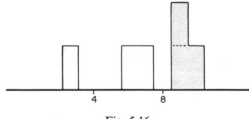

Fig. 5.16

know that half of them were placed on the plank below 8 and half of them above 8. *In other words, half of the terms are below 8 and half of them are above* 8.

It is entirely possible that the original distribution looks like one or the other of the two parts of Fig. 5.17.

In either event our inference is correct. It happens that one-sixth of the area is below 4, so we may conclude that one-sixth of the terms in the distribution are less than 4. In general, we may interpret the proportion of area over an interval as the proportion of the terms in the distribution that fall in that interval. This interpretation does not depend upon our knowing how many terms there are in the whole distribution.

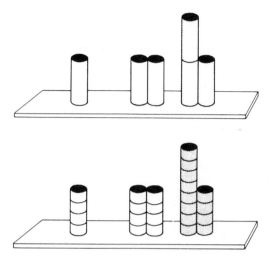

Fig. 5.17

5.8 Applications. Bar graphs and histograms are extensively used to represent virtually all kinds of data. Histograms are used more frequently, but when two distributions are represented on the same axes it is often advantageous to use frequency polygons, since they are less likely to coincide over sections of the graph.

As we shall see, it is a relatively easy matter to compute the various measures already discussed once a mass of data has been "poured" into the intervals of a group frequency table. The experienced researcher can conjure in his mind the shape of a distribution by looking at the frequency table, and often he plunges into computing the various measures without stopping to draw a graph.

PROBLEM SET A

5.1 Determine the midpoints and boundaries of the following intervals: (a) 85–87, (b) 1–5, (c) 101–109, (d) 72–73.

5.2 The weights in pounds of the 40 football players on a professional team are 217, 200, 199, 229, 182, 244, 192, 204, 223, 255, 185, 169, 192, 212, 194, 228, 236, 187, 212, 207, 233, 212, 200, 197, 167, 267, 227, 216, 200, 237, 181, 190, 185, 220, 193, 238, 220, 200, 176, and 218. Construct a group frequency table showing the distribution of these weights, using 10-pound intervals. Also, prepare a column indicating the boundaries of each interval.

5.3 Draw the histogram and frequency polygon of the distribution you prepared in Problem 5.2.

5.4 Construct a group frequency table for the data in Problem 5.2 using only approximately half as many intervals as you used in the original problem. Construct the histogram and frequency polygon and compare these with the results in Problem 5.3.

5.5 Construct a group frequency table for the data in Problem 5.2 using approximately twice as many intervals as you used in the original problem. Construct the histogram and frequency polygon and compare these with the results in Problems 5.3 and 5.4.

5.6 The following are the weights of players on a college football squad: 185, 213, 186, 195, 167, 185, 185, 199, 163, 227, 194, 234, 199, 212, 188, 235, 204, 178, 192, 213, 233, 217, 204, 184, 181, 200, 214, 223, 153, 169, 232, 200, 203, 221, 216, 188, 200, 210, 216, 199, 192, 168, 198, 166, 206, 162, 172, 188. Construct a group frequency table using 10-pound intervals. Also, prepare a column indicating the boundaries of each interval.

5.7 Draw the histogram and frequency polygon of the distribution you prepared in Problem 5.6.

5.8 Construct a group frequency table for the data in Problem 5.6 using only approximately half as many intervals as you used in the original problem. Construct the histogram and frequency polygon and compare these with the results in Problem 5.7.

5.9 Construct a group frequency table for the data in Problem 5.6 using approximately twice as many intervals as you used in the original problem. Construct the histogram and frequency polygon and compare these with the results in Problems 5.7 and 5.8.

5.10 Construct a group frequency table, histogram, and frequency polygon for the following numbers: 143, 171, 154, 139, 166, 192, 219, 162, 147, 116, 165, 121, 157, 147, 84, 188, 151, 173, 140, 177, 162, 141, 179, 167, 135, 140, 155, 176, 140, 152, 207, 123, 137, 151, 88, 173, 127, 161, 163, 164, 130, 182, 175, 158, 145, 117, 151, 109, 183, 169, 172, 146, 152, 154, 158, 107, 124, 163, 147, 177, 101, 113, 180, 142, 134.

PROBLEM SET B

5.11 Determine the midpoints and boundaries of the following intervals: (a) 4–6, (b) 190–199, (c) 1–8, (d) 63–67.

5.12 The IQ scores of 25 students in a bright third-grade class are 119, 125, 102, 112, 111, 124, 133, 103, 115, 147, 114, 146, 113, 105, 117, 111, 115, 136, 103, 115, 116, 103, 102, 135, and 119. Using six-point intervals, construct a group frequency table of these IQ's. Also, prepare a column indicating the boundaries of each interval.

5.13 Draw the histogram and frequency polygon of the distribution you prepared in Problem 5.12.

5.14 Construct a group frequency table for the data in Problem 5.12 using only approximately half as many intervals as you used in the original problem. Construct the histogram and frequency polygon and compare these with the results in Problem 5.13.

5.15 Construct a group frequency table for the data in Problem 5.12 using approximately twice as many intervals as you used in the original problem. Construct the histogram and frequency polygon and compare these with the results in Problems 5.13 and 5.14.

5.16 The social studies achievement-test scores of a group of seventh-grade students are 54, 60, 46, 38, 54, 68, 50, 42, 47, 58, 40, 62, 61, 56, 64, 47, 34, 53, 44, 69, 48, 68, 60, 57, 74, 46, 51, 71, 65, 56, 60, 43, 59, 52, and 58. Construct a group frequency table showing the distribution of these scores. Also, prepare a column indicating the boundaries of each interval.

5.17 Draw the histogram and frequency polygon of the distribution you prepared in Problem 5.16.

5.18 Construct a group frequency table for the data in Problem 5.16 using only approximately half as many intervals as you used in the original problem. Construct the histogram and frequency polygon and compare these with the results in Problem 5.17.

5.19 Construct a group frequency table for the data in Problem 5.16 using approximately twice as many intervals as you used in the original problem. Construct the histogram and frequency polygon and compare these with the results in Problems 5.17 and 5.18.

5.20 Construct a group frequency table, histogram, and frequency polygon for the following numbers: 180, 169, 173, 148, 164, 155, 177, 133, 148, 89, 193, 135, 197, 152, 140, 154, 142, 216, 182, 156, 137, 143, 151, 155, 187, 171, 142, 111, 126, 162, 172, 173, 163, 180, 138, 176, 187, 170, 207, 141, 155, 114, 181, 160, 131, 152, 133, 118, 125, 126, 179, 145, 169, 148, 127, 203, 165, 199, 170, 152, 180, 119, 133, 105, 120.

six

___

computing
various measures
from grouped data

6.1 Introduction. In Chapters 2, 3, and 4, we discovered the meaning of certain measures of average value, variability and location, and we learned how to compute these measures for distributions of given data.   In Chapter 5 we saw the desirability of grouping a large mass of data into intervals to form a grouped frequency distribution and we learned how to represent such a distribution graphically by the use of histograms and frequency polygons.   In the present chapter we shall see how to combine these previous ideas and compute measures for grouped data.   Our previous graphical analysis will prove useful, and we shall also consider an additional graphical procedure.

6.2 The Median and Percentiles from a Histogram. We recall that the median is a number that is larger than or equal to half of the terms in a distribution and smaller than or equal to half of them.   It is also the fiftieth percentile.   We saw that the median need not be the same number as a term in the distribution.   It is extremely important to keep these facts in mind when discussing the median of a distribution of grouped data where, as we have seen, the individual terms of the original distribution have lost their identity.

For reference we reproduce here (Table 6.1) the group frequency table and histogram for the children's spelling scores.   So far we know only that the *median* is larger than half of these 1,000 scores and smaller than half of them.   To find it, we must find a number on the horizontal axis of Fig. 6.1

that is larger than 500 of the scores. Looking at the frequency table and adding the frequencies, we have 18 in the lowest interval, $18 + 32 = 50$ in the

TABLE 6.1 GROUP FREQUENCY TABLE

INTERVALS	BOUNDARIES	MIDPOINTS	FREQUENCIES
92–100	91.5–100.5	96	60
83–91	82.5–91.5	87	140
74–82	73.5–82.5	78	160
65–73	64.5–73.5	69	120
56–64	55.5–64.5	60	140
47–55	46.5–55.5	51	80
38–46	37.5–46.5	42	119
29–37	28.5–37.5	33	81
20–28	19.5–28.5	24	50
11–19	10.5–19.5	15	32
2–10	1.5–10.5	6	18
			1,000

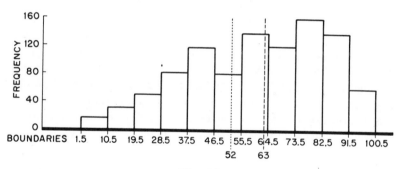

Fig. 6.1

lowest two intervals combined, $50 + 50 = 100$ in the lowest three intervals combined. Continuing in this way we find 181 in the lowest four intervals, 300 in the lowest five, 380 in the lowest six, and 520 in the lowest seven intervals. In other words, there are 380 scores less than the boundary 55.5 and there are 520 scores less than the boundary 64.5. The median must be a number larger than 55.5 and smaller than 64.5, but closer to 64.5.

The area above the interval from 55.5 to 64.5 represents the 140 scores that were grouped into that interval by rounding them off to the value 60. In order to find the median we must find a point on the horizontal axis at which to split the bar by a vertical line so that 120 scores out of the 140 are to the left of that line. This will put 500 scores to the left of the line and our point on the axis will then be at the median. In drawing in such a line we must put 120/140 or 12/14 or 6/7 of the area of the bar to the left of the line 69 and the remaining 1/7 to the right. It should be noted that the fact that the top of the bar is horizontal or parallel to the base has an important implication for us at this point. It means that in considering the histogram and the median we are assuming that the 140 scores (which lost their individual identities when they were all rounded to the value 60) are distributed evenly throughout the interval.* With this consideration in mind a vertical line has been drawn in with dashes on the histogram. It crosses the horizontal axis at about the score value of 63. This means that the median of the frequency distribution is about 63. Note that half of the total area of bars on the histogram is to the left of the dashed line and half is to the right.

The histogram can be used in the same way to obtain percentiles for a grouped frequency distribution. We recall that a percentile is also a term in the distribution or is a score value intermediate to two terms. For example, the thirty-fifth percentile is a score that exceeds 35 per cent of the scores (including half of those at that score). If we wish to find the thirty-fifth percentile for the spelling distribution we must find a number on the horizontal axis such that 35 per cent of the scores are smaller than that number. Or stated differently, now we wish to draw in a vertical line such that 35 per cent of the area is to the left of that line and 65 per cent is to the right of it.

Note first that 35 per cent of 1,000 scores represents 350 scores. Looking back to where we added successive interval frequencies, we found 300 scores in the lowest five intervals combined and 380 in the lowest six intervals combined. This means that the 35th percentile is between 46.5 and 55.5. We must put 50/80 or 5/8 of the area of the bar for that interval to the left of the line we draw and 3/8 to its right. Such a line has been drawn in with dots in Fig. 6.1. It crosses the horizontal axis at about the value 52. This means that the thirty-fifth percentile of the frequency distribution is about 52. Note that 35 per cent of the total area of bars on the histogram is to the left of the dotted line and 65 per cent is to the right. The value of any percentile can be found from the histogram by the same procedure as we have used here. The reader will recognize that the median is just another name for the fiftieth percentile and that its estimation from the histogram followed the same pattern.

6.3 Computation of the Median and Other Percentiles. The method for computation of any percentile, including the median, for a grouped frequency distribution is simply a numerical version of the procedure that we used in the preceding section. We shall go over the two examples again in some detail and refine the method so that reference to the histogram is unnecessary.

First, it will be convenient to introduce what is called a "cumulative less than" column in our frequency table. We shall refer to this simply as the "less than" column. In this column we record the sums we obtained when

___
* It is standard statistical procedure to make this assumption in the computation of the median.

we added the frequencies of successive intervals in our previous work. This has been done in Table 6.2 in the last column. The column is constructed from the bottom up. There are 18 scores less than 10.5, $18+32=50$ scores

TABLE 6.2 "CUMULATIVE LESS THAN" FREQUENCIES FOR SPELLING SCORES

BOUNDARIES	FREQUENCIES	LESS THAN
91.5–100.5	60	1,000
82.5–91.5	140	940
73.5–82.5	160	800
64.5–73.5	120	640
55.5–64.5	140	520
46.5–55.5	80	380
37.5–46.5	119	300
28.5–37.5	81	181
19.5–28.5	50	100
10.5–19.5	32	50
1.5–10.5	18	18

less than 19.5, $50+50=100$ scores less than 28.5, and so on. In other words, the "less than" refers to the *upper* boundary of each interval.

Now in seeking the median we want that point where the "less than" reading is 500. The reason is that 500 represents half of our total of 1,000 cases. Our table shows that this point will be between 55.5 and 64.5. But how far must we go starting from 55.5 and proceeding towards 64.5 before reaching this point? At 55.5 we have 380 scores, so we need 500 minus 380 or 120 scores out of the 140 scores in the interval from 55.5 to 64.5. In the previous section we discussed splitting the bar of the histogram so that 120/140 or 6/7 of its area was to the left of the dashed line that we drew in. But we noticed there the importance of the fact that the tops of the bars in the histogram are parallel to the horizontal axis. We are actually just finding the point on the horizontal scale (Fig. 6.2) that is 6/7 of the distance from 55.5

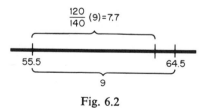

Fig. 6.2

to 64.5. Since the total distance is 9, we want to go to the point that corresponds to the number that is 55.5 plus 6/7 of 9. That is, the median is $55.5+(6/7)(9)=55.5+54/7=55.5+7.7=63.2$. To the nearest whole number this is 63, which we had estimated from the histogram. This means that 500 spelling scores are less than 63.2 and 500 are greater.

Now for the thirty-fifth percentile we must find that point where the "less than" reading is 35 per cent of 1,000, or 350. Our "less than" column in

Table 6.2 shows that it will be in the interval from 46.5 to 55.5; we must then determine the stopping point as we proceed from 46.5 to 55.5 along the horizontal scale (Fig. 6.3). We have 300 scores up to 46.5, and we need 350 minus 300 or 50 scores out of the 80 scores that have been grouped into the interval that we are considering. This means that we need to cover

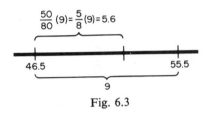

Fig. 6.3

50/80 or 5/8 of the 9 units of distance from 46.5 to 55.5. If we do, we will arrive at the point that represents the thirty-fifth percentile, often designated as P_{35}. Hence,

$$P_{35} = 46.5 + (5/8)(9) = 46.5 + 45/8 = 46.5 + 5.6 = 52.1.$$

To the nearest whole number this is 52, which we had also estimated from the histogram. This means that 35 per cent of the scores are less than 52.1 and 65 per cent are greater.

For those who prefer symbolic statements, we could summarize our method for finding percentiles by the formula:

(6.1) $$P_n = L + (s/f)(i).$$

Here we are finding the nth percentile, P_n. We refer to the "less than" column and find that P_n falls in the interval having L for its lower boundary. We still need s scores in order to arrive at the point on the horizontal scale (Fig. 6.4) that corresponds to the percentile. The distance from the lower

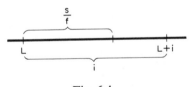

Fig. 6.4

boundary L to the upper boundary $L+i$ is i and there are f scores in the interval from L to $L+i$. For instance, reference to the preceding example gives us $P_{35} = 46.5 + (5/8)(9)$, which we had before.

6.4 Quartiles and Deciles. Two terms frequently used in educational literature are *quartile* and *decile*. The first quartile, or Q_1, is the value corresponding to that point on the scale of scores such that 25 per cent (or one-quarter) of the scores are smaller than it and 75 per cent are greater. In other words it is another name for P_{25}. The second quartile, or Q_2, is another name for the median or P_{50}. Similarly, the third quartile, or Q_3, is another name for P_{75}. A student whose spelling score is among the lowest one-fourth is properly spoken of as being "in the lowest quarter" of the distribution but a student is not *at* the first or lower quartile unless his score falls at that point on the scale of scores such that exactly 25 per cent of the scores are less than his. Similarly, a student is *at* the upper quartile, Q_3, if his score exceeds exactly 75 per cent of the scores; but a pupil is "in the upper quarter" if his score has a percentile rank anywhere in the interval from 75 to 100.

In the same way, deciles refer to tenths of a population. The first decile, or D_1, is the value corresponding to that point on the scale of scores such that ten per cent of the scores are smaller than it. It should be clear that $D_1=P_{10}$, $D_2=P_{20}$, and so on.

6.5 The Cumulative Curve. We shall consider now the "less than" column that we added in Table 6.2 for the computation of percentiles. The cumulative "less than" graph for the spelling scores is shown in Fig. 6.5.

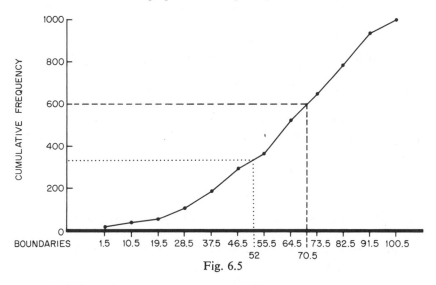

Fig. 6.5

Remembering that the "less than" refers to the upper boundary of each interval, we have plotted a point for each value in the "less than" column of Table 6.2 and we have placed these points directly above the upper boundaries of the corresponding intervals. In addition we added a point at the extreme left of the curve to bring the curve down to the base line and indicate that there are zero scores less than 1.5, the lower boundary of the lowest interval.

The cumulative "less than" curve for a grouped frequency distribution may readily be used to obtain both percentiles and percentile ranks. Suppose

73

we want the sixtieth percentile, P_{60}. Since there are 1,000 scores, we go up the vertical scale to the point corresponding to 60 per cent of 1,000, or 600 scores. The dashed line is drawn parallel to the base until it hits the curve. Then it is dropped perpendicular to the base and seems to strike the scale of scores at about 70.5. Hence, $P_{60} = 70.5$. Had we started with a score of 70 or 71 and desired its percentile rank, we would have drawn the parts of the line in the reverse order and would have estimated the frequency on the vertical scale to be about 600. This would represent 600/1,000 of the scores, and when divided this gives 60 per cent, or a percentile rank of 60.

The dotted line in Fig. 6.5 shows the estimate of P_{35} to be 52 or the percentile rank of 52 to be 35. This is the example that we have done twice previously. As the examples show, the cumulative "less than" curve (often called an *ogive*) can be very useful if several percentiles or percentile ranks must be estimated for the same frequency distribution.

6.6 The Mode for Grouped Data. There is little to be said about the determination of the mode for data that have been placed in a grouped frequency distribution. The reader will recall from Chapter 2 that a distribution of actual scores may have two or more modes—that is, the mode is not always uniquely determined. The same sort of situation prevails for grouped data. First, we have to decide what number to identify as the mode. It is customary to identify the interval having the greatest frequency as the *modal interval* and to call its midpoint the *crude mode*. Thus, in the spelling scores example (Table 6.1) the modal interval is the interval 74–82, since its frequency, 160, is greater than that of any other interval in the table. The crude mode is 78, the midpoint of the modal interval. Had there been another interval with frequency 160, we would have had two modal intervals and two crude modes. The crude mode is not used in further computations for grouped frequency distributions.

6.7 The Mean for Grouped Data. In Chapter 2 we learned how to calculate the mean of a distribution by direct application of its definition. We also discussed some important properties of the mean and saw how we could "cash in" on one of these properties in order to shorten the calculation of the mean when the data involved large numbers. This was the property that states that if a constant is subtracted from each term in a distribution, the mean is decreased by the same constant. We also saw that dividing each term by a constant has the effect of dividing the mean by that same constant. In the present section we shall see how to calculate the mean from a grouped frequency table. Again we shall first do this by direct application of the definition and then we shall see how to shorten the work by making use of the two properties of the mean that were just mentioned.

We have seen that a simple method of rounding off terms is to put them in a grouped frequency table. When this is done, the terms in each particular interval get rounded off to the value of the midpoint of that interval. For instance, according to Table 6.1, the spelling scores of 18 children are in the interval from 2 through 10. The midpoint of this particular interval is 6. When we compute the mean for this grouped frequency distribution we must have one value to use for all of the 18 scores in this interval. The logical choice is the value at the midpoint, so we proceed as if 18 children had spelling scores of exactly 6. Similarly, 32 terms in the distribution are in the interval from 11 through 19, so we proceed as if there are 32 terms that have the value 15, and so on. The reader may well object at this stage that we are not being consistent because earlier in this chapter, when we computed percentiles for grouped data, we made the assumption that the terms in an interval are evenly distributed throughout that interval. This assumption was based on the observation that the tops of the bars in the histogram are horizontal. A little reflection, however, should remove the reader's objection, for if the terms in an interval are, indeed, distributed evenly throughout the interval, then their mean will be at the midpoint of that interval. It is this very midpoint that we are now choosing when faced with the necessity of using one value to represent all of the terms in the interval.

When we round off terms, we are in effect creating a new distribution. We admit only certain possible values, those that are the midpoints of the successive intervals. The frequencies of the intervals become the frequencies of these values. The process of rounding off entails increasing slightly the values of some of the terms and decreasing slightly the values of others. The reason is that the terms that happen to fall in a particular interval may include some that are above its midpoint and others that are below it; but we proceed as if all the terms have the value of the midpoint. Note that the width of the intervals tells exactly how much rounding off we are doing. The wider the interval, the more we may be changing the terms from their original values.

We shall compute the mean of a distribution by the group method. Consider the group frequency distribution of the spelling marks of the 1,000 fourth grade children (Table 6.1). Our first step is to obtain the sum of these 1,000 spelling marks or terms in the distribution. Remember that we are now considering the distribution of rounded-off terms. The highest 60 terms in this grouped distribution each have the same value, 96. Thus the sum of these highest 60 terms is $(60)(96) = 5,760$. The next highest 140 terms have each been rounded off to the value of 87. Thus the sum of these next 140 terms is $(140)(87) = 12,180$. In the same way we can find the sum of the terms in each of the eleven intervals. To get the sum of the rounded-off terms in any particular interval, we multiply the midpoint of that interval by the number of terms included, *i.e.*, the frequency of the interval.

We shall now add to Table 6.1 a column indicating the sum of the terms in each interval. This is done in Table 6.3. In this table we use the letter X to refer to the midpoints of the intervals and the letter F to refer to the frequencies of the intervals. Thus the sum of the terms in each interval is denoted by FX.

75

TABLE 6.3   COMPUTATION OF MEAN

(1) INTERVALS	(2) MIDPOINTS X	(3) FREQUENCIES F	(4) SUM OF TERMS FX
92–100	96	60	5,760
83–91	87	140	12,180
74–82	78	160	12,480
65–73	69	120	8,280
56–64	60	140	8,400
47–55	51	80	4,080
38–46	42	119	4,998
29–37	33	81	2,673
20–28	24	50	1,200
11–19	15	32	480
2–10	6	18	108
		$\sum F = 1{,}000$	$\sum FX = 60{,}639$

Column (4) tells us the sums of the terms in each interval. To get the sum of all the terms in the distribution, we add up all of the values in Column (4). The sum of these values is denoted by $\sum FX$ and is found to be 60,639. In other words, the sum of the rounded-off spelling marks of all the 1,000 children is 60,639.

Remember that when the terms are listed individually and we consider each term as an individual value of X, then the sum of the terms is denoted $\sum X$. Now that we are using the group frequency method, the sum of the terms is $\sum FX$, where the X values are the midpoints of intervals rather than individual terms. Thus the two symbols $\sum X$ (used in Section 2.5) and $\sum FX$ (used in this section) are equivalent except for errors caused by the grouping. The difference in notation merely means that we have summed up all of the terms in the distribution by different methods.

The frequencies of the different intervals are indicated in Column (3) of Table 6.3. In practice, we add up these frequencies to determine how many terms there are in the distribution (test scores). Here the sum of the frequencies, denoted $\sum F$, is 1,000. In other words, we are working with 1,000 test scores. It should be clear that $\sum F = N$.

It follows that the formula for the mean of a distribution, computed by the group method, is

(6.2)
$$\mu = \frac{\sum FX}{\sum F}$$

Thus for the data given in Table 6.3,

$$\mu = \frac{\sum FX}{\sum F} = \frac{60{,}639}{1{,}000} = 60.639 \quad \text{or} \quad 60.6.$$

Although the computation just completed by Formula (6.2) was perfectly straightforward, the reader may have been chagrined by the fairly large numbers involved in it. He may recall that on page 20 we were faced with a similar situation and we made the work simpler by applying Theorem 2.3 and subtracting a constant from each score. That constant was then added back to the mean of the new scores. We saw that the choice of a constant to be subtracted had no effect on the final result. It was advantageous to choose a constant that appeared to be as near the mean as possible, since this led to calculations with smaller numbers. Now let us see how we can apply this same method to our grouped data on spelling scores. Since we are working only with the midpoints, we will need to subtract the same constant from each midpoint.

The midpoints and frequencies have been entered in columns (1) and (2) of Table 6.4. We shall choose 60 as the constant to subtract from each midpoint. This value is the midpoint of one of the middle intervals. Our choice is dictated by the large frequency in the interval with 60 as its midpoint. Now the new score for the value 60 itself becomes zero. The fact that a large frequency, namely 140, will now be multiplied by zero will reduce our computations considerably. The value 60 has been subtracted from each of the midpoints and the set of new midpoints thus obtained appears in Column (3) of Table 6.4. These new midpoints have been labeled X'. Note that $X' = X - 60$.

TABLE 6.4 COMPUTATION OF MEAN BY SHORTER METHOD

(1) MIDPOINTS X	(2) FREQUENCIES F	(3) NEW MIDPOINTS X'	(4) SUM OF NEW TERMS FX'
96	60	36	2,160
87	140	27	3,780
78	160	18	2,880
69	120	9	1,080/ + 9,900
60	140	0	0
51	80	− 9	− 720
42	119	− 18	− 2,142
33	81	− 27	− 2,187
24	50	− 36	− 1,800
15	32	− 45	− 1,440
6	18	− 54	− 972/ − 9,261
	$\sum F = 1,000$		$\sum FX' = $ + 639

Now we must find the mean, using the X' midpoint values, and then add back the 60 to that new mean. We proceed as in the previous computation, except that we now use the X' values. Column (4) of Table 6.4 shows the product FX' for each interval. We need to remember that the product of a positive number by negative number is a negative number. The sums of the positive and negative values in Column (4) are shown separately. The sum of all the terms after subtraction of the constant is denoted by $\sum FX'$ and for

CHAPTER 6

the data in Table 6.4 it is found to be 639. Then the mean of the new (X') scores is $\sum FX'/\sum F$. We shall denote this by μ'. Thus we have $\mu' = \sum FX'/\sum F = 639/1{,}000 = .639$. Now we must add back the 60 to this mean of the new scores, i.e., $\mu = \mu' + 60$. This gives us $\mu = 60.639$ or 60.6, exactly the same as when we computed the mean without subtracting a constant from each midpoint score. It is always true that this short-cut method will give a result identical with that obtained by the longer method when both are applied to the same grouped distribution. It will be instructive for the student to do the computation of Table 6.4 over again, this time using 51 as the constant to be subtracted from each midpoint value. Although the intervening numbers will be different, the final result should again be the same.

The reader may have noticed that all of the X' values in Column (3) of Table 6.4 are multiples of 9. This occurs because the distance between successive midpoints is 9. The fact that these numbers are all multiples of 9 enables us to devise a further short cut in the computation by applying Theorem 2.4. We shall divide each of these numbers by 9 and then find the mean of the new numbers. The theorem tells us that dividing each number by 9 will have the effect of dividing the mean by 9. Hence, we shall then multiply back the new mean by 9. This will give us only the mean of the X' values, so we will still need to add 60 to get the mean of the original spelling scores.

In Table 6.5, the X, X', and F columns from Table 6.4 have been recorded as Columns (1), (2), and (3), respectively. Column (4) contains the results of dividing each X' value by 9. These values have been denoted by U. We then proceed as in the previous computations except that we now use the U values. Column (5) shows the product FU and gives the sum of the U scores for each interval. The sum of all of the 1,000 U scores for the distribution is the sum of Column (5) and is denoted by $\sum FU$. This is found to be $+71$, as shown in the table. Then the mean of these newest or U scores is $\sum FU/\sum F$. This may be denoted by μ''. For the data in the table we then have

$$\mu'' = \sum FU/\sum F = 71/1{,}000 = .071.$$

Now we must multiply back by 9 so that $\mu' = 9(.071) = .639$, the mean of the X' scores, as before. Finally, $\mu = .639 + 60 = 60.639$ or 60.6, the mean of the original grouped spelling scores and the same result as obtained before.

Now, admittedly, if we had to go through all of the above computations our method of using U scores would not be much of a short cut from our original computation of the mean for the grouped spelling scores. However, a little reflection will show the reader that the U values in Column (4) of Table 6.5 could be written in at once without the intervening step of the X' scores in Column (2). We simply choose 60 as the original midpoint score to be replaced by a new or U score of zero. This particular midpoint (60 in our example) becomes our reference point. We then enter 0 in the U column opposite 60 and successively add 1 to get the U scores for the intervals having midpoints larger than 60. This gives 1, 2, 3, . . . and so on for these

78

TABLE 6.5   COMPUTATION OF MEAN USING
CODED SCORES

(1) X	(2) X'	(3) F	(4) U	(5) FU
96	36	60	4	240
87	27	140	3	420
78	18	160	2	320
69	9	120	1	120/ + 1,100
60	0	140	0	0
51	− 9	80	−1	− 80
42	− 18	119	−2	− 238
33	− 27	81	−3	− 243
24	− 36	50	−4	− 200
15	− 45	32	−5	− 160
6	− 54	18	−6	− 108/ − 1,029
		$\sum F = 1,000$		$\sum FU = + \quad 71$

U scores.   Likewise, we successively subtract 1 to get the U scores for the intervals having midpoints smaller than 60.   This gives -1, -2, -3, . . . and so on for these U scores.   The U scores or *coded scores*, as they are often called, can thus be entered directly in Column (4) of Table 6.5.   It must be remembered that we are subtracting 60 from each midpoint value and then dividing the result by 9 when we write the U scores.   Then once we have found the mean of the U scores (denoted by μ''), the mean of the original scores must be found by first multiplying back by the 9 and then adding back the 60.   Symbolically $\mu = 9\mu'' + 60$ or $\mu = 9(\sum FU / \sum F) + 60$.   To generalize this for distributions other than spelling scores, let us denote the reference point (60 in our example) by R and the distance between successive midpoints (9 in our example) by i (since it is the same as the length of the interval from boundary to boundary).   The general formula is then

(6.3)
$$\mu = i \frac{\sum FU}{\sum F} + R.$$

As an example of direct coding of scores, let us calculate the mean for the IQ data shown in Table 6.6.   Here we have entered the midpoints, X, in

TABLE 6.6   CALCULATION OF MEAN OF IQ
SCORES USING CODED SCORES

(1) IQ INTERVAL	(2) F	(3) X	(4) U	(5) FU
121–130	20	125.5	2	40
111–120	40	115.5	1	40/ + 80
101–110	77	105.5	0	0
91–100	55	95.5	−1	− 55
81–90	40	85.5	−2	− 80
71–80	18	75.5	−3	− 54/ − 189
	$\sum F = 250$			$\sum FU = -109$

79

Column (3); the coded scores are shown in Column (4), which indicates that we have decided to use 105.5 as the reference point. Note that in this case $i = 10$, so that the entry of coded scores in Column (4) really amounts to subtracting 105.5 from each midpoint or X value and then dividing the result by 10. The reader should verify this for the individual midpoints. We then compute Column (5) by multiplying the F and U values for each row. We find that $\sum FU = -109$. Then using Formula (6.3) we have

$$\mu = i(\sum FU / \sum F) + R = 10(-109/250) + 105.5$$
$$= 10(-.436) + 105.5$$
$$= -4.36 + 105.5 = 101.14 \quad \text{or} \quad 101.1$$

The reader should note that the $-.436$ represents the mean of the U or coded scores and that we then multiplied back by the 10 and added back the 105.5. It will be instructive for the reader to do the work for Table 6.6 over again, this time using 95.5 as the reference point. Note that the final result is the same (101.14) although the intervening numbers are different.

6.8 The Variance and the Standard Deviation for Grouped Data. The reader should recall the definition of the variance (as given in Chapter 3), and that the standard deviation was defined as the positive square root of the variance. It was noted that one useful formula for finding the variance is the following:

(6.4)
$$\sigma^2 = \frac{\sum X^2}{N} - \left(\frac{\sum X}{N}\right)^2$$

In Formula (6.4), σ^2 stands for the variance, $\sum X$ stands for the sum of the terms, $\sum X^2$ stands for the sum of the squares of the terms, and N stands for the total number of terms involved.

Several important properties of the variance and the standard deviation were also discussed in Chapter 3. One of these was that the subtraction of a constant from each term in a distribution has no effect on either the variance or the standard deviation. Another was that division of each term in a distribution by a constant has the effect of dividing the standard deviation by the absolute value of that constant and of dividing the variance by the square of that constant. In the present section we shall follow the same plan that we used in the preceding section. We shall first see how to carry out the computation for grouped distributions using a modification of Formula (6.4). Then we shall see how to shorten the work by making use of the two properties that we have just mentioned.

Once again we shall use a single value to represent all of the scores in each interval of the distribution. The logical choice is again the midpoint, for the same reasons discussed in the preceding section. We are, therefore, assuming that all of the scores in each interval are located at the midpoint of that interval. For example, the 18 spelling scores in the interval from 2 through 10 in the distribution of Table 6.1 are all considered to have the value 6, the midpoint of that interval.*

Now to illustrate the computation by means of a modification of Formula (6.4), let us again use the distribution of children's spelling scores. We will need the data in Table 6.3, and these have been reproduced in Table 6.7. We will also need the values of X^2 in order to find σ^2. As before, we shall use X to represent the midpoints of the intervals; the sum of the terms in each interval is given by the corresponding entry in the FX column. The sum of all of these values in Column (4) is $\sum FX = 60,639$. Column (5) shows the square of each X value in Column (2) and Column (6) shows the value of the sum of the squares of the terms in each interval. For example, each term in the 2–10 interval (last row) is 6 and there are 18 such terms. We have $X = 6$, so $X^2 = 36$ and the sum of all these 18 terms is $FX^2 = (18)(36) = 648$. In the first row $X = 96$, so $X^2 = 9,216$ and the sum of the 60 squares is $FX^2 = (60)(9,216) = 552,960$. The Column (6) numbers for the other rows may be obtained in like fashion—by multiplying the corresponding numbers in Columns (3) and (5). Some readers may recognize that $FX^2 = (FX)(X)$ and hence that for the last row $FX^2 = (108)(6) = 648$. This shows that Column (5) is unnecessary and that the values in Column (6) may be obtained by multiplying the corresponding values in Columns (2) and (4). As a further illustration of this, note that for the first row $(FX)(X) = (X)(FX) = (96)(5,760) = 552,960 = FX^2$.

TABLE 6.7   COMPUTATION OF THE VARIANCE

(1) INTERVALS	(2) MIDPOINTS X	(3) FREQUENCIES F	(4) SUM OF TERMS FX	(5) SQUARES X^2	(6) SUM OF SQUARES FX^2
92–100	96	60	5,760	9,216	552,960
83–91	87	140	12,180	7,569	1,059,660
74–82	78	160	12,480	6,084	973,440
65–73	69	120	8,280	4,761	571,320
56–64	60	140	8,400	3,600	504,000
47–55	51	80	4,080	2,601	208,080
38–46	42	119	4,998	1,764	209,916
29–37	33	81	2,673	1,089	88,209
20–28	24	50	1,200	576	28,800
11–19	15	32	480	225	7,200
2–10	6	18	108	36	648
		$\sum F = 1,000$	$\sum FX = 60,639$		$\sum FX^2 = 4,204,233$

The sum of the squares of all 1,000 terms in the distribution is then the sum of Column (6) and is denoted by $\sum FX^2$. Hence, $\sum FX^2 = 4,204,233$ in

---

* A correction factor is sometimes introduced to make up for the error occasioned by the rounding off done when computing the variance as described. However, the correction itself is usually trivial when the distribution contains a large number of cases, and we need not consider it here.

our example. The formula for the variance of a distribution as computed by the group method is then

$$(6.5) \qquad \sigma^2 = \frac{\sum FX^2}{N} - \left(\frac{\sum FX}{N}\right)^2.$$

Thus for the data given in the table we have

$$\sigma^2 = (4{,}204{,}233/1{,}000) - (60{,}639/1{,}000)^2 = 4{,}204.23 - (60.64)^2$$
$$= 4{,}204.23 - 3{,}677.21 = 527.0.$$

Therefore the variance is 527 and the standard deviation $\sigma = \sqrt{527} = 22.96$ or 23 as we find from the table of square roots in Appendix I.

By now the reader may well be dismayed at the prospect of having to carry out computations that involve squaring and adding such large numbers. However, just as before, the short cut of using coded scores may be employed. To begin with, we may subtract some particular constant from each score in the distribution—that is, from each midpoint value listed. Theorem 3.1 tells us that to do so will have no effect whatsoever on the variability of the distribution, so that we shall obtain the same values of σ and σ^2 as if we had not bothered to subtract the constant. It follows that having subtracted a constant from each midpoint will not even necessitate our making a re-adjustment by adding it back to compute the standard deviation or the variance.

The reader will recall that the final step in setting up the short cut for the mean was to divide the new midpoint values by the width of the interval in order to obtain the final coded or U scores. We shall do precisely the same thing when computing the variance. Theorem 3.2 tells us that this will have the effect of dividing the standard deviation by that same amount and hence of dividing the variance by the square of that amount. We will then have to "multiply back" in order to get the final values for these measures.

To see the application of this coded score method to the spelling score data, we have reproduced four columns of Table 6.5 as the first four columns of Table 6.8. These columns show the original midpoint values, X; the frequencies, F; the coded scores, U; and the sums of the coded scores for each row, FU. Remember that the U scores were obtained by subtracting 60 from the corresponding X scores and then dividing by 9.

Two new columns have been added to the table: Column (5) shows the square of the coded score for each row, and Column (6) gives the sum of these squared scores for each row. In obtaining Column (5) keep in mind that the product of two negative numbers is a positive number and, in particular, the square of a negative number is a positive number. The values in Column (6) may be obtained by multiplying the corresponding entries in Columns (2) and (5). Thus in the first row $FU^2 = (F)(U^2) = (60)(16) = 960$, in the second row $FU^2 = (140)(9) = 1{,}260$, and so on. The reader may also recognize that $FU^2 = (FU)(U) = (U)(FU)$ and hence the Column (5) values are unnecessary and the Column (6) values may be obtained by multiplying

TABLE 6.8 COMPUTATION OF VARIANCE AND
STANDARD DEVIATION USING CODED SCORES

(1) X	(2) F	(3) U	(4) FU	(5) U^2	(6) FU^2
96	60	4	240	16	960
87	140	3	420	9	1,260
78	160	2	320	4	640
69	120	1	120	1	120
60	140	0	0	0	0
51	80	-1	-80	1	80
42	119	-2	-238	4	476
33	81	-3	-243	9	729
24	50	-4	-200	16	800
15	32	-5	-160	25	800
6	18	-6	-108	36	648
	$\Sigma F = 1,000$		$\Sigma FU = 71$		$\Sigma FU^2 = 6,513$

the corresponding values in Columns (3) and (4). Thus in the first row $FU^2 = U(FU) = (4)(240) = 960$, in the second row $FU^2 = (3)(420) = 1,260$, and so on. Again we must remember that the product of two negative numbers is a positive number; for instance, in the last row $FU^2 = U(FU) = (-6)(-108) = 648$.

The sum of the squares of all 1,000 terms in the coded or U score distribution is the sum of Column (6) and is denoted by ΣFU^2. Hence, $\Sigma FU^2 = 6,513$ in our example. We now proceed as in the previous computation except that we use the U values. Let σ_U^2 denote the variance of the U values. Then we have

$$(6.6) \qquad \sigma_U^2 = \frac{\Sigma FU^2}{N} - \left(\frac{\Sigma FU}{N}\right)^2.$$

Thus for the data in Table 6.8 we have

$$\sigma_U^2 = \frac{6,513}{1,000} - \left(\frac{71}{1,000}\right)^2 = 6.513 - (.071)^2 = 6.513 - .005 = 6.508.$$

Then the standard deviation of the U values $\sigma_U = \sqrt{6.508} = 2.55$ according to the table of square roots.

As stated earlier, Theorems 3.1 and 3.2 tell us that the only adjustment needed to get the values of σ^2 and σ for the original distribution is to multiply back to undo the effect of having divided by 9. By Theorem 3.2 we must multiply σ_U by 9 to get σ. We have, then, $\sigma = 9\sigma_U = 9(2.55) = 22.95 = 23$, the same result that we got with uncoded scores. By squaring σ, we get the value of σ^2. $\sigma^2 = (22.95)^2 = 527$. Note that this value of the variance is what we also obtained when we worked with uncoded scores. It will be instructive for the student to do the computation over again, this time using 51 as the reference point for the coded scores. This should again illustrate that the choice of reference point has no effect on the final result.

83

For those who prefer a completely symbolic statement, we can combine Formula (6.6) with the multiplying back and obtain

$$(6.7) \qquad \sigma = i\sqrt{\frac{\sum FU^2}{N} - \left(\frac{\sum FU}{N}\right)^2}$$

where we have again used i for the length of the interval in the distribution.

It should be noted that Table 6.8 (even with Column (5) omitted) contains all data needed for computation of the mean as well as the variance and standard deviation of the spelling marks using coded scores. The reader will have noted that $(\sum FU)/N$ appears in both computations. This quantity is the mean of the coded scores. For grouped frequency distributions the coded scores approach is generally the fastest and easiest to apply when these measures are needed.

PROBLEM SET A

Distribution of Achievement-Test Scores

Interval	Frequency
96–100	3
91–95	6
86–90	9
81–85	14
76–80	8
71–75	5
66–70	4
61–65	1

6.1 For the distribution above, compute the median, thirty-fifth percentile, third quartile, and fourth decile.

6.2 For the distribution above, draw the cumulative curve.

6.3 For the distribution above, determine the crude mode and compute the mean and variance. Also, compute the mean and variance using coded scores by subtracting a constant of 63.

Distribution of Efficiency Ratings

Interval	Frequency
70–72	2
67–69	8
64–66	32
61–63	26
58–60	10
55–57	2

6.4 For the distribution above, compute the median, twentieth percentile, first quartile, and sixth decile.

6.5 For the distribution above, draw the cumulative curve.

6.6 For the distribution above, determine the crude mode and compute the mean and variance. Also, compute the mean and variance using coded scores by subtracting a constant of 50.

Distribution of Quiz Scores

Interval	Frequency
70–75	2
64–69	5
58–63	8
52–57	7
46–51	7
40–45	4
34–39	2

6.7 For the distribution above, compute the median, twelfth percentile, third quartile, and ninth decile.

6.8 For the distribution above, draw the cumulative curve.

6.9 For the distribution above, determine the crude mode and compute the mean and variance. Also, compute the mean and variance using coded scores by subtracting a constant of 30.5.

Distribution of Tuition Cost per Student

Interval	Frequency
$400.00–499.99	3
300.00–399.99	6
200.00–299.99	10
100.00–199.99	5
0–99.99	1

6.10 For the distribution above, compute the median, thirty-third percentile, first quartile, and seventh decile.

6.11 For the distribution above, draw the cumulative curve.

6.12 For the distribution above, determine the crude mode and compute the mean and variance. Also, compute the mean and variance using coded scores by subtracting a constant of $50.

PROBLEM SET B

Distribution of IQ Scores

Interval	Frequency
115–119	3
110–114	5
105–109	12
100–104	30
95–99	24
90–94	18
85–89	6
80–84	2

6.13 For the distribution above, compute the median, seventieth percentile, first quartile, and second decile.

6.14 For the distribution above, draw the cumulative curve.

6.15 For the distribution above, determine the crude mode and compute the mean and variance. Also, compute the mean and variance using coded scores by subtracting a constant of 82.

Distribution of Spelling-Test Scores

Interval	Frequency
67–69	10
64–66	21
61–63	30
58–60	19
55–57	11
52–54	6
49–51	3

6.16 For the distribution above, compute the median, ninety-second percentile, third quartile, and eighth decile.

6.17 For the distribution above, draw the cumulative curve.

6.18 For the distribution above, determine the crude mode and compute the mean and variance. Also, compute the mean and variance using coded scores by subtracting a constant of 50.

Distribution of Test Scores

Interval	Frequency
96–100	23
91–95	19
86–90	11
81–85	6
76–80	15
71–75	22
66–70	7

6.19 For the distribution above, compute the median, fifty-eighth percentile, first quartile, and second decile.

6.20 For the distribution above, draw the cumulative curve.

6.21 For the distribution above, determine the crude mode and compute the mean and variance. Also, compute the mean and variance using coded scores by subtracting a constant of 103.

Distribution of Weekly Salaries

Interval	Frequency
$55.00–59.99	2
50.00–54.99	2
45.00–49.99	6
40.00–44.99	7

35.00–39.99	3
30.00–34.99	3
25.00–29.99	2

6.22 For the distribution above, compute the median, eighty-fourth percentile, third quartile, and first decile.

6.23 For the distribution above, draw the cumulative curve.

6.24 For the distribution above, determine the crude mode and compute the mean and variance. Also, compute the mean and variance using coded scores by subtracting a constant of $22.50.

seven

review: determinants of the shape of a distribution

7.1 Review. In later chapters the reader will need full grasp of the concepts already presented. Accordingly we shall pause briefly to review some of the more important ideas.

Earlier we defined the mean or average of a distribution of terms and showed that it could be thought of as a balance point or center. However, the mean tells us nothing about how scattered the terms are. The standard deviation provides this information. That is, a small standard deviation tells us that the terms are similar. A large standard deviation tells us that they are quite different from each other.

The standard deviation is thought of as a kind of yardstick, and measurements are always made from the mean or balance point. Consider the distribution of ages of boys living on a particular street. Suppose the ages of the boys (rounded off to the nearest year) are as follows:

Bob	3	Ned	11
Tom	5	Sam	13
Bill	6	Al	14
Dick	7	Jack	14
Jim	7	John	20

The mean of these ages is 10 and the standard deviation is 5. The histogram of the distribution is shown in Fig. 7.1. The names of the boys are written in the rectangles representing their ages.

The standard deviation may be thought of as a ruler five units long. The z score of a term tells how many ruler lengths that term is above or below the

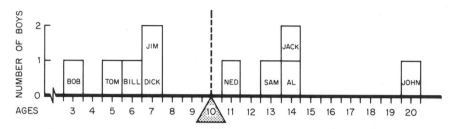

Fig. 7.1

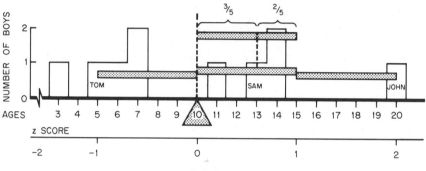

Fig. 7.2

mean. For instance, John's age is 20. Starting from the mean, which is 10, it takes precisely *two* of these yardsticks placed end to end to extend to John's score (Fig. 7.2). We say that John's age is two standard deviations above the mean. In statistical language, John's z score is 2.0.

In the same way, Tom's age, 5, is one standard deviation below the mean (Fig. 7.2).

Another way of saying that Tom's age is one standard deviation below the mean is to say that Tom's z score is -1.0.

Sam's age is 3/5 of a standard deviation above the mean; in other words, Sam's z score is $+.6$ (Fig. 7.2).

The z score of any term in a distribution is the number of standard deviations that particular term is above or below the mean. The distribution might be drawn over, using z score units as a base and omitting the original scores (the ages) entirely. Both types of units, z scores and the original scores, are indicated in Fig. 7.3.

The percentile rank of a term in a distribution is the sum of the per cent of terms below it and one-half of the per cent of the distribution that the term comprises.

There are ten terms in our distribution of ages. The term representing Sam's age is higher than 60 per cent of the terms in the distribution. Sam's age is itself a term that comprises 10 per cent of the distribution. Thus Sam's age has a percentile rank of 65.

89

In Fig. 7.4a we have split the rectangle representing Sam's age in half by drawing a vertical midline. The area to the left of this line is shaded. The

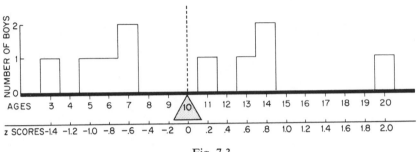

Fig. 7.3

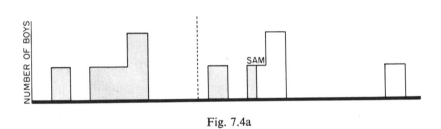

Fig. 7.4a

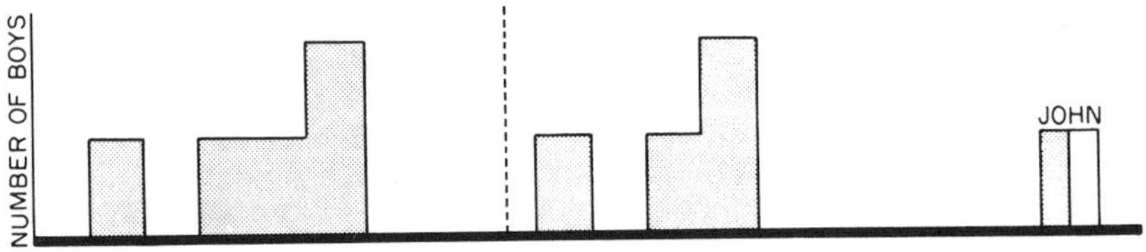

Fig. 7.4b

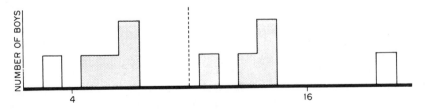

Fig. 7.5

per cent of the total area that is shaded (65 per cent) is the percentile rank of Sam's age.

Similarly John's age has a percentile rank of 95, since 95 per cent of the area is to the left of the midline of the rectangle standing for John's age (Fig. 7.4b).

We shall have frequent occasion to consider the area of a sector of a graph. For instance, in Fig. 7.5, 80 per cent of the area (the shaded portion) is between 4 and 16, reflecting the fact that 80 per cent of the terms in the distribution are between 4 and 16.

The terms that are either less than 4 or greater than 16 comprise 20 per cent of all the terms. The area of part of a graph is often interpreted as the proportion of terms contained in that part.

7.2 Data That Account for the Shape of a Graph. The graphs of distributions take many shapes. Some of these shapes arise repeatedly. That is, they fit many different types of data. Other shapes arise quite infrequently.

We shall now discuss certain aspects of data that suffice to determine how their graph will look. To set the stage, suppose that we are given some information about the distribution of ages of citizens of Daytonville and asked to determine the shape of the graph of this distribution. We are told that one person's age has a z score of -2.0 and that other ages have z scores of -1.0, 0, 1.0 and 2.0. If we attempt to draw the graph of the distribution using this information, our first move might be to construct a horizontal axis and then to note where the five z scores appear. But once this is done, we can proceed no further. We do not yet have enough information to determine the shape of the distribution.

Now suppose we are given the following five additional facts:

1. The age with a z score of -2.0 has a percentile rank of 0.
2. The age with a z score of -1.0 has a percentile rank of 40.
3. The age with a z score of 0 has a percentile rank of 70.
4. The age with a z score of 1.0 has a percentile rank of 80.
5. The age with a z score of 2.0 has the highest percentile rank in the distribution.

This information contributes a great deal. We can now build a histogram from it. The base of the first rectangle is the segment going from $z = -2.0$ to $z = -1.0$ (Fig. 7.6).

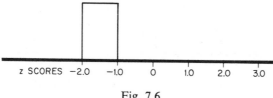

z SCORES   −2.0   −1.0   0   1.0   2.0   3.0

Fig. 7.6

There are no scores to the left of $z = -2$ because the percentile rank of $z = -2$ is 0.

Since 40 per cent of the ages in the distribution have z scores between -2.0 and -1.0, we know that the area of the rectangle we have drawn comprises 40 per cent of the area of the entire graph.

We know also that 70 per cent of all the ages must have z scores less than 0. Thus 30 per cent of the area of the entire graph must be in the rectangle based on the interval $z = -1.0$ to $z = 0$. This rectangle must be $\frac{3}{4}$ as high as the first one.

By this same reasoning, 10 per cent of the area must be in the next rectangle and 20 per cent in the last rectangle. We obtain the graph in Fig. 7.7.

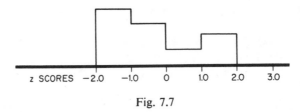

Fig. 7.7

We have not been given sufficient information to reconstruct the graph with a high degree of accuracy. However, given an elaborate table, one can construct a detailed graph.

The point is that a graph is determined by a table giving z scores and the percentile ranks that go with them. That is, from a detailed set of pairings of z scores and percentile ranks, we can draw the shape of the graph. The pairings of z scores and percentile ranks tell us how the area is divided up over different portions of the base line.

We might say that a table of pairings like the one given is a description of a picture. The picture, of course, is the graph.

If two graphs differ in shape, the tables that correspond to them must differ. If two graphs have identical shapes, the tables that correspond to them must be identical. For instance, the graph of the distribution of men's weights in City A is shown in Fig. 7.8.

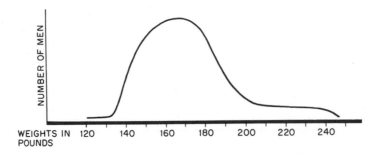

Fig. 7.8

The graph of people's incomes in City B happens to have an identical shape (Fig. 7.9).

Since the two curves are identical, a single table of z scores and their corresponding percentile ranks describes both distributions. Thus the weight in City A, which has a z score of 0, has a percentile rank of 60. Similarly, in

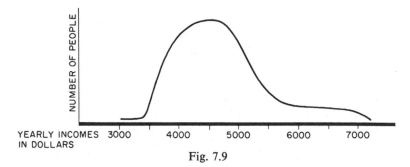

Fig. 7.9

City B the person whose income has a z score of 0 earns more than 60 per cent of his competitors. The percentile ranks that go with each particular z score are the same in each distribution, reflecting the fact that the shapes of the two graphs are the same.

We have seen how to construct a graph from this information. The purpose of this section was simply to show that there is an equivalence between *a graph* on the one hand and *a table of z scores and their percentile ranks* on the other. Ordinarily it might not occur to the reader that a set of pairings of z scores and percentile ranks corresponds to a particular graph.

Incidentally, the person drawing the graph decides in advance how much space he is willing to give to it. For instance, successive z score units might have been set much closer to each other than they were in Fig. 7.7. The graph would have looked more compact but essentially it would not have been different. The reader should not be fooled into thinking that two graphs are different when their dissimilarities are wholly due to differences in spacing of units along one or both axes.

PROBLEM SET A

For each of Problems 7.1 through 7.4, sketch a histogram of a distribution that satisfies the conditions given in the table.

7.1	z Score	Percentile Rank
	−3.0	0
	−2.0	15
	−1.0	35
	0.0	55
	1.0	75
	2.0	95
	3.0	100

93

7.2

z Score	Percentile Rank
−4.0	0
−3.0	5
−2.0	10
−1.0	20
0.0	40
1.0	85
2.0	95
3.0	100

7.3

z Score	Percentile Rank
−4.0	0
−3.0	5
−2.0	15
−1.0	30
0.0	40
1.0	50
2.0	90
3.0	95
4.0	100

7.4

z Score	Percentile Rank
−3.0	0
−2.0	5
−1.0	30
0.0	50
1.0	75
2.0	95
3.0	100

7.5 A distribution contains 10 scores arranged in order of magnitude, $X_1, X_2, \ldots$ X_{10}, with X_1 the smallest and no two the same. Find the percentile ranks of scores X_2, X_5, and X_9.

7.6 A distribution contains 20 scores arranged in order of magnitude, $X_1, X_2, \ldots,$ X_{20}, with X_1 the smallest and no two the same. Find the percentile ranks of scores X_1, X_8, X_{12}, and X_{19}.

7.7 The 10 scores in a distribution are 1, 2, 3, 4, 5, 6, 7, 8, 9, and 10. Find the z scores and percentile ranks of scores 4, 7, 9, and 10.

7.8 (a) In Problem 7.7 replace the score of 10 by 100 and find the z scores and percentile ranks of scores 4, 7, 9, and 100. Compare with results for the original data. (b) Repeat, replacing the score of 100 by 1000.

7.9 Find P_{10}, P_{30}, P_{50}, P_{80}, and corresponding z scores for the distribution below.

Interval	Frequency
159.5–169.5	2
149.5–159.5	1
139.5–149.5	1
129.5–139.5	5

119.5–129.5	5
109.5–119.5	40
99.5–109.5	61
89.5–99.5	38
79.5–89.5	17

7.10 Find P_{10}, P_{30}, P_{50}, P_{80}, and corresponding z scores for the distribution below.

Interval	Frequency
139.5–149.5	1
129.5–139.5	0
119.5–129.5	9
109.5–119.5	37
99.5–109.5	51
89.5–99.5	47
79.5–89.5	21

7.11 (a) Find the mean and standard deviation of the IQ scores presented below.
(b) Find P_5, P_{15}, P_{30}, P_{50}, P_{75}, P_{95}, and corresponding z scores.

IQ	Frequency
130–139	1
120–129	12
110–119	68
100–109	150
90–99	101
80–89	45
70–79	1

7.12 (a) Find the mean and standard deviation of the IQ scores presented below.
(b) Find P_5, P_{15}, P_{30}, P_{50}, P_{75}, P_{95}, and corresponding z scores.

IQ	Frequency
120–129	1
110–119	10
100–109	59
90–99	167
80–89	241
70–79	182
60–69	73

PROBLEM SET B

For each of Problems 7.13 through 7.15, sketch a histogram that satisfies the conditions given in the table.

7.13

z Score	Percentile Rank
−2.0	0
−1.0	10
0.0	15
1.0	20
2.0	25
3.0	95
4.0	100

7.14

z Score	Percentile Rank
−3.0	0
−2.0	10
−1.0	50
0.0	60
1.0	85
2.0	90
3.0	100

7.15

z Score	Percentile Rank
−3.0	0
−2.0	10
−1.0	40
0.0	60
1.0	80
2.0	90
3.0	100

7.16 A distribution contains 10 scores arranged in order of magnitude, $X_1, X_2, \ldots,$ X_{10}, with X_1 the smallest and no two the same. Find the percentile rank of scores X_3, X_6, and X_8.

7.17 A distribution contains 20 scores arranged in order of magnitude, $X_1, X_2, \ldots,$ X_{20}, with X_1 the smallest and no two the same. Find the percentile rank of scores X_2, X_6, X_{11}, and X_{17}.

7.18 The 10 scores in a distribution are 1, 2, 3, 4, 5, 6, 7, 8, 9, and 10. Find the z score and the percentile rank of scores 2, 5, 8, and 10.

7.19 (a) In Problem 7.18 replace the score of 10 by 100 and find the z score and percentile rank of scores 2, 5, 8, and 100. Compare with results for the original data. (b) Repeat, replacing the score of 100 by 1000.

7.20 Find P_{20}, P_{40}, P_{60}, P_{90}, and corresponding z scores for the distribution below.

Interval	Frequency
159.5–169.5	2
149.5–159.5	1
139.5–149.5	1
129.5–139.5	5
119.5–129.5	5
109.5–119.5	40
99.5–109.5	61
89.5–99.5	38
79.5–89.5	17

7.21 Find P_{20}, P_{40}, P_{60}, P_{90}, and corresponding z scores for the distribution below.

Interval	Frequency
139.5–149.5	1
129.5–139.5	0
119.5–129.5	9
109.5–119.5	37
99.5–109.5	51
89.5–99.5	47
79.5–89.5	21

7.22 (a) Find the mean and standard deviation of the IQ scores presented below.
(b) Find P_5, P_{15}, P_{30}, P_{50}, P_{75}, P_{95}, and corresponding z scores.

IQ	Frequency
150–159	2
140–149	9
130–139	27
120–129	73
110–119	176
100–109	208
90–99	190
80–89	87
70–79	43
60–69	9

7.23 (a) Find the mean and standard deviation of the IQ scores presented below.
(b) Find P_5, P_{15}, P_{30}, P_{50}, P_{75}, P_{95}, and corresponding z scores.

IQ	Frequency
140–149	3
130–139	8
120–129	17
110–119	63
100–109	227
90–99	301
80–89	265
70–79	111
60–69	3

7.24 (a) Find the mean and standard deviation of the IQ scores presented below.
(b) Find P_5, P_{15}, P_{30}, P_{50}, P_{75}, P_{95}, and corresponding z scores.

IQ	Frequency
110–119	4
100–109	11
90–99	26
80–89	54
70–79	39
60–69	7

eight

the normal distribution

8.1 Theoretical Distributions. We have seen that as we add more and more terms to a distribution we may often depict its shape with increasing accuracy by means of a curve. Under some conditions, as we continue collecting terms, the effect of new terms on the shape of the distribution becomes increasingly negligible. After a while, the shape of the distribution may be said to "settle down" and it becomes possible to consider properties of the distribution without specifying the number of terms included in it. We need specify only that the number of terms is "large enough."

We shall use the phrase *an infinitude of terms* to mean so many terms that the effect of adding more by the same process need not concern us. A *theoretical distribution* is one composed of an infinitude of terms. In other words, a theoretical distribution is one that has "settled down" and we may talk about its properties and its shape as if they were fixed. In a theoretical distribution, each term comprises a negligible per cent of the distribution and we say that the percentile rank of a term is merely the per cent of terms below it.

8.2 The Normal Distribution and the Normal Curve. We shall now give our attention to a particular theoretical distribution called the *normal distribution* and the graph that depicts it, the *normal curve*. The normal distribution is extraordinarily important from both a theoretical and a practical standpoint. Much of the development of statistics—indeed, perhaps most of it—has sprung from consideration of the normal distribution.

The normal distribution is of prominent importance in virtually all fields where data are gathered.

The graph in Fig. 8.1 is the normal curve. *z* score values are plotted along the base line. Consecutive *z* score units are equal distances apart.

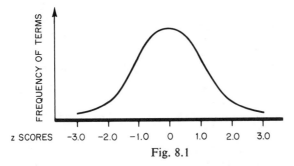

Fig. 8.1

The exact shape of the normal curve depends somewhat upon how far apart we decide to set the *z* score units along the base line. For instance, the normal curve looks somewhat different when we set the *z* score units a quarter-inch apart (Fig. 8.2). The spacing of units along the vertical axis is the same in Fig. 8.2 as in Fig. 8.1.

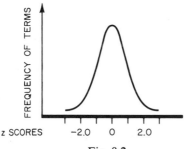

Fig. 8.2

Several properties of the normal distribution are reflected in the normal curve, no matter how it is drawn. For instance, the normal curve is symmetric and its highest point is over the value $z=0$, facts that the reader may verify by looking at either Fig. 8.1 or 8.2. We shall point out more of these properties in Section 8.3.

The normal distribution and the normal curve are precisely and completely defined by a mathematical formula. Unfortunately, we cannot define the normal distribution in the usual way without presupposing a knowledge of calculus. Instead we shall present a somewhat imprecise but roughly equivalent definition of the normal distribution, to which we shall repeatedly refer.

We saw in Section 7.2 that a table listing *z* score values along with the percentile ranks of the terms with those *z* score values constitutes a numerical description of a distribution. We were able to draw the graph of a distribution from such a table with no further information.

99

There exists a particular set of relationships between the z scores of terms and their percentile ranks such that, when these relationships hold, the distribution is said to be a normal distribution.

The particular set of relationships that defines the normal distribution is given in Appendix II. This table is known as a normal distribution table. A number of z scores ranging from -4 to $+4$ are listed. When we know the z score of a term in the normal distribution and would like to know what fraction of the total number of terms is below the given term, this normal distribution table tells us. For instance, according to the table, in the normal distribution the term with a z score of -1.0 is bigger than approximately $159/1,000$ of the terms in the distribution. In other words, the term with a z score of -1.0 is at about the 16th percentile. The term with a z score of 0 is bigger than half of the terms in a normal distribution; thus it is at the 50th percentile. The term with a z score of 1.5 is bigger than approximately $933/1,000$ of the terms in the distribution. Thus the term with a z score of 1.5 is roughly at the 93rd percentile.

The normal distribution may be roughly defined as the distribution in which the relationships listed in Appendix II hold. The statement that a distribution is normal means that the relationships between the z scores of the terms and their percentile ranks in the distribution are as indicated in the table. In short, if the relationships are as indicated in the table, the distribution is normal at least to the extent that the table, being incomplete, is able to identify the distribution. Admittedly we have given an approximate characterization of the normal distribution rather than a precise definition of it.

Appendix II is necessarily incomplete and imprecise. Obviously, no table can be extensive enough to include all possible z scores; even the most elaborate tables provide only an approximate definition. The proportions in the table go one digit beyond percentile ranks but they are still not precise. In any event, our approximate definition will prove to be serviceable for characterization of the normal distribution, as we shall see.

Given the information in the table, the reader can proceed to draw an approximate graph of the normal distribution. Actually he would end up with a histogram rather than a smooth curve, but if he were carefully to "smooth out" the tops of the bars of the histogram after drawing it, he would get a figure virtually identical with the normal curve. Whether his picture would resemble the normal curve in Fig. 8.1 or the one in Fig. 8.2 would depend upon how he decided to space the z score units along the horizontal axis.

We have defined the normal distribution by referring to a table of relationships rather than by referring to the normal curve. The curve of the distribution characterized by the table is the normal curve whether it looks relatively flat, as in Fig. 8.1, or peaked, as in Fig. 8.2. Appendix II is our best substitute for the formula that perfectly defines the normal distribution and the normal curve. A more extensive table would naturally define the normal distribution more precisely.

8.3 Four Properties of the Normal Distribution. We shall now consider four properties of the normal distribution and note their influences on the normal curve (as drawn in Fig. 8.3). We shall refer to the vertical center line in Fig. 8.3 as the *vertical axis*. The vertical axis is at the value $z=0$ where the curve reaches its maximum point.

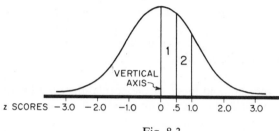

Fig. 8.3

1. One property of the normal distribution is that the terms tend to cluster around the point $z=0$. That is, as the curve moves in either direction away from the vertical axis, it drops, revealing the fact that there are fewer and fewer terms. For instance, there are more terms with z scores between 0 and .5 than there are terms with z scores between .5 and 1.0. In Fig. 8.3 the area in sector 1 (which signifies the proportion of terms with z scores between 0 and .5) is larger than the area in sector 2 (which signifies the proportion of terms with z scores between .5 and 1.0). The further we go from the vertical axis in either direction, the fewer terms we find.

2. A second property is that the normal curve is symmetrical around the vertical axis. The height of the curve over any z score value is exactly the same as its height over the negative of that value. For example, the curve is the same height over the point $z=1.0$ as it is over the point $z=-1.0$. The curve is the same height over the point $z=-2.57$ as it is over the point $z=2.57$.

It follows that half of the terms are to the left of the vertical axis and have negative z scores and half of them are to the right of the vertical axis and have positive z scores. Also there are the same proportion of terms in any sector as in its corresponding sector on the opposite side of the vertical axis. For instance, it turns out that about 15 per cent of the terms have z scores between .5 and 1.0. The same per cent of the terms have z scores between $-.5$ and -1.0 (Fig. 8.4).

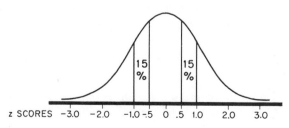

Fig. 8.4

In Section 8.4 we shall make use of the fact that each sector contains the same proportion of terms as its symmetrically opposite sector. We shall use this fact to draw up a simplified and convenient table of the normal distribution.

3. The sizes of the terms in the normal distribution are not bounded in either direction. In theory there exist terms with z scores of all magnitudes. There are terms with z scores as small as $-1,000$ and z scores as large as 1,000.* Such terms are incredibly rare and in practice we do not consider them. However, we take cognizance of them in the normal curve by not dropping the curve completely down to the base line no matter how far out we extend it in either direction.

4. The mean, median, and mode of the normal distribution all have the same value. The term that has a z score equal to 0 has all three properties (see Fig. 8.3). The term with the z score of 0 is at the mean since it is the balance point of the distribution. The same term is the mode since the normal curve is highest where $z=0$, indicating that more terms have the z score of 0 than any other z score value. Finally, the property of symmetry tells us that exactly half of the area is to the left of the vertical axis, where $z=0$, and half of the area is to the right of it. Thus half of the terms have z scores less than 0 and half of them have z scores larger than 0. Accordingly the term with the z score of 0 is also the median term.

8.4 Making Inferences about Terms from the Normal Distribution Table. Because the normal distribution is symmetric, we can put the information in Appendix II into a more convenient form. We need only describe the relationships for half of the distribution and the reader can use the table whether he is working with positive or negative z score values. The table in Appendix III gives the proportion of terms that have z scores between $z=0$ and various z score values. Different values of z scores are listed, with the first two digits in the left-hand column and the third digit at the top of one of the other columns. Suppose we wish to find in Appendix III the proportion of terms that have z scores between $z=0$ and $z=.70$. We look for the entry that is in the row for $z=.7$ and that is also in the column for .00. We do this because $.70=.7+.00$. We find this entry to be .2580. Loosely speaking, slightly more than one-quarter of the terms in the normal distribution have z scores between 0 and .70.

Examine Fig. 8.5. We have just stated that the sector bounded by the vertical axis and $z=.70$ contains .2580 of the area under the normal curve. Appendix III also tells us the proportion of terms that have z scores between $z=0$ and $z=-.70$. Due to the symmetry of the curve, the same proportion of terms are in this sector as in the sector we have considered. That is, the

---

* Thus the formula for the normal distribution divulges to the mathematician the exact proportion of terms in the normal distribution which have z scores between 1,000 and 1,001.

proportion of terms with z scores between $z = -.70$ and $z = 0$ is .2580. The sector bounded by $z = -.70$ and the vertical axis is also delineated in Fig. 8.5.

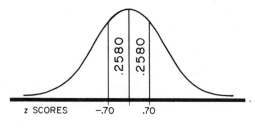

Fig. 8.5

Now consider a different kind of problem. Suppose we wish to determine the proportion of terms with z scores between two values, neither of which is $z = 0$. We must either add or subtract sectors in such a way as to isolate the sector we are after. For instance, suppose we wish to determine the proportion of terms in the normal distribution that have z scores between $z = -1.96$ and $z = 1.96$. We first read in Appendix III that the proportion of terms with z scores between $z = -1.96$ and $z = 0$ is .4750. This is found in the row for $z = 1.9$ and the column with .06 at the top, because $1.96 = 1.9 + .06$. That is, the proportion of area in the sector bounded by $z = -1.96$ and the vertical axis is .4750 (Fig. 8.6). Similarly the proportion of terms with z scores between 0 and 1.96 is .4750.

The proportion of terms in the two sectors together is the proportion of terms with z scores between -1.96 and 1.96. Therefore, the proportion of terms with z scores between -1.96 and 1.96 is $.4750 + .4750 = .9500$ (Fig. 8.7).

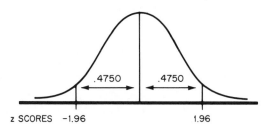

Fig. 8.6

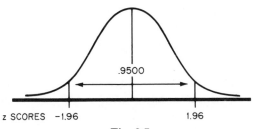

Fig. 8.7

103

We have just added the areas in two separate sectors to get the area in a larger sector not directly given in Appendix III. However, we sometimes find it necessary to subtract the area of one sector from that of another in order to isolate the area of a particular sector (which is the proportion of terms in that sector).

For instance, suppose the problem is to determine the proportion of terms in the normal distribution that have z scores between $z=.50$ and $z=1.0$. We compute the area of the specified sector by subtracting the area of one sector from that of another. According to Appendix III, the proportion of area in the sector bounded by $z=0$ and $z=1.0$ is .3413 (Fig. 8.8a).

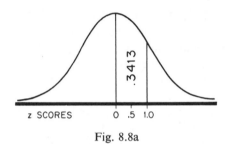

z SCORES

Fig. 8.8a

The proportion of area in the sector bounded by $z=0$ and $z=.50$ is .1915 (Fig. 8.8b).

The proportion of area in the sector $z=.5$ to $z=1.0$ may be found by subtracting the area of the second sector from that of the first. Thus the proportion of area in the sector we are after is $.3413-.1915=.1498$. In other words, .1498 of the terms in a normal distribution have z scores between $z=.50$ and $z=1.0$.

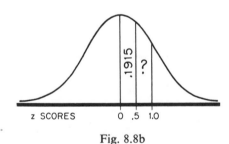

z SCORES

Fig. 8.8b

The proportion of terms in various sectors may be determined by the methods described. Appendix III gives enough information so that by adding sectors or subtracting them, as the situation requires, the reader can "isolate" the sector he is after and determine the proportion of terms contained in it.

We shall now make some inferences about terms in a normal distribution, using Appendix III. The distribution of IQ scores of American males is normal.   Its mean is 100 and its standard deviation is 15 points (when scores are obtained from the Wechsler Adult Intelligence Scale).*

Remember how we get z scores (Section 4.2).   The IQ score of 115 is one standard deviation above the mean and thus has a corresponding z score of 1.0; the IQ score of 130 has a corresponding z score of 2.0; the IQ score of 85 corresponds to a z score of -1.0, and so on.

To determine the position of an IQ score in the distribution, we first convert the IQ score into a z score.   Then we use either Appendix II or Appendix III to determine the proportion of terms in the normal distribution which have smaller z scores.   We use Appendix III in this section, since we shall make exclusive use of it in the chapters that follow.

To begin with, suppose we are to determine the percentile rank of the IQ score of 115.   The IQ score of 115 corresponds to a z score of 1.0.   Our problem becomes that of determining the percentile rank of the z score of 1.0. Note that 50 per cent of the terms have z scores less than zero (Fig. 8.9).

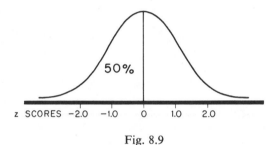

Fig. 8.9

According to Appendix III, roughly 34 per cent of the terms have z scores between 0 and 1.0 (Fig. 8.10).

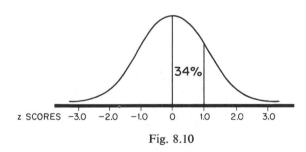

Fig. 8.10

When added, the two sectors of the distribution comprise roughly 84 per cent of it (Fig. 8.11).   Thus the term with a z score of 1.0 is higher than

* In practice, we consider a distribution normal even when it does not follow the table at the extreme tails.   Virtually all actual distributions are bounded. For instance, the distribution of IQ scores has no terms below zero, and no score can be above 157, which is the upper limit of the Wechsler Adult Intelligence Scale.   So long as a distribution "behaves normally" between $z = -3.0$ and $z = 3.0$ we shall consider it normal.

approximately 84 per cent of the terms in the distribution. In effect, we have said that the IQ of 115 is higher than approximately 84 per cent of the IQ's in the entire distribution.

Perhaps the reader can guess how to use Appendix III to locate the position of an IQ score that is below the mean. For instance, our task might be to determine the percentile rank of the IQ of 70. As usual, the first step is to convert the IQ score into a z score. The IQ score of 70 corresponds to a z score of -2.0. Appendix III tells us that roughly 48 per cent of the terms in a normal distribution have z scores between 0 and -2.0 (the same as between 0 and 2.0). Thus the term with the z score of -2.0 is surpassed by roughly 48 per cent + 50 per cent of the terms in the distribution (see Fig. 8.12). To put it another way, the term with a z score of -2.0 is at the second percentile. Thus the person with an IQ of 70 surpasses only about 2 per cent of the people in the entire distribution.

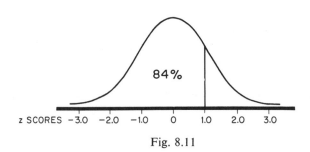

Fig. 8.11

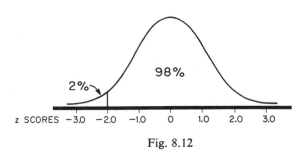

Fig. 8.12

Somewhat more complex problems relating terms in a normal distribution arise but, if what has been said is clear, the reader should have no trouble handling them. The first step is always to convert the original scores (or terms of another kind) into z scores. Once the conversion has been made, one of the procedures already discussed may be used. For instance, the problem might be to determine the proportion of people who have IQ's between 90 and 110. This problem becomes that of determining the proportion of terms in the normal distribution which have z scores between $-.67$ and

+.67. It is advisable to draw the normal curve for each problem and to delineate the relevant sectors on it before proceeding. It usually becomes apparent which sectors must be added or subtracted to reach the solution.

Once we are told that a distribution is normal and are given its mean and standard deviation we can draw the graph of the distribution indicating the original scores (or more generally, the original terms) along the base line, along with the z scores. For instance, the distribution of IQ's has a mean of 100 and a standard deviation of 15 points, so that the normal curve of IQ's looks like that in Fig. 8.13.

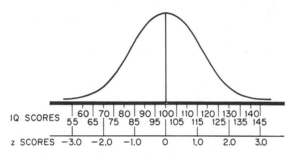

Fig. 8.13

Note that the IQ of 100 is the mean, the median, and the modal IQ of the distribution. The IQ's tend to cluster around 100; in other words, we find fewer and fewer IQ's as we move further and further from 100 in either direction. The symmetry of the distribution tells us that as many people have IQ's of 70 (that is, 69.5 to 70.5) as of 130 (that is, 129.5 to 130.5), the same number of people have IQ's of 95 as those who have IQ's of 105, and so on.

8.5 The Distribution of Shots. We shall now consider a completely different kind of situation in which the normal distribution is found. In Chapter 9 are specified conditions, found in nature, that "force data" into the normal distribution shape. The purpose of this section is simply to give the reader more experience in working with the normal distribution and a greater appreciation of its generality.

Suppose a rifle-shooting tournament is held in which the contestants are presented with the task of hitting a telephone pole a given distance away. The rules are that each marksman is allowed a single shot. Each marksman has to wait in a nearby building until it is his turn to shoot so that he cannot witness his predecessors at the firing line. An electronic device is set up to record how far each bullet is from the pole when it passes it. For instance, the device may record that one bullet passes seven feet to the right of the pole and another three feet to the left of it. The score given to each marksman indicates where his bullet passes the pole; for example, the marksman whose

107

shot goes seven feet to the right of the pole gets a score of $+7$. The one who fires three feet to the left of the pole gets -3. Any one who hits the pole gets a score of zero, which guarantees him at least a tie for the prize. We shall assume that the conditions are such that a "miss" is as apt to occur on one side of the pole as on the other.

The contest is starting and it is our job to record the outcome. The first marksman steps to the line and fires. Our device tells us that his bullet passes 3 feet to the right of the pole. His score is $+3$ and we record it by placing a tiny rectangle on a horizontal axis that we have drawn specially for the occasion (Fig. 8.14).

The sound of the shot is a signal for the next person to emerge from the building and step to the firing line. Bang! His shot goes five feet to the left of the pole and again we record it (Fig. 8.15). After several shots our picture looks like Fig. 8.16. We may suppose that the contest goes on indefinitely but it is time for us to leave. Our picture has come to look like Fig. 8.17.* Note that virtually as many shots went to the left of the telephone pole as to the right of it and that hits are more numerous near the pole than farther away on either side. The approximate representation by a curve becomes more and more accurate as more shots are fired. As you may have guessed,

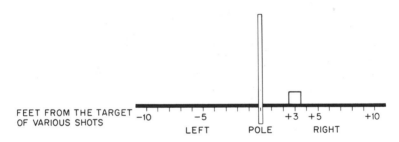

Fig. 8.14

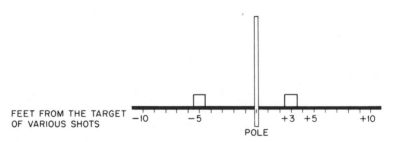

Fig. 8.15

* In this illustration the scores are rounded off to the nearest unit (foot).

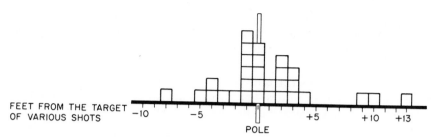

FEET FROM THE TARGET
OF VARIOUS SHOTS   −10   −5   +5   +10   +13
                                              POLE

Fig. 8.16

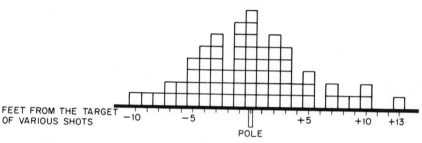

FEET FROM THE TARGET
OF VARIOUS SHOTS   −10   −5   +5   +10   +13
                                              POLE

Fig. 8.17

this curve is the normal curve and the distribution of shots is the normal distribution.*   The standard deviation of the distribution reflects the abilities of the marksmen as a group.   A small standard deviation would indicate that the shots tend to cluster around the telephone pole and a large standard deviation would indicate that they are relatively scattered.   To be specific, we shall now suppose that the standard deviation of the shots is four feet. The distribution of shots looks like Fig. 8.18.

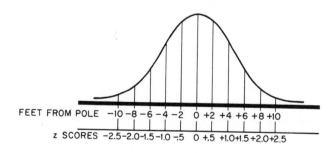

FEET FROM POLE −10 −8 −6 −4 −2   0 +2 +4 +6 +8 +10

z SCORES −2.5 −2.0 −1.5 −1.0 −.5   0 +.5 +1.0 +1.5 +2.0 +2.5

Fig. 8.18

---

* We shall consider the principle "explaining" the normality of the distribution of shots in Chapter 9.

Appendix III tells us that in a normal distribution about 19 per cent of the terms lie within a half of a standard deviation to the right of the mean. Thus about 19 per cent of the shots went between the telephone pole and an imaginary vertical line a half of a standard deviation (or 2 feet) to the right of the pole. Similarly 19 per cent of the shots went between the telephone pole and an imaginary vertical line 2 feet to the left of the pole. Altogether we might say that 38 per cent of the shots went within 2 feet of the telephone pole.

Altogether about 68 per cent of the terms in the normal distribution are within one standard deviation of the mean on one side or the other. Thus about 68 per cent of the shots landed within four feet of the telephone pole on one side or the other. The reader may verify that about 95 per cent of the shots landed within eight feet of the pole. Note that we need know only the values of the mean and standard deviation of a normal distribution in order to make inferences about any of the individual terms, using Appendix III.

8.6 Determining Scores from Percentile Ranks. Another type of problem that we can solve using the table of normal curve areas in Appendix III is essentially the opposite of those previously illustrated in this chapter.

Suppose that we wish to know what z score has a percentile rank of 35. This means that 35 per cent of the area is to be to the left of that score and 15 per cent is to be in the sector between the desired z score and $z = 0$ (Fig. 8.19).

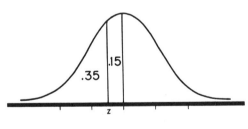

Fig. 8.19

We look in the body of the table in Appendix III this time and our answer will be a z score read from the margins. The proportion given in the body of the table that is nearest to .15 is .1517. This corresponds to a z score of .39. We obtain .39 by adding the heading of the column in which .1517 appears to the z value at the left of its row. That is, $0.3 + .09 = 0.39$. Therefore, since we know that the desired z score is below the mean, the answer is $z = -.39$.

To illustrate the application of this last type of problem, let us suppose that only the top 10 per cent of the applicants for a position are to be interviewed. The test that is given to determine the top 10 per cent results in a normal distribution with a mean of 80 and a standard deviation of 5. What is the cut-off score? Now we must again refer to the body of the table in Appendix III. This time we seek the z score for a percentile rank of 90, from which we will find the actual score that corresponds to this z score. Fig. 8.20

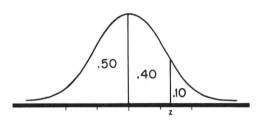

Fig. 8.20

illustrates the situation. Since 90 per cent of the area is to be to the left of the desired z score, we must have 40 per cent of the area between that score and the mean at $z=0$. Therefore we look in the body of the table for the number closest to .40. It is .3997 and corresponds to a z score of 1.28 as read from the margins. This means 1.28 standard deviations. Since one standard deviation in our given distribution is 5 score units, we must multiply 1.28 by 5 to get to our desired cut-off point. The result tells us that the cut-off point is 6.40 score units above the mean. Hence it is $80+6.40$ or 86.4.

8.7 Applications. The researcher makes use of the fact that the positions of terms with various values in the normal distribution can be determined from only the values of the mean and the standard deviation of the distribution. In effect, he can provide an elaborate description of the normally distributed observations he has made simply by stating these two values—the mean and standard deviation. Thus educators and psychologists often obtain tests that have been given to a large number of people as part of what is called a standardization process. The distribution of obtained scores is often approximately normal and the computed values of the mean and standard deviation of this distribution are published. The educator or psychologist can use these two values to determine precisely the place of an individual score in the distribution. His procedure is to convert the individual score into a z score and then to locate this z score in the normal distribution, using the table in Appendix III.

PROBLEM SET A

8.1 Find the percentile ranks of the terms in a normally distributed population which have the following z scores: (a) -2.0, (b) -1.5, (c) $+1.0$, (d) $+1.8$.

8.2 Find what per cent of the terms in a normally distributed population have z scores between (a) 0 and $+1.0$, (b) -1.0 and $+1.0$, (c) -0.5 and $+1.0$, (d) -2.0 and $+1.5$.

8.3 Find what per cent of the terms in a normally distributed population have z scores between (a) -1.64 and -1.03, (b) -2.15 and -1.96, (c) $+0.74$ and $+1.62$, (d) $+1.62$ and $+2.94$.

8.4 Find the z scores of the terms in a normally distributed population which have the following percentile ranks: (a) 20, (b) 45, (c) 65, (d) 80.

8.5 Find the z scores of the terms in a normally distributed population which have the following percentile ranks: (a) 15, (b) 40, (c) 55, (d) 75.

8.6 The distribution of scores on a mechanical aptitude test is normal, with a mean of 72 and a standard deviation of 8. Find the percentile rank for each of the following scores: (a) 52, (b) 67, (c) 72, (d) 75, (e) 83.

8.7 For the test described in Problem 8.6, find (a) the cut-off point for the top 10 per cent of the scores, (b) the cut-off point for the lowest 10 per cent of the scores, (c) an interval, centered at the mean, that includes the middle 50 per cent of the scores.

8.8 A set of measurements is normally distributed with a mean of 75 and a standard deviation of 15. Find (a) P_{30}, (b) P_{45}, (c) P_{50}, (d) P_{70}, (e) P_{85}.

8.9 For a distribution of 1,000 test scores, the mean is found to be 500 and the standard deviation is 100. Assume that the distribution is normal and answer the following questions: (a) What per cent of the students had scores less than 750? (b) What is the percentile rank of a score of 475? (c) What is the score exceeded by 60 per cent of the students? (d) Find an interval, centered at the mean, that includes 60 per cent of the scores. (e) There are only 10 scores less than what score?

8.10 The time lengths of films produced by a French film company are found to be normally distributed with a mean length of 92 minutes and a standard deviation of 23 minutes. Draw the normal curve depicting the distribution of film lengths. Find what per cent of films last (a) less than 65 minutes, (b) more than 77 minutes, (c) less than 112 minutes, (d) more than 120 minutes, (e) between 69 and 115 minutes, (f) between 92 and 100 minutes, (g) between 100 and 108 minutes.

8.11 The mean grade of a class of home economics students was 7.2 and the standard deviation was 1.2. If the grades were normally distributed, find (a) the highest grade for the lowest $33\frac{1}{3}$ per cent of the class, (b) the lowest grade for the highest 15 per cent of the class.

8.12 The mean wage in a certain industry is $4.75 per hour with a standard deviation of $0.50. If the wages follow a normal distribution, what per cent of the employees receive wages between $4.00 and $5.50 per hour?

8.13 The mean life of a certain appliance is 36 months with a standard deviation of 6 months. Assuming a normal distribution, what per cent of these appliances should be expected to last from 27 to 41 months?

8.14 In a normal distribution of scores with a mean of 100 and a standard deviation of 50, there are 165 scores larger than 200. How many scores are to be expected between 170 and 210?

PROBLEM SET B

8.15 Find the percentile ranks of terms in a normally distributed population which have the following z scores: (a) -1.9, (b) -0.8, (c) 0, (d) $+0.5$.

8.16 Find what per cent of the terms in a normally distributed population have z scores between (a) 0 and $+0.5$, (b) -1.2 and $+1.2$, (c) -0.8 and $+1.3$, (d) -1.7 and $+0.9$.

8.17 Find what per cent of the terms in a normally distributed population have z scores between (a) -2.19 and -1.72, (b) -2.11 and $+1.69$, (c) $+0.51$ and $+1.37$, (d) $+1.32$ and $+2.41$.

8.18 Find the z scores of the terms in a normally distributed population which have the following percentile ranks: (a) 30, (b) 50, (c) 70, (d) 90.

8.19 Find the z scores of the terms in a normally distributed population which have the following percentile ranks: (a) 25, (b) 35, (c) 85, (d) 95.

8.20 The final-exam scores of high school juniors in a large city are normally distributed with a mean of 82 and a standard deviation of 7. Find the percentile rank for each of the following scores: (a) 61, (b) 75, (c) 82, (d) 90, (e) 95, (f) 103.

8.21 For the scores described in Problem 8.20, find (a) the score exceeded by the top 15 per cent of the students, (b) the score exceeded by 95 per cent of the students, (c) an interval, centered at the mean, that includes the middle 40 per cent of the students.

8.22 A set of scores is normally distributed with a mean of 85 and a standard deviation of 5. Find the values of (a) Q_1, (b) Q_2, (c) Q_3, (d) P_{40}, (e) P_{60}.

8.23 The mean of an approximately normal distribution of test scores of 3,000 students is found to be 250 and the standard deviation is 50. Find (a) the percentile rank of a score of 280, (b) the number of scores less than 175, (c) the per cent of scores between 160 and 260, (d) an interval, centered at the mean, that includes 40 per cent of the scores, (e) the cut-off point for the top 15 per cent of the scores, (f) the cut-off point for the top 10 scores.

8.24 Suppose that the weights of U.S. adult males are normally distributed with a mean of 180 and a standard deviation of 12 pounds. (a) What per cent of the men weigh 190 pounds or more? (b) What per cent of the men weigh less than 160 pounds? (c) What per cent of the men differ from the mean by at least 35 pounds? (d) Find an interval, centered at the mean, that includes the middle 50 per cent of the weights. (e) What weight is exceeded by only 5 per cent of the weights?

8.25 The mean grade on an examination was 75 and the standard deviation was 10. The top 10 per cent of the class will receive A's. What is the lowest grade a student can score and still receive an A?

8.26 A certain machine manufactures steel balls. The diameters of the balls are normally distributed with a mean of 0.3456 inches and standard deviation 0.0078 inches. Find the per cent of balls with diameters (a) between 0.3400 and 0.3500 inches, (b) greater than 0.3560 inches, (c) less than 0.3232 inches.

8.27 Certain commodities are graded by weight and normally distributed; 20 per cent are called standard, 50 per cent large, 20 per cent super, and 10 per cent colossal. If the mean weight is 0.92 ounces with a standard deviation of 0.08 ounces, find the limits for the weight of the supers.

8.28 The test scores of a group of students are normally distributed with a mean of 74.8. If 17 per cent of the scores are at least 79.5, what is the standard deviation of these test scores?

distributions of sample sums and sample means

9.1 The Central Limit Theorem. The reader may wonder how it is that the curve that depicts the incidence of bullets around a target also describes the distribution of IQ's in the United States. The heights of mature trees are also distributed normally and might have been used as illustrative observations. In fact, we are scarcely restricted to subject matter when looking for illustrations of the normal distribution.

This chapter introduces a highly theoretical discussion leading to an explanation of why the normal distribution arises in so many contexts. For the sake of discussion, we are going to pretend that it is our job instead of nature's to make up the values of a multitude of terms. In making up these values, we shall use the same procedure that nature often employs when she produces a distribution of normally distributed terms.

To clarify its details we shall discuss the procedure as if we are carrying it out ourselves. In Section 9.2 we will see how nature makes use of the procedure. As scientists our role is that of observing the normal distributions that she produces. We will postpone examining the relevance of the procedure for the time being and begin simply by discussing its technical aspects.

Assume that to begin with we have access to an infinitely large population of terms and that we can choose at random as many terms from this population as we wish. *To proceed we need know nothing about the shape of the distribution of terms in this infinite population.* We assume only that it has some unknown mean and standard deviation and that we have access to its terms. Theorem 9.1 describes the procedure for gathering terms from the infinite population and combining them in such a way as to create a normal distribution.

Theorem 9.1 (The Central Limit Theorem) *Suppose that a multitude of equal-sized random samples are gathered from the same infinite population. The sum of each sample is computed and the sums of the different samples are put together to form a new distribution. It follows that the new distribution is normal. (An assumption is that the random samples yielding the sums are large enough.*)*

Note the assumption that we have access to the terms in an infinite population about which we know nothing. The procedure is to gather equalized random samples from this population, and *to contribute the sum of each sample as a single term in a new distribution.* Theorem 9.1 states that under these conditions the new distribution will be normal even when the original distribution is not normal.

The following illustration should make Theorem 9.1 clear. Picture a giant wooden crate filled with an inexhaustible supply of tiny slips of paper. On each slip a number is written. For simplicity, assume that only the numbers 0, 1, and 2 are included and that each of these numbers appears on roughly one-third of the slips. (Strictly speaking, we need not make either of these assumptions.) Remember that the supply of slips of each denomination is infinite. That is, there are so many that we can withdraw any number of slips of any denomination and still we need not be concerned with having changed the balance among those remaining in the crate.

The wooden crate has a small opening in its corner through which a slip can be released. It also contains an apparatus that shuffles the slips so thoroughly that every one of them stands an equal chance of being released when we decide to draw one. To draw a slip we press a button that automatically shuffles the slips before releasing one. Thus we can obtain a slip at will, but when we do, every slip in the crate is an equally likely candidate to be the one that is released. Accordingly, our observations are independent and our sample is random.

We are now ready to gather equal-sized samples. We decide to gather samples of 200 slips (*i.e.*, terms) each.

We begin by taking 200 slips, one after the other. The number on the first one is a 2; the number on the second one is a 0; on the third, a 2; and so on. The total of all the numbers on the 200 slips (comprising the first sample) turns out to be 210. In other words, the sum of the terms in the first sample is 210. This sum becomes the first term in the new distribution about to be created. We denote the value 210 on our "scorecard."

The sum of the next 200 slips turns out to be 194. The process is repeated for a multitude of samples of 200 slips each. Theorem 9.1 tells us that the distribution of sums of samples approaches the normal distribution as more and more sums are contributed. Fig. 9.1 shows one stage of the process.

---

* Nearly always the samples are large enough if they contain at least 30 terms each. The proof of Theorem 9.1 is beyond the scope of this book.

The reader should make sure it is clear how Theorem 9.1 predicts the normality of the distribution in Fig. 9.1. The surprising thing about Theorem

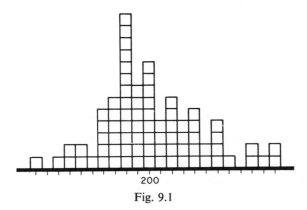

200

Fig. 9.1

9.1 is that nothing is stated about the population from which the samples are drawn. The distribution of the sums of the samples is normal regardless of the shape of the distribution from which the samples are drawn. However, when the "parent" population is extremely unbalanced, it is necessary to take giant-sized samples to ensure that the sums of these samples will be normally distributed.

9.2 How Nature Uses Theorem 9.1   Theorem 9.1 explains why the normal distribution arises in so many different contexts. The method we used in sampling with the slips of paper seems to be a favorite procedure of nature. In each situation presented, the numbers that comprise a normal distribution may be viewed as sums of equal-sized samples of independent observations.

Consider again the bullets that fell into a normal distribution surrounding the telephone pole. The place where each bullet struck was, in effect, the sum of a number of influences. For instance, the way a particular marksman stood may have directed his bullet two feet to the right of the target $(+2)$. A wind may have veered his bullet a foot to the left (-1). A cloud passing overhead changing the light may have had no effect whatsoever (0).* The marksman's proneness might be to fire "first shots" a foot to the left (-1), and an involuntary tension may have jerked his aim two feet to the right $(+2)$. A shadow may have shifted his aim one-and-a-half feet to the right $(+1\frac{1}{2})$.

These and other influences operate simultaneously on a particular marksman and they are different for different marksmen. A marksman's score,

_____

* This influence would have to be included if its effect on any other shots was different from zero.

117

indicating where his bullet finally went, is the sum of a sample of influences. To be concrete, let us say that each marksman's score was the sum of 70 major influences. Therefore, the score given to each shot may be viewed as the sum of a sample of 70 terms and corresponds to the sum of the numbers on 70 slips. Thus the scores obtained by the different marksmen may be viewed as the sums of different equal-sized samples. The distribution of shots (or sample sums) was normal, as would have been predicted by Theorem 9.1.

The normality of people's intelligence levels can be explained by the theorem. Each person's intelligence level may be viewed as a sum of small contributions from various sources. An individual's actual intelligence is the sum of contributions from hereditary factors and other factors like nutrition, opportunity, emotional makeup, and interest. For example, the contribution to an individual's intelligence from his grandmother might correspond to one IQ point. The contribution to his intelligence from his nutrition might be worth two points. According to this view, the intelligence level of each person is the sum of a vast number of contributions. Thus the distribution of intelligence levels of a multitude of people is normal by Theorem 9.1.

The distribution of the heights of trees is normal for essentially the same reason. The height of a tree may also be viewed as the sum of a vast number of small contributions. Its heredity and environment both contribute to its growth in numerous ways. Similarly, whenever the terms in a population are each subject to the same vast number of small and relatively independent influences, then the distribution of the terms is normal.*

9.3   The Central Limit Theorem for Sample Means. Let us now consider a variation of Theorem 9.1. To begin with an illustration, we shall use the crate analogy again. Suppose as before that we draw a multitude of random samples (of 200 numbered slips each) from the crate. In Section 9.1 we made up a distribution out of the sums of these various samples. Theorem 9.1 stated that the distribution of these sample sums is normal.

This time suppose that we find the mean of each sample instead of its sum. Now the various sample means are put together to form a new distribution. Theorem 9.2 states that the distribution of sample means is normal.

Theorem 9.2   (The Central Limit Theorem for Means) *Suppose that a multitude of equal-sized random samples are gathered from the same infinite population. The mean of each sample is computed and the means of the*

---

* When there are a large number of terms in each sample, the distribution of sample sums is normal even though not all the observations (or influences) are mutually independent. In other words, the demand for absolute independence becomes diminished when the samples yielding their sums are large. Undoubtedly many of the influences on the marksman's aim were somewhat related when he fired, and many of the influences on intelligence are interrelated. Yet the distribution of shots and that of intelligence levels are each normal.

different samples are put together to form a new distribution. It follows that the new distribution is normal. (An assumption is that the random samples yielding the means are large enough.) *

Theorem 9.2 follows directly from Theorem 9.1. The mean of each sample is its sum divided by the number of terms included in the sample. Remember that each sample must have an identical number of terms. (This number was 200 in the crate analogy.)

Let us say that there are N terms per sample. Thus if the sums of five samples happen to be 200, 400, 400, 800, and 1,200, then the corresponding sample means are $200/N$, $400/N$, $400/N$, $800/N$, and $1,200/N$. Specifically, where $N = 200$, the means of the five samples are:

$$\frac{200}{200} \quad \frac{400}{200} \quad \frac{400}{200} \quad \frac{800}{200} \quad \frac{1,200}{200}$$

or, more simply, 1, 2, 2, 4, 6. One might say the set of means has been obtained from the set of sums by "proportional shrinking." Thus the distribution of means is a small-scale version of the distribution of sums. The graph of the sums of the five samples is shown in Fig. 9.2, and the graph of their means is shown in Fig. 9.3.

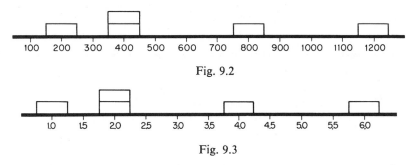

Fig. 9.2

Fig. 9.3

According to Theorem 9.1, as more and more samples are gathered, the distribution of samples' sums approaches the normal distribution, as shown in Fig. 9.4.

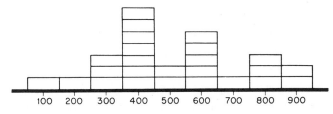

Fig. 9.4

---

* See the footnote on page 116.

The distribution of sample means continues to mirror precisely the distribution of sample sums. Thus the distribution of sample means also becomes normal (Fig. 9.5).

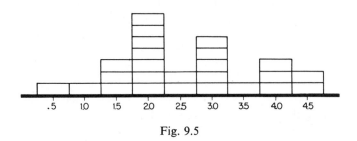

Fig. 9.5

We have demonstrated Theorem 9.2 but we have not proven it rigorously. The reader who has mathematical facility might deduce Theorem 9.2 from Theorem 9.1. He would have to show that dividing all the terms in a distribution by the same constant does not change either the z scores or the percentile ranks of any of the terms. Thus the operation of dividing all the the terms by N leaves the pairings of z scores and percentile ranks unchanged. The set of pairings fully defines the shape of a distribution. Since the distribution of sample sums is normal by Theorem 9.1, the distribution of sample means is also normal.

9.4 Properties of the Distribution of Sample Means. Our main focus will now be on the distribution of sample means. It is helpful to view this distribution as a secondary distribution. Specifically, starting with any population, suppose that equal-sized random samples are drawn from it and that the means of these samples become the numbers or elements that form a new distribution. It is the new distribution thus formed that is called the distribution of sample means.

From now on it will be assumed that this new distribution is infinite. That is, an infinite number of samples are drawn from the original population so that an infinite number of sample means comprise the new distribution.

There exist important relationships between the mean and standard deviation of the original population and the mean and standard deviation of the distribution of sample means created from this population. The mean (or balance point) of the distribution of sample means is the same as the mean of the original population.

Theorem 9.3 *Suppose an infinitude of equal-sized random samples is drawn from the same population and the means of these samples are put together to form a new distribution. The mean of the new distribution (composed of sample means) has the same value as the mean of the original population.*

We call such a new distribution "the sampling distribution of means." A sampling distribution is a theoretical distribution of some particular value, like a mean, as it would be obtained from an infinite number of samples.

Thus the symbol μ, used to denote the mean of the original population, may also be used to denote the mean of the distribution of sample means. The standard deviation of the distribution of sample means shall be called the standard deviation of means.*   It is related to the standard deviation of the original population.

Theorem 9.4   *Suppose an infinitude of equal-sized random samples is drawn from the same population.   Denote the number of terms in each of these samples by N.   The distribution of the means of these samples has a standard deviation (called the standard deviation of means or the standard error of the mean).   The standard deviation of means equals the standard deviation of the original population divided by the square root of N.*

The symbol for the standard deviation of the means is σ_M.   Thus, Theorem 9.4 states that:

$$(9.1) \qquad\qquad \sigma_M = \frac{\sigma}{\sqrt{N}}$$

where σ is the standard deviation of the population from which the samples are drawn and N is the size of the samples.

Consider the population of IQ scores of adult males in New York City. The mean of this population is 100. ($\mu = 100$). The standard deviation of this population is 15 points ($\sigma = 15$). Suppose we gather an infinitude of random samples of 10 scores each from this population and we put together the means of these samples to comprise a distribution.

According to Theorem 9.3, the distribution of sample means has a mean (or balance point) at the value 100.   Theorem 9.4 enables us to find the standard deviation of the distribution of sample means.

$$\sigma_M = \frac{\sigma}{\sqrt{N}} = \frac{15}{\sqrt{10}} = 4.7$$

Fig. 9.6 shows the graph of the distribution of individual IQ scores.   The graph of the means of random samples of 10 scores each is also shown.

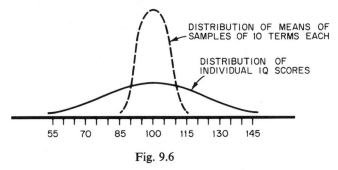

DISTRIBUTION OF MEANS OF SAMPLES OF 10 TERMS EACH

DISTRIBUTION OF INDIVIDUAL IQ SCORES

55   70   85   100   115   130   145

Fig. 9.6

---

* This standard deviation of means is often called the *standard error of the mean.*

The graph of the distribution of sample means is the more compact, reflecting the fact that the standard deviation of means is 4.7 points whereas the standard deviation of the original population is 15 points. Viewed in another way, the sample means cluster around the balance point more than do the individual IQ's. Thus the means of different samples resemble each other more closely than do the individual IQ scores.

The resemblance between the means of samples may be ascribed to a "balancing process" within each sample. Within any sample there are apt to be both high scores and low scores. Usually these scores tend to "average each other out" so that the sample mean turns out to be somewhere near 100, the value of the population mean. Thus it would be very rare to find a sample with a mean as low as 80 or as high as 120. The members of the sample would have to perform either consistently poorly or consistently well, and, since the members in each sample are randomly chosen, this is unlikely. Whereas individual scores are apt to differ markedly from the population mean, the sample means are less apt to spread out.

We may view the situation another way. Theorem 9.4 implies that the standard deviation of a distribution of means of samples is always smaller than the standard deviation of the original distribution from which the samples are drawn. Thus, the means of samples always resemble each other more closely (and cluster closer to the balance point) than do the individual terms in the population from which the samples are drawn.

The size of the samples yielding the means determines how closely the means will cluster around the balance point. The larger the samples are, the more closely their means cluster together. We can see in Formula 9.1 that, as the value of N increases, the value of σ_M decreases. In other words, when the number of terms per sample is large, the standard deviation of the sample means becomes small.

Three distributions are shown in Fig. 9.7. The first (Curve A) is the distribution of individual IQ scores with a mean of 100 and a standard deviation of 15 points. The second (Curve B) is the distribution of means of samples of 10 terms each drawn from the IQ population. This distribution of sample means has a mean of 100 and a standard deviation of 4.7 points. It is clearly more compact than the distribution of individuals' scores. Finally the distribution of the means of samples of 100 scores each is also shown (Curve C). The mean of this distribution is 100 but its standard deviation is only 1.5 points. In other words, the means of samples of 100 scores each tend to be very much like each other.

9.5   Locating a Sample Mean in the Distribution of Sample Means.   Suppose that a vast number of equal-sized random samples are drawn from the same population and that the means of these samples form a normal distribution. We have seen how to compute the z score and the percentile rank of a single term in a normal distribution. We shall now consider how to compute the z score and the percentile rank of a single sample mean in the normal distribution of sample means. (Note that we are assuming that the samples are large enough to make the distribution of their means normal.)

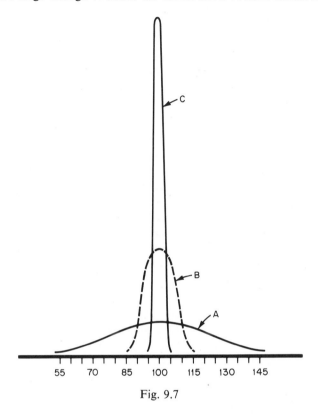

Fig. 9.7

To begin with, suppose we know the values of the mean and standard deviation of the population of individual terms (from which the samples are drawn). We have access to only one of the samples and we compute its mean. Without seeing any of the other samples, we are to determine the z score and the percentile rank of our particular sample mean in the distribution of sample means.

Before illustrating the method, we shall outline it briefly. Remember that we are given only the mean and the standard deviation of a population. We are told that many equal-sized random samples have been drawn from this population. We know the mean of one of these samples and wish to determine its z score and its percentile rank in the distribution of sample means.

Remember, Theorem 9.2 tells us that the distribution of sample means is normal. The first step is to compute the mean and standard deviation of the distribution of sample means. Our particular sample mean is a single term in this normal distribution of sample means. We compute the z score of our sample mean in its distribution by a variation of the standard method used to

123

compute the z score of a term. Since we are locating the sample mean in the distribution of sample means, we use as a yardstick the standard deviation of means. The z score of our particular sample mean is the number of yardsticks it is above or below the balance point of the distribution of sample means.

At this point, we have computed the z score of our particular sample mean in the normal distribution of sample means. It should be clear that the sample mean is simply a term in a normal distribution. Now that we know the z score of this term we are able to find its percentile rank by referring to the normal distribution table.

We shall now illustrate in detail the method of finding the z score and the percentile rank of a sample mean in its distribution of sample means. Once again we begin with the population of individual IQ's, which we will say has a mean of 100 ($\mu = 100$) and a standard deviation of 15 points ($\sigma = 15$). An infinite number of random samples of 25 cases are each drawn from this population ($N = 25$) and their means comprise a distribution. However, we have access to only one sample and have found that its mean is 106. We shall denote the mean of a sample by $\bar{X}$ so that $\bar{X} = 106$. The problem is to find the z score and the percentile rank of the sample mean of 106 in the distribution of sample means.

The distribution of sample means which has been created is normal (by Theorem 9.2) and has a mean of 100. By Formula 9.1, the standard deviation of the distribution of sample means turns out to be 3 units.

$$\sigma_M = \frac{\sigma}{\sqrt{N}} = \frac{15}{\sqrt{25}} = 3$$

Therefore the distribution of sample means looks like Fig. 9.8.

The sample mean of 106 is located in Fig. 9.8. Our next step is to find the z score for the sample mean of 106. Remember how we found the z score of an individual's IQ—we used the standard deviation (of 15 points) as a

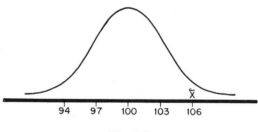

94 97 100 103 106

Fig. 9.8

yardstick; we then determined how far the individual's score was from the balance point and divided that distance by the length of the yardstick.

Our attention is now focused on the distribution of sample means. Therefore, we will use an analogous yardstick, the standard deviation of means.

The z score of a particular sample mean is obtained by first finding the distance from the balance point to the particular sample mean. We then divide this distance by the length of the yardstick we are using. The quotient we get, which tells how many yardsticks the particular sample mean is from the balance point, is the z score of the particular sample mean.

The sample mean of 106 is 6 points above the balance point of its distribution. The standard deviation of means is 3 points. Thus the sample mean is above the balance point a distance that is twice the length of the standard deviation of means, as illustrated in Fig. 9.9. In other words, the z score of the sample mean is 2.

Finally we can determine the percentile rank of the sample mean of 106 from its z score. Appendix III tells us that, in a normal distribution, the term with a z score of 2 has a percentile rank of 98. Thus the sample mean of 106 is higher than approximately 98 per cent of all the sample means.

Before considering a second problem, it is worthwhile restating the main principles in what has been said. The supposition is that a vast number of equal-sized random samples are drawn from the same population. The means of these random samples form a normal distribution, called the

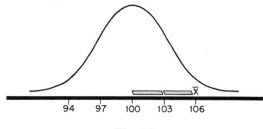

Fig. 9.9

distribution of sample means. Our knowledge of the original population enables us to determine the values of the mean and standard deviation of the distribution of sample means. Once we know the values of the mean and standard deviation of the distribution of sample means we can readily compute the z score of any particular sample mean. We are then able to determine the percentile rank of that sample mean from its z score.

Incidentally, from now on when we speak of the z score or the percentile rank of a sample mean, it will be assumed that the particular sample mean is being considered as a member of the theoretical distribution of equal-sized random samples drawn from the same population.

To illustrate the procedures again, suppose that the population of IQ's among men of army age has a mean of 100 and a standard deviation of 15 points. After induction, the vast number of draftees are randomly assigned to squadrons containing 400 soldiers each.

The mean IQ of the men in a particular squadron, Squadron B, is 99.25. Our problem is to determine the relative standing of Squadron B among all the squadrons. Specifically, we ask what per cent of all the squadrons have (mean) IQ's below 99.25.

We are given that the distribution of individual IQ's has a mean of 100 and a standard deviation of 15 points. The squadrons are each random samples of 400 IQ's drawn from this population. Therefore, the distribution of squadron means has a balance point at 100 (Theorem 9.3). The standard deviation of means is found by Formula 9.1.

$$\sigma_M = \frac{\sigma}{\sqrt{N}}$$

$$\sigma_M = \frac{15}{\sqrt{400}} = \frac{3}{4}$$

In sum, the distribution of squadron means has a mean of 100 and a standard deviation of 3/4. (The relevant yardstick is 3/4.) The distribution of squadron means is shown in Fig. 9.10.

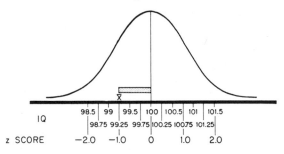

Fig. 9.10

The mean of Squadron B, 99.25, is located in the illustration. The z score of the mean of Squadron B may be determined from two facts:

1. The mean of Squadron B is .75 points below the balance point.
2. The standard deviation of means, which is the relevant yardstick, is 3/4 of a point.

The z score of the mean of Squadron B is $\frac{-.75}{.75} = -1$. The distribution of squadron means is normal (since it is a distribution of large equal-sized random samples from the same population).

Appendix III tells us that in a normal distribution the term with a z score of -1 is larger than approximately 16 per cent of the other terms. Thus the mean of Squadron B is at the 16th percentile. In effect we have compared a particular sample with the other samples by finding the z score of its mean and then converting this z score into a percentile rank. In the chapters to follow, we shall have many occasions to locate a particular sample mean in the distribution of sample means.

PROBLEM SET A

9.1 A musical aptitude test was given to all third-grade classes in New York City. The mean score on this test was 75 and the standard deviation of children's scores was 15 points. Suppose that the classes are composed of 36 children each, so that in effect there are a vast number of samples of 36 scores each. (a) In what per cent of the classes would we expect the class mean to be as high as 8C? (b) In what per cent would the mean be as low as 72?

9.2 Repeat Problem 9.1 using random samples of 50 scores each.

9.3 Repeat Problem 9.1 using random samples of 25 scores each. Compare with the results of Problems 9.1 and 9.2.

9.4 A certain population has a mean of 90 and a standard deviation of 20. Give the mean and the standard deviation of a distribution of sample means for a large number of random samples of size (a) 50, (b) 200, (c) 800.

9.5 For each part of Problem 9.4, find what per cent of the random samples of the specified size would be expected to have means below 88.

9.6 For each part of Problem 9.4, find what per cent of the random samples of the specified size would be expected to have means in the interval from 88.5 to 91.5.

9.7 For each part of Problem 9.4, find the score that we would expect the means of only 30 per cent of the random samples of the specified size to exceed.

9.8 For each part of Problem 9.4, find an interval, centered at the mean, that would be expected to include the means of 50 per cent of the random samples of the specified size.

9.9 The distribution of scores on a test of driving skill is found to have a mean of 80 and a standard deviation of 9. (a) Find the expected mean and standard deviation for a distribution of means of random samples of size 100. (b) Find the per cent of random samples of size 100 that would be expected to have a mean above 87. (c) Find the per cent of random samples of size 30 that would be expected to have a mean below 82. (d) Find an interval, centered at the mean, that would be expected to include the means of 50 per cent of random samples of size 50.

9.10 Suppose that the weights of U.S. adult males are normally distributed with a mean of 180 and a standard deviation of 12 pounds. (a) What per cent of samples of size 500 have a mean as small as 168 pounds? (b) What per cent of the samples of size 50 have a mean as large as 174 pounds? (c) Find the weight that we would expect to be exceeded by the means of 80 per cent of the random samples of size 100. (d) Find an interval, centered at the mean, that would be expected to include the means of 50 per cent of the samples of size 30.

9.11 Transistors manufactured by a certain process have a mean lifetime of 600 hours and a standard deviation of 45 hours. Samples of 30 transistors are taken from this group. What per cent will have a mean lifetime between 580 and 620 hours?

9.12 Repeat Problem 9.11 using samples of 75 transistors. Compare your answer with the answer to 9.11 and explain any difference.

PROBLEM SET B

9.13 The distribution of scores obtained by high school seniors on a language aptitude test has a mean of 80 and a standard deviation of 10. This distribution is normal. Suppose that random samples of 50 scores each are taken from the original distribution of scores and the means of these samples are calculated. (a) What would be the mean and standard deviation of this new distribution? (b) In what per cent of the samples would we expect the sample mean to be lower than 78? (c) In what per cent of the samples would we expect the sample mean to be higher than 83?

9.14 Repeat Problem 9.13 using random samples of 30 scores each.

9.15 Repeat Problem 9.13 using random samples of 400 scores each. Compare with the results of Problems 9.13 and 9.14.

9.16 A certain population has a mean of 90 and a standard deviation of 20. Give the mean and the standard deviation of a distribution of sample means for a large number of random samples of size (a) 100, (b) 400, (c) 1,600.

9.17 For each part of Problem 9.16, find what per cent of the random samples of the specified size would be expected to have means above 91.

9.18 For each part of Problem 9.16, find what per cent of the random samples of the specified size would be expected to have means in the interval from 88.0 to 92.5.

9.19 For each part of Problem 9.16, find the score that we would expect the means of 60 per cent of the random samples of the specified size to exceed.

9.20 For each part of Problem 9.16, find an interval, centered at the mean, that would be expected to include the means of 40 per cent of the random samples of the specified size.

9.21 Suppose that the heights of American soldiers are normally distributed with a mean of 5 feet $7\frac{1}{2}$ inches and a standard deviation of $3\frac{1}{2}$ inches. (a) What per cent of samples of size 350 have a mean as great as 5 feet 8 inches? (b) What per cent of samples of size 50 have a mean as small as 5 feet 6 inches? (c) Find the height that we would expect to be exceeded by the means of only 10 per cent of the samples of size 100.

9.22 Consider a population consisting of only the scores 3, 6, 8, 11, and 15. (a) Find the mean and standard deviation of these scores. (b) Find all the possible samples of size two (that is, 3 and 6, 6 and 3, 3 and 3, etc.) which can be drawn from this population, *with replacement*, and list these (there are 25 such samples). (c) Find the means of all the samples of size two. (d) Find the mean of the sample means. (e) Find the standard deviation of the sample means. Compare the mean of all the samples of size two and the standard deviation with the results in (a).

9.23 Repeat Problem 9.22 when the sampling is done without replacement—only twenty samples of size two.

9.24 Consider a population consisting of 2, 7, 12, and 17. Take all the possible samples of two, *without replacement*. Find (a) the population mean, (b) the population standard deviation, (c) the means of the samples of size two and the mean of these means, and (d) the standard deviation of the samples' means. (e) Recalculate (c) and (d) from (a) and (b) by appropriate formulas.

ten

probability and sampling from the normal distribution

10.1 Probability. Probability is a topic that has long fascinated philosophers and mathematicians. The concept of probability is subtle; as a matter of fact, a century after two mathematicians named Pascal and Fermat founded the science of probability, no one knew quite how to define it. Probability seemed to imply that things "partly happen and partly don't," whereas in the world things either happen completely or they don't happen at all.

Fortunately, the science of probability was fruitful enough to survive until a satisfactory rationale was developed for it. The science of probability has received considerable attention, and today there are several ways of defining probability which are adequate for most purposes. Probability theory has flowered in the twentieth century and it now enriches many applied fields.

On some level we all sense the meaning of probability, even though we do not have a handy definition of it. Children seem to have some understanding of probability, because they show surprise at improbable happenings. Even white rats manifest a sense of probability: they act very differently when their rewards or punishments are made more or less probable by psychologists.

Probability must be viewed in the context of an experiment or procedure that permits definable outcomes. For instance, the flip of a coin is an experiment that permits two outcomes, a head and a tail. A race is an experiment in which we may specify as many outcomes as there are possible winners. If we are interested in the exact order in which the participants finish, we must consider each possible order of finishing as an outcome that may occur.

The weather on a Sunday afternoon is in this sense also an experiment. If we are interested in whether it will rain we may specify two outcomes, rain and no rain. Remember that the outcomes of an experiment must be defined and recognizable and that the experiment must inevitably lead to some outcome. The outcomes will be defined in such a way that no two of them can possibly occur together.

The drawing of a slip from the wooden crate, as described in the preceding chapter, is an experiment. If the numbers on the slips are 0, 1, and 2, then these numbers are the three possible outcomes. The drawing of a *sample* of slips to determine its mean may also be viewed as a single experiment. The mean of the sample may be considered the outcome of this experiment. In the same way, selecting a person at random and determining his height is an experiment that theoretically permits an infinitude of outcomes corresponding to the possible heights that may be obtained. Choosing a random sample of people and finding the mean height in this sample is also an experiment with an infinitude of possible outcomes.

The word *probability* is always used with reference to the outcome of an experiment. The exact nature of the experiment and the specified outcome must be made clear before the probability of the outcome can be found. The probability of an outcome occurring is a *fraction* or *proportion*.

Definition 10.1   *The probability of an outcome is the proportion of times that the outcome would occur if the experiment were repeated indefinitely.*

Suppose the experiment is the flip of a coin, which has two possible outcomes, a head and a tail. The probability that a head will turn up is defined as the proportion of times that heads would appear if the coin was flipped over and over again an infinitude of times.

Picture someone flipping a coin into the air over and over again. It is our job to record the number of heads and the number of flips of the coin. We construct a fraction that has as numerator the number of heads already obtained and as denominator the total number of flips of the coin. To be concrete, suppose that a head comes up four times in the first ten flips. At this point the fraction is 4/10. The proportion of heads obtained is .40. The flipping continues and after 100 flips suppose that the number of heads is 53. The fraction is now 53/100 and the proportion is .53. The repetitions of the experiment go on and on; that is, the flipping of the coin continues indefinitely.

Table 10.1 provides a record of the number of heads obtained after various numbers of flips of the coin. The different proportions of heads are also indicated.

TABLE 10.1   HEADS OBTAINED IN COIN EXPERIMENT

Heads	4	53	139	237	480
Flips	10	100	300	500	1,000
Proportion of Heads	.40	.53	.46	.47	.48

Early in the game, the proportion of heads obtained is greatly influenced by the outcome of each flip. For instance, the appearance of one more head during the first ten flips would have raised the proportion of heads after ten flips from .40 to .50. However, as the number of flips increases, the importance of a few flips becomes less and less. The proportion of heads gains a kind of stability, and after a while even long runs of heads or tails scarcely influence it.

Mathematicians like to say that the proportion of heads *settles down* at some value, and this value is defined as the probability that a head will turn up. If the coin is a fair one, the value is .50. If the coin is biased in favor of heads, the fraction is more than .50. If the coin is biased against heads, the fraction is less than .50.

Let us suppose the coin being investigated has a tail on each side. Now as the repetitions of the experiment continue, the numerator—indicating the number of heads obtained—remains equal to zero. The denominator—indicating the number of repetitions of the experiment—keeps increasing. No matter when we stop, the proportion of heads stays equal to zero. Similarly, *whenever any outcome is impossible, its probability is zero.*

On the other hand, suppose the coin has a head on both sides. The number of times a head turns up will always equal the number of repetitions of the experiment. Thus the numerator of the fraction will always equal the denominator. Now the probability of getting a head equals one. In general, *whenever an outcome is certain, its probability is one.*

The probability of any outcome is a number between zero and one. That is, the probability of any outcome ranges between impossibility and certainty. Incidentally, if we add up the probabilities of all the possible outcomes of an experiment, the total we get equals one (certainty). Thus if there are two possible outcomes and the probability of one of them is .37, then the probability of the other must be .63. (1.00 − .37 = .63).

10.2 The Word Probability in Everyday Language. Many theorists believe that our definition of probability covers the way the word is used in common parlance. The belief is that whenever we say "probably," we are applying knowledge of what occurs in the long run to a specific instance.

The statement "There is a 70 per cent chance of measurable precipitation tomorrow" is a typical weather-forecast statement. It sounds as if it were about tomorrow and no day but tomorrow. However, in actuality the statement refers to a whole collection of days, and tomorrow is only one member of this set. The statement implies that on 70 per cent of the occasions when there has been a weather situation like today's there has been measurable precipitation on the following day.

It is worthwhile illustrating the importance of clearly defining an experiment or process leading to an outcome before trying to determine the probability of that outcome. Two people may conjecture about the same outcome, but if they view the procedure that leads to it differently, their expectations are apt to be different.

For instance, suppose Person A knows only that his friend was hit by an automobile and brought to the hospital three months ago. Let us be diabolical and think of the accident as an experiment. Consider one particular outcome—that the victim was seriously injured. Suppose it is known that one-half of hospitalized auto-accident victims are seriously injured. Person A, who knows that his friend was hospitalized following an auto accident, would make his prediction. He would classify his friend as a member of the group of people hit by automobiles and brought to the hospital. This would lead him to say that the probability is one-half that his friend was seriously injured.

However, suppose we know (but Person A doesn't) that the victim left the hospital one week after being admitted. Furthermore we know that one-fifth of the individuals hit by automobiles and detained in hospitals for one week are seriously injured. On the basis of what we know, it is correct for us to state that the probability is one-fifth that the victim was seriously injured.

Person A has arrived at his probability statement correctly and so have we. The two statements differ because they are based on different information. The moral is that we must be careful to describe in detail our experiment or process when we discuss probability. If we omit information, we cannot expect another person to attach the same probability value to an outcome that we do.*

10.3 Probability and the Sampling Experiment. One kind of experiment in which we make probability statements is of primary concern to researchers. It involves selecting a random sample consisting of one or more terms from a population. The variable of interest is one that may assume different values in different samples. For instance, the values of the variable may be the means of the samples. Its value in the particular sample chosen is considered the outcome of the experiment.

Each sample value or outcome has some probability of occurring. Once a particular value is specified, we can determine the probability that the sample drawn will have this value. Our focus now is on how to determine the probability of getting any particular outcome when an experiment is performed. In this section, we will assume that the sample drawn consists of one term. That is, the experiment involves choosing a single term at random from the population. The problem is to determine the probability that the term chosen has some specified value.

---

* For an interesting discussion of probability, see "It's More Probable Than You Think," by Martin Gardner in the November 1967 issue of *The Reader's Digest.*

An illustrative experiment was conducted not long ago on a farm. The farmer, a man named Hiram, owned 100 chickens—70 of the Plymouth Rock variety and 30 Rhode Island Reds. Hiram was working in the barn and could not see the hawk that soared high overhead. Lazily, the hawk descended and circled until he was at an angle where the barn momentarily concealed him from the chicks. An instant later he bolted forward and was gone. Ninety-nine chicks were left scurrying about the yard.

Their chattering brought Hiram out of the barn, and he quickly realized what had happened. One thing made him curious though. Was it a Plymouth Rock chick or a Rhode Island Red that was at this moment sailing through the heavens? Probably it was a Plymouth Rock chick, the farmer thought. There were 70 of them and only 30 Rhode Island Reds. A hawk can be easily seen when he attacks and he has no time to choose his prey during his momentary plunge.

Hiram's problem may be formulated this way. We shall assume as he did that the hawk made a random selection of one chick from the population of 100. The problem is to determine the probability that the victim was a Plymouth Rock chick and the probability that it was a Rhode Island Red. Note that the variable in this experiment is the discrete variable, type of chicken, and the two possible outcomes are the two values this variable may assume.

The probability of an outcome of an experiment is a statement (in the form of a proportion) of what would happen if the experiment were to be repeated indefinitely. The experiment described cannot be repeated in actuality, but we can conceive of it being repeated, say once a day, and that is enough for us to reach a solution.

Suppose that every day for an indefinitely long period the *same* 100 chickens could be placed in the same yard. Each day the hawk swoops down to make his random selection of one of them. The fact that each single chick is an equally likely candidate to be chosen means that in the long run each chick is chosen on one out of every 100 days. Thus on 70 out of every 100 days, the abducted chicken is a Plymouth Rock chicken. We are considering as one possible outcome of the *sampling experiment, the abduction of a Plymouth Rock chicken.* The probability of this outcome is 70/100.

By the same reasoning, on 30 out of every 100 days in the long run the abducted chicken would be a Rhode Island Red. Thus the probability that a Rhode Island Red chicken gets taken in a single experiment is .30.

The point is that the members of the Plymouth Rock group comprise a certain proportion of the population (70/100) and this proportion tells us the probability that in a single experiment a member of the Plymouth Rock group will be selected. The group of Rhode Island Reds also comprises a proportion of the population (30/100) and that proportion tells the probability that a Rhode Island Red will be selected in a single experiment.

In general, suppose the experiment of selecting a random sample is performed. We wish to know the probability that the sample has some particular value or characteristic. The proportion of samples in the population with that value or characteristic tells us the probability that the single sample chosen will have that value.

For example, suppose a single United States inhabitant is chosen at random; consider the outcome that the person chosen lives in the State of New York. Suppose there are about eighteen million New Yorkers and about one hundred eighty million United States inhabitants all together. Then New Yorkers would comprise about one-tenth of the population and the probability that the person chosen will be a New Yorker is about 1/10.

By the same reasoning, suppose there are roughly a quarter of a million medical doctors residing in America. Then about one person out of every 720 is a medical doctor. Therefore the probability that the person chosen at random will be a medical doctor is about 1/720.

So far we have discussed only the situation in which the sample consists of one term. The generalization to samples made up of more than one term will be made later on. It is not difficult, as we shall see.

10.4 Probability and the Normal Distribution. Our knowledge about the normal distribution enables us to make probability statements in connection with it. Consider the experiment of drawing a single term at random from a population of normally distributed terms. Suppose that a prediction about this term is made. For instance, it might be conjectured that the term that is picked will have a z score larger than 0 or that it will have a z score between 0 and 1. Our knowledge about the normal distribution enables us to determine the probability that a prediction of this kind will come true.

Remember how to determine the probability that a randomly drawn sample of one or more cases will have a particular characteristic: A certain proportion of samples in the population have the characteristic; this proportion tells us the probability that a single sample chosen will have the particular characteristic.

Now we are proposing to draw a sample of one case from a normal distribution. Consider the outcome that the term drawn will have a z score larger than 0. In a normal distribution the probability of this outcome equals the proportion of terms in the population that have z scores larger than 0. Therefore, the probability is 1/2 that the term drawn will have a z score larger than 0.

Instead we may specify that the term drawn will have a z score between -1 and $+1$. To determine the probability of this outcome, we must find out the proportion of terms in a normal distribution which have z scores between -1 and $+1$. Appendix III shows that in a normal distribution about .68 of the terms have z scores between -1 and $+1$. Therefore the probability is roughly .68 that a randomly chosen term will have a z score in that interval. The area representing these z scores has been indicated in Fig. 10.1.

135

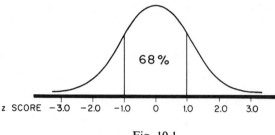

z SCORE −3.0 −2.0 −1.0 0 1.0 2.0 3.0

Fig. 10.1

Another outcome is that the term chosen will have a z score that is not between −1 and +1. That is, it may be specified that the term chosen will have a z score that is either less than −1 or greater than +1. For this outcome to occur, the term must come from one of the "tails" of the distribution depicted in Fig. 10.1. These two tails considered together include about 32 per cent of the terms. Therefore, the probability is 32/100 that a randomly chosen term from a normal distribution will have a z score not in the interval −1 to +1.

About 95 per cent of the terms in a normal distribution have z scores between −2 and +2. Therefore the probability is about 95/100 that a randomly chosen term will have a z score between −2 and +2.

Only 5 per cent of the terms in a normal distribution have z scores that are either less than −2 or greater than +2. In other words, about 5 per cent of the terms are two or more z score units from the balance point in either direction. These terms are included in the tails in Fig. 10.2. The probability that a randomly chosen term will have a z score that is either less than −2 or greater than +2 is about 5/100.

About $2\frac{1}{2}$ per cent of the terms in a normal distribution have z scores larger than +2. That is, about $2\frac{1}{2}$ per cent of the terms are in the right tail of the graph in Fig. 10.2. Therefore the probability is about .025 that a term chosen at random from a normally distributed population will have a z score greater than +2.

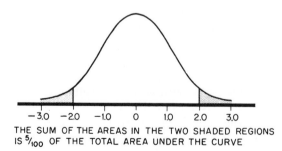

−3.0 −2.0 −1.0 0 1.0 2.0 3.0

THE SUM OF THE AREAS IN THE TWO SHADED REGIONS IS 5/100 OF THE TOTAL AREA UNDER THE CURVE

Fig. 10.2

The material just discussed may now be applied to several problems. Suppose the heights of the men in a certain large city are normally distributed with a mean of 5 feet 8 inches and a standard deviation of 3 inches. One of these men is to be randomly selected from the population. What is the probability that this man is between 5 feet 8 inches and 5 feet 11 inches?

The first step in solving this problem is to convert the specified interval into a z score interval. The value 5 feet 8 inches corresponds to a z score of 0. The value 5 feet 11 inches has a z score of 1.0. Thus the original problem leads us to ask what is the probability that a term randomly chosen from a normal distribution will have a z score value between 0 and 1.0. Appendix III tells us that the probability of this happening is 34/100. Thus the probability that the man who is chosen is between 5 feet 8 inches and 5 feet 11 inches is 34/100.

One might ask what is the probability that a randomly chosen man will be over 5 feet 11 inches. After we translate into z score values, our problem leads us to determine the probability that a randomly chosen term has a z score value greater than 1. Appendix III tells us that this probability is 16/100.

One last problem is to determine the probability that the man who is chosen will be either shorter than 5 feet 2 inches or taller than 6 feet 2 inches. After translating into z score values, we are led to determine the probability that a randomly chosen term will have a z score that is either less than -2 or greater than $+2$. The probability of this happening is about 5/100. Thus the probability is 5/100 that the man who is chosen is either less than 5 feet 2 inches or taller than 6 feet 2 inches.

The turning point in solving problems like those given is to note that the interval specified in the problem defines a corresponding z score interval. The first step is to find this z score interval and the second step is to determine the proportion of terms included in it. This proportion tells us the probability that the term chosen will be a member of the interval specified in the problem.

10.5 Probability Statements about Sample Means. In scientific experiments, usually a sample consisting of many cases is drawn at random from a population. The mean of this sample is found. Suppose as before that a particular interval is specified when the experiment is performed. We can now ask what is the probability that the sample mean will be in the specified interval. The method of solving a problem of this kind is similar to the method used when a single term is drawn. The only difference is that we now refer to the distribution of sample means and not to the distribution of individual terms. The reason is that we are dealing with a sample mean instead of an individual term.

Consider the same population as in the last problem. The distribution of men's heights in this population has a mean of 5 feet 8 inches and a standard deviation of 3 inches. This time a squadron of 100 men is randomly drawn from the population and the mean height of these men is found. When the experiment is performed, what is the probability that the mean height in the squadron is between 5 feet 7 inches and 5 feet 9 inches?

Our reasoning goes this way. We were asked the probability that the mean of a sample of 100 cases would lie in a specified interval. That is, if a vast number of similar samples were drawn, in the long run what proportion of them would have means inside the specified interval? To answer the question, we imagine that an infinitude of random samples of 100 cases each are drawn from the given population. We can determine what the distribution of the means of these samples looks like. For one thing, the distribution of sample means is normal.* Its standard deviation (*i.e.*, the standard deviation of means) is $3/\sqrt{100} = .3$ inches. Therefore the distribution of sample means, which we have conjured up to solve the problem, looks like Fig. 10.3.

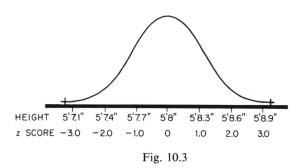

| HEIGHT | 5'7.1" | 5'7.4" | 5'7.7" | 5'8" | 5'8.3" | 5'8.6" | 5'8.9" |
| z SCORE | −3.0 | −2.0 | −1.0 | 0 | 1.0 | 2.0 | 3.0 |

Fig. 10.3

Our job is to determine the probability that the mean of a particular randomly chosen sample will be between 5 feet 7 inches and 5 feet 9 inches. In the distribution of sample means the value 5 feet 7 inches has a *z* score of −3.3. The value 5 feet 9 inches has a *z* score of +3.3 inches. Thus the original problem leads us to ask what proportion of terms (sample means) in the distribution of sample means have *z* scores between −3.3 and +3.3.

Appendix III provides an answer. One can see that virtually all the sample means are in the specified interval. Thus the probability that a randomly chosen sample will be in the specified interval is almost 1. The probability is roughly 999/1,000.

Note that a sample mean of 5 feet 8.3 inches has a *z* score of 1.0. Appendix III tells us that approximately 34 per cent of the sample means in the distribution are between 5 feet 8 inches and 5 feet 8.3 inches. The probability is about 34/100 that the mean height in a randomly chosen squadron of men will be between 5 feet 8 inches and 5 feet 8.3 inches.

---

* The normality is owing to Theorem 9.2. When considering the mean of a large-sized sample, it is not necessary to assume normality in the population.

Now consider this problem. Suppose that the average length of films produced during the last decade is 100 minutes and that the standard deviation of film lengths is 24 minutes. If 36 different films are randomly chosen, what is the probability that the mean length of a film in this sample is either less than 92 minutes or greater than 108 minutes?

To solve this problem we imagine that a multitude of similar samples are drawn. That is, different samples of 36 films each are gathered and the mean length of each sample is found. The means of different samples are put into a distribution.

This distribution of sample means is normal. Its mean is 100 and its standard deviation turns out to be 4 minutes. It looks like Fig. 10.4.

The value 92 corresponds to a z score of -2 and the value 108 corresponds to a z score of $+2$ in this distribution. We were asked the probability that a sample mean chosen at random is either less than 92 or greater than 108 minutes. An equivalent question is what is the probability that a randomly chosen sample mean has a z score either less than -2 or greater than $+2$.

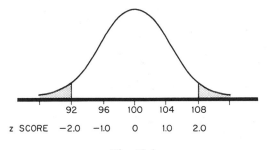

Fig. 10.4

The answer is about 5/100. Thus the probability is about 5/100 that the mean length in a sample of 36 randomly chosen films is either less than 92 minutes or greater than 108 minutes.

The question might have been asked in a different form. One might have asked what is the probability that the mean length of films in the sample chosen is *at least 8 minutes away from 100 minutes* (the mean length of all the films produced). The answer is that the probability is about 5/100 that the mean length in a sample of 36 films will differ from the average film length by as much as 8 minutes. In other words, only 5 samples out of 100 will have means that differ from the population mean by as much as 8 minutes.

PROBLEM SET A

10.1 Raffle tickets numbered from 1 through 728 have been sold. The draw consists of taking one of these tickets randomly from a crate containing all of them. Express answers to the following questions in decimals. What is the probability that (a) the number 628 will be chosen? (b) the number 545 will be chosen? (c) the number will be less than 225? (d) the number will be more than 600? (e) the number 839 will be chosen? (f) a number in the seven hundreds will be chosen? (g) the number will be either less than 125 or greater than 725? (h) the number will be either less than 50 or greater than 700?

10.2 Scores on a dental aptitude test are normally distributed with a mean of 65 and a standard deviation of 15. One of the vast number of subjects who took the test is to be chosen at random. What is the probability that his score will be (a) above 88? (b) below 88? (c) between 58 and 88? (d) either below 58 or above 88? (e) between 57.5 and 72.5? (f) over 95?

10.3 The distribution of IQ's in a particular army camp is normal, with a mean of 105 and a standard deviation of 12.5. Consider a sample of 30 IQ scores as a random sample chosen from the entire population of IQ scores. (a) What is the probability that the mean of this sample will be over 110? (b) Suppose the mean of this sample turns out to be as high as 125. What speculation might be made about this finding?

10.4 The distribution of scores on a mechanical aptitude test is normal, with a mean of 72 and a standard deviation of 8. Find the probability that a sample score chosen at random from this population is (a) over 65, (b) between 65 and 76, (c) below 76, (d) below 65, (e) either under 65 or over 76.

10.5 The burning times of candles produced by the Flasho Candle Company are found to be normally distributed with a mean time of 92 minutes and a standard deviation of 23 minutes. Find the probability that a single candle chosen at random from this population will burn (a) over 120 minutes, (b) between 60 and 120 minutes, (c) under 40 minutes, (d) neither under 50 minutes nor over 100 minutes.

10.6 From a distribution of 1,000 test scores, the mean is found to be 500 and the standard deviation is 100. Assume the distribution to be normal. Find the probability that a single score chosen at random from this population is (a) more than 800, (b) between 450 and 800, (c) less than 450, (d) not more than 700.

10.7 Suppose that the weights of U.S. adult males are normally distributed with a mean of 180 and a standard deviation of 12 pounds. Find the probability that a single weight chosen at random from this population is (a) 200 pounds or more, (b) 150 pounds or less, (c) between 150 and 200 pounds, (d) neither under 160 pounds nor over 180 pounds.

10.8 Consider a random sample of size 50 chosen from the population described in Problem 10.7. Find the probability that the mean of this sample is (a) 200 pounds or more, (b) 185 pounds or more, (c) under 178 pounds, (d) between 177 and 183 pounds.

10.9 The distribution of scores on a test of driving skill is found to have a mean of 80 and a standard deviation of 9. Consider a random sample of size 100 drawn from this population. Find the probability that the mean of this sample is (a) over 86, (b) under 82, (c) between 83 and 87.

10.10 Repeat Problem 10.9 for a random sample of 81.

10.11 Repeat Problem 10.9 for a random sample of 64.

10.12 Repeat Problem 10.9 for a random sample of 49.

PROBLEM SET B

10.13 What is the probability that a term chosen at random from a normally distributed population will have a z score that is (a) greater than 1.5? (b) between $-.75$ and .75? (c) less than -1.5? (d) between -1.25 and 1.45?

10.14 Suppose a random sample of 105 people is given the test described in Problem 10.3. What is the probability that the mean score found for the sample will be (a) over 70? (b) over 80? (c) below 50? (d) below 71?

10.15 The mean height of Texas men is 5 feet $9\frac{1}{2}$ inches and the standard deviation of heights is $3\frac{1}{4}$ inches. Eighty-one Texas males are chosen at random and their heights are taken. What is the probability that the mean of this sample of heights will be (a) over 5 feet 10 inches? (b) under 5 feet 7 inches? (c) either under 5 feet 8 inches or over 5 feet 10 inches?

10.16 The spelling-test scores of all fifth-grade children in a large school district are normally distributed with a mean of 79 and a standard deviation of 10. Find the probability that a single score chosen at random from this population is (a) over 95, (b) between 80 and 95, (c) not between 80 and 95.

10.17 The achievement-test scores of high school students in a large city are normally distributed with a mean of 82 and a standard deviation of 7. Find the probability that a single score chosen at random from this population is (a) over 95, (b) between 80 and 95, (c) not between 80 and 95.

10.18 The mean of an approximately normal distribution of test scores of 3,000 students is found to be 250 and the standard deviation is 50. Find the probability that a single score chosen at random from this population is (a) less than 75, (b) between 175 and 325, (c) more than 175, (d) not less than 125.

10.19 The distribution of scores obtained by high school seniors on a language aptitude test has a mean of 80 and a standard deviation of 10. Consider a random sample of size 50 drawn from this population. Find the probability that the mean of this sample is (a) under 71, (b) over 76, (c) between 77 and 82, (d) not over 83.

10.20 Repeat Problem 10.19 for a random sample of 30.

10.21 Repeat Problem 10.19 for a random sample of 400.

10.22 A musical aptitude test was given to all third-grade classes in New York City. The mean score on this test was 75 and the standard deviation of the scores was 15. Consider a random sample of size 36 drawn from this population. Find the probability that the mean of this sample is (a) over 75, (b) under 71, (c) between 70 and 75, (d) not over 74.

10.23 Repeat Problem 10.22 for a random sample of 50.

10.24 Repeat Problem 10.22 for a random sample of 25.

eleven

decision making and risk

11.1 Introduction. We are constantly faced with the necessity of making decisions, large and small, and acting in accordance with them. Sometimes the inferences that lead to these decisions are made unconsciously and the decisions that spring from them resemble reflex actions. This is true, for instance, where a traveler steps to the side of the road upon hearing the noise of a vehicle behind him. It is often true where one decides which tie or sweater to wear on a given day or which elective courses to take in a given semester. However, sometimes the evidence considered is quite well defined, and prolonged consideration is needed before a conclusion can be reached and action taken.

Very important decisions are made every day by the management of large corporations and by other persons in positions of responsibility. In many cases a knowledge of probability and sampling procedures plays an important role in the making of these decisions. In such cases inferences are made based on certain available evidence. The purpose of this chapter is to explain some of the theory behind the making of such inferences and in particular some of the risks taken when they are made.

Two prime considerations are present whenever an action is to be made. First, one must consider the probability that the inference that prompted the action is correct. Second, one must weigh the possible gains and losses. Statistics enable us to find the probability values attached to different kinds of inferences. However, it usually tells us nothing regarding how costly a mistake will be if a decision turns out to be incorrect.

When there is a choice of alternatives, it can sometimes be surmised that a mistake in one direction will be roughly as costly as a mistake in the other. Then it is reasonable to choose the action with the higher probability of being correct. However, sometimes when there are alternative actions, a mistake in one direction threatens to be much more costly than a mistake in the other. Then the potentially costly action should be taken only when it has a considerably higher probability of being correct.

It follows that the same probability value might indicate a particular action in one situation and not in another. For instance, a stock market investor who already has diversified holdings is apt to purchase a commodity that has a 7/10 probability of increasing its value. However, suppose that a man is on trial for murder and the evidence against him is such that 7 out of every 10 defendants in his position are guilty in the long run. The probability is 7/10 that this defendant is guilty, but under these circumstances a jury would most likely vote for acquittal.

The reason has to do with the stakes that are involved. The penalty for a faulty conviction, the life of an innocent man, threatens to be much greater than the penalty for a mistaken acquittal. Because of this imbalance, the decision would most likely be to exonerate the defendant despite the 7/10 probability that he is guilty. The jury, hopefully, would reach the verdict with the lesser probability of being correct.

A make-believe incident dramatizes the importance of the stakes of the game in decision making. Consider the plight of Tom, a boy stranded in a strange town. The bus back home costs a dollar and Tom has only 99 cents. He is too proud to beg or borrow and now he is both angry and sad as he ambles toward the edge of the city.

Tom has always liked to flip coins for small sums. Soon he approaches a boy about his own age. "I'm gonna flip a coin and you take your choice," he says. "Guess either heads or tails when the coin is in the air. The winner gets a penny."

The boy, however, refuses to play and so do several others whom Tom approaches.

At this point our hero is about to hurl his 99 cents into a nearby lake when a daring plan enters his mind. The last boy he challenged looked hesitant before refusing his offer. Tom accosts him once more.

"Same deal as before," Tom says. "But this time if you win you get 99 cents and if you lose you pay me just one cent."

The deal is made. Tom's 99 cents and the boy's penny are put under a rock. Now the coin is in the air.

"Heads," the boy cries, but the coin falls tails. Tom scrambles to pick up the money—99 cents of his own and the penny he has won.

"Let's try it once more," the boy asks. But by now Tom has already scampered down the road on his way to the bus terminal.

The decision that Tom made was to risk 99 cents in order to possibly win a penny if the other boy did not guess the outcome of the coin correctly. The

143

venture would ordinarily be a foolish one, but the stakes of the game made it worthwhile. The probability was 1/2 that Tom would win the bet, but from his point of view Tom was actually getting incredible odds in his favor. The loss of 99 cents would not have changed his stranded status. The conquest of one penny meant that he was no longer stranded.

Note that the gamble was wise for the other boy too. For him the stakes of the game were favorable. He could win 99 cents and stood to lose only one cent and he had the same chance to win as to lose.

It often happens that a gambling or business venture is favorable to both parties, since their stakes are likely to be different. To elaborate would bring us into the worlds of gambling and finance. The purpose here is merely to sensitize the reader to the importance of stakes. Their influence is ever present, even in the realm of science.

In experimental work, there are stakes in avoiding different kinds of errors. For instance, there may be a heavy penalty for accepting the thesis that a drug is beneficial when it is not. Thus there is a stake in not making this error. On the other hand, the risk of failing to discover a useful drug because of exaggerated skepticism is also to be avoided. We shall see in the pages to follow how an experimenter runs various risks of error which he cannot avoid entirely. In fact, the more he reduces the chance of one kind of error, the more he increases the chance of another. The experimenter must consider carefully the penalties that may accrue to him from making each kind of error and then choose his procedure. Often the approach that reduces the risk of making one kind of error entails increasing the risk of making another kind—but one that would be less costly.

11.2 Hypothesis Testing. In any branch of science, it becomes necessary to evaluate hypotheses that arise. Sometimes these hypotheses, or conjectures about the nature of phenomena, are suggested by experimental findings. Sometimes they grow out of practical work in a field.

A hypothesis has the status of a bill presented to the legislature. We are the decision-making body and we must evaluate the hypothesis to determine whether to accept or reject it. Our procedure is to carry out what is called a *hypothesis-testing experiment*. This experiment is a formalized procedure for gathering and interpreting evidence relevant to deciding whether to accept or reject a hypothesis. We shall discuss this procedure and later we shall present different applications of it, designed to illustrate tests of different kinds of hypotheses.

To clarify some issues that arise when a hypothesis is tested, consider the simple hypothesis that a particular coin is fair. This hypothesis states that the probability is 1/2 that the coin will fall heads when flipped. Imagine that an experimenter takes the coin and walks to the front of a huge auditorium filled with scientists. It is decided that the experimenter will flip the coin into the air 100 times in plain view of all. The outcome of each flip will be announced, and when the experiment is completed the assemblage will decide either to accept or reject the hypothesis that the coin is fair.

The experiment begins and the audience is silent. The coin is flipped into the air over and over again. After each flip, the coin is shaken up in a little box from which it is thrown the next time. Thus each flip (or observation) is independent. As the experiment proceeds, a giant scoreboard records the number of heads and tails.

When the experiment is over, the scoreboard reads:

HEADS: 55   TAILS: 45

The question is now put to the learned audience: "Judging by what you have seen, do you accept or reject the hypothesis that the coin is fair?"

Dr. R, a young researcher, rises from his front-row seat. "Reject the hypothesis," he says. "The evidence contradicts the hypothesis that the coin is fair. The number of heads obtained exceeded the number of tails by enough margin to indicate that the coin is biased."

Unexpectedly, Dr. A, an elderly man, stands up at the back of the auditorium. "On the contrary, accept the hypothesis that the coin is fair. One cannot condemn a coin that turns up 55 heads and 45 tails. A fair coin may turn up in that ratio. Our evidence is consistent with the hypothesis that the coin is fair."

Dr. R is becoming angry. "What would it take to get you to reject the hypothesis?" he asks. "I mean what kind of disparity between heads and tails would convince you that a coin is unfair?"

Dr. A replies, "At least 90 heads as opposed to 10 tails or 90 tails as opposed to 10 heads." He flicks on a cigarette lighter and holds it to his pipe. "Remember that I am a much older man than you and have seen much more. A fair coin when tested may frequently show a great disparity between heads and tails." (Dr. R is becoming quite angry, but he dares not interrupt the speaker, who is often referred to as the dean of American scientists.) Dr. A continues: "If our policy is to reject a coin that shows a 55 to 45 ratio or more, this policy will lead us to condemn many fair coins that by chance happen to show up that way."

Dr. A is choosing his words slowly, because he is poised and because he senses that they are infuriating Dr. R. "I am sorry. You make it too easy to condemn a fair coin. Your plan will too often lead us to reject the hypothesis simply because some chance phenomenon occurs during an experiment."

Young Dr. R is now at the bursting point. "Why, that's ridiculous. You demand that a coin show a 90 to 10 ratio before you'll say that it is biased. Your over-credulity will make it almost impossible ever to reject a hypothesis about a coin. Granted, we will seldom reject a fair coin. But we will find it too hard to reject even a lopsided coin. Even lopsided coins do not usually show a disparity as great as 90 to 10. Following your plan, we will repeatedly make the mistake of accepting the hypothesis that a coin is fair when in actuality it is lopsided."

The arguments of Dr. A and Dr. R are both inherently sound, but the stakes of the two men differ. To focus on the difference, suppose that the

145

experiment is repeated with a different coin each day over a long time period. Naturally it is not known whether the coin used on a particular day is fair or biased.

Dr. A's rule is to accept the hypothesis that a coin is fair unless the ratio in an experiment is as unequal as 90 to 10. Presumably, Dr. A abhors the error of condemning a fair coin. He has gone out of his way to minimize the chance of making this kind of error. He runs little risk of condemning a fair coin, since a fair coin will almost never show a ratio as unequal as 90 to 10. However, the price he pays is to lower greatly the power of detection of the experiment. As statisticians say, he has made it extremely difficult to reject the hypothesis. A large number of biased coins will not show a disparity as great as 90 to 10 and consequently these coins will go undetected. When Dr. A happens to be testing a biased coin, he runs considerable risk of mistakenly accepting the hypothesis that the coin is fair. Dr. A may be said to have a predilection to accept hypotheses. In fact, A is short for "Accept." As a consequence, when the hypothesis is true, he will seldom make the error of rejecting it. However, when the hypothesis is false, Dr. A will often make the error of accepting it. Dr. A is like the person who says "You can do anything you want but I have faith in your innocence." He refrains from condemning. He has the virtue of accepting true hypotheses even when coincidental factors throw them in a bad light. But he has the failing of being slow to revise an opinion even when considerable evidence contradicts it.

On the other hand, let us say that Dr. R's rule is to accept the hypothesis that a coin is fair unless the ratio is as unequal as 55 to 45. That is, he will accept the hypothesis only if the coin performs within narrow limits. Otherwise he rejects the hypothesis. Dr. R views the experiment like a government security check conducted during the threat of national calamity. He deems it essential to have high power of detection of biased coins. His plan provides for minimum risk of accepting a false hypothesis. The price is to increase the number of times that fair coins will be rejected.

Dr. R may be described as having a readiness to reject hypotheses. In fact, R is short for "Reject." When the hypothesis happens to be false, he seldom makes the error of accepting it. When the hypothesis happens to be true, he will often make the error of rejecting it. Dr. A and Dr. R evaluate the context of the coin experiment differently. They disagree on the costs of making different kinds of errors. Thus they specify different cut-off points to indicate when to accept or reject the hypothesis in the experiment.

The ideal hypothesis-testing experiment obviously would always lead an experimenter to the correct decision. When a hypothesis happened to be true, the decision would be to accept it. When a hypothesis happened to be false, the decision would be to reject it. Knowing that he was conducting an ideal experiment, the experimenter could always bank on his decision. He would know that when he accepted a hypothesis, it was in reality true and that when he rejected a hypothesis, it was in reality false. Of course this ideal experiment can never actually be conducted, because in any practical experiment there are inevitable risks of error.

In practice, two possibilities must be considered when a decision is made. Suppose the decision is to reject the hypothesis:

(a) One possibility is that this decision is correct. In reality the hypothesis is false and deserved to be rejected.

(b) The other is that the decision to reject was in error. The hypothesis is true but·the finding from the sample gathered deluded the experimenter into rejecting his hypothesis. An error of this kind is called a *type-one error*.

Definition 11.1 *When a hypothesis is true but the experimenter mistakenly rejects it, he is said to be committing a type-one error.*

Suppose, on the other hand, the decision is to accept the hypothesis:

(a) One possibility is that this decision is correct. In reality the hypothesis is true and deserved to be accepted.

(b) The other is that the decision to accept was in error. The hypothesis is false, but the experimenter made the error of accepting it. An error of this kind is called a *type-two error*.

Definition 11.2 *When a hypothesis is false but the experimenter mistakenly accepts it, he is said to be making a type-two error.*

11.3 The Long-Run Outcomes with a Fair and a Biased Coin. Quite naturally, when any given experiment is carried out, the experimenter who has decided about accepting or rejecting the hypothesis does not know for certain whether his decision is correct. One cannot properly compare for accuracy Dr. A's decision to accept the coin hypothesis with Dr. R's decision to reject it, making reference only to what happened in the single experiment described. The coin, so far as we know, may have been either fair or biased, so that either one of them may have been correct.

As usual, the way to gain full insight is to imagine that the experiment described is carried out each day over a long time period. To begin with, let us suppose that a fair coin is used in carrying out the same experiment each day for a great many days. That is, each day a fair coin is flipped into the air 100 times and the number of heads is recorded. The fact that the coin is fair is being arbitrarily assumed by us, although of course this fact is not known to either Dr. A or Dr. R. In this way we can see what would happen were Dr. R to continue to use his rule and were Dr. A to continue to use his.

On any given day, whether it be the first or any other, the coin is flipped into the air exactly 100 times. (Each experiment consists of 100 flips and we shall consider an experiment as a sample for which $N = 100$.) On each given day the fair coin yields some number of heads, which is then recorded, and this total number is perceived by both scientists. The obtained number of heads is the only fact that becomes known to the two scientists, each of whom is left to make his own decision about whether the coin is fair.

Suppose that on the first day the coin shows 52 heads. This outcome may be indicated by placing a marker on a horizontal axis (see Fig. 11.1). Incidentally, on this particular day the score of 52 heads would lead Dr. A and Dr. R each independently to the correct conclusion that the coin is fair.

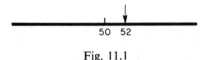

Fig. 11.1

On the next day the outcome might be 47 heads and the diagram would look like Fig. 11.2.

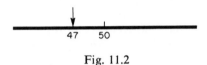

Fig. 11.2

Now suppose that the same experiment is performed on a vast number of successive days. The number of heads obtained on a given day would as a rule be close to 50, although on occasion it would be far removed. (We shall use the value X to refer to the number of heads obtained on a given day, so that for the first day $X=52$ and for the second day $X=47$.) In the long run, the different numbers of heads (or different values of X) would form a normal distribution. This distribution would have a mean of 50 and a standard deviation of 5 and would look like Fig. 11.3.

The rules whereby we determined that this distribution is normal and found its mean and standard deviation are worth discussing briefly. A variable like the flip of a coin, which must show one of two outcomes in a

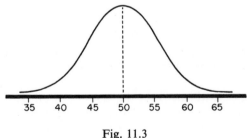

Fig. 11.3

single observation, is called a *dichotomous variable*. The occurrence of interest to us is the turning up of heads. When we assume the coin is fair, we are assuming that the probability that a head will turn up on a flip is 1/2. That is, we are considering that $P=1/2$ where P stands for the probability that a head will turn up. In general, when we talk about a dichotomous variable we single out one of the two possible occurrences and use the letter P to stand for its probability. What is shown in Fig. 11.3 is the distribution of outcomes of successive experiments, each of which is a sample of 100 flips. This distribution was determined by references to the following theorem:

Theorem 11.1 *Suppose for a dichotomous variable an occurrence of interest has a probability of happening, designated by P. A vast number of equal-sized samples, each of which consists of N observations, is collected. In each sample the outcome (designated by X) is defined as the frequency of the occurrence of interest in the particular sample. The distribution of outcomes (or values of X) obtained from an infinite number of such samples approximates the normal distribution. The mean is NP and its standard deviation is*

$$\mu = NP \quad \text{and} \quad \sigma = \sqrt{NP(1 - P)}.^*$$

We have been considering what happens when $P=.5$ and when $N=100$. Theorem 11.1 states that the long-run distribution of outcomes or values of X is normally distributed. Its mean is $NP=(100)(.5)=50$. Its standard deviation is $\sqrt{NP(1-P)}=\sqrt{(100)(.5)(1-.5)}=5$.

Now let us leave the technical discussion and get back to the main line of argument. We are interested in seeing what would happen with the rules of Dr. A and Dr. R if a fair coin were used repeatedly for all experiments. It must be emphasized that both men are testing the hypothesis that the coin is fair. The only distinction is in their readiness to reject that hypothesis.

First consider Dr. A's rule, which is to accept the hypothesis if the number of heads obtained in the experiment is between 10 and 90. According to this rule, only if the number of heads is 10 or less, or 90 or greater, will Dr. A

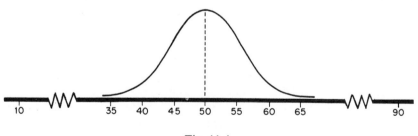

Fig. 11.4

reject the hypothesis that the coin is fair. Dr. A's rule is shown in Fig. 11.4.

This illustration shows that when a coin is fair it will virtually always show a number of heads between 10 and 90 in an experiment. Since the

---

* The value of the mean and also the value of $N(1-P)$ should be at least 5 for the approximation to be satisfactory.

acceptance region in Dr. A's rule covers eight standard deviations on each side of the mean, a fair coin will almost never deviate from 50 so far as to lead Dr. A to reject his hypothesis mistakenly. In other words, he will hardly ever commit a type-one error.

Now let us consider the decision rule that Dr. R has advocated, which is to accept the hypothesis only if a coin shows heads more than 45 but less than 55 times. Fig. 11.5 shows what would happen if a fair coin were used repeatedly, this time in connection with Dr. R's decision rule.

We can compute the probability that a fair coin will show a finding in the acceptance region using Theorem 11.1. The endpoints of the acceptance region are 45 and 55, and each of these numbers differs from the mean of 50

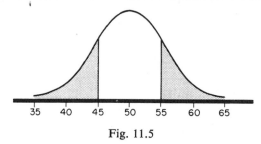

Fig. 11.5

by one standard deviation of 5.* Theorem 11.1 and Appendix III tell us that the acceptance region contains about 68 per cent of the total area and that the shaded areas of Fig. 11.5 contain together about 32 per cent of the total area. In other words, the probability is approximately 68/100 that a fair coin will produce a sample in which the number of heads is in the acceptance region.

Of importance is the fact that the probability is about 32/100 that a fair coin will produce a sample in which the number of heads is in one of the rejection areas designated by Dr. R. This is the probability that even though the hypothesis is true, the finding will be such that Dr. R will mistakenly reject the hypothesis. In other words, Dr. R has set up a rule such that when a coin is fair, the probability of his making a type-one error is about 32/100.

So far we have considered what happens in the long run when the coin is fair. We have shown that Dr. A would over the long run virtually never reject the hypothesis mistakenly or commit a type-one error. Dr. R would over the long run mistakenly reject the hypothesis about 32 per cent of the time, so that when a coin is fair the probability of his committing a type-one error is about 32/100.

Now suppose that the coin is biased—that is, its probability of turning up heads on a given flip is some fraction other than ½. Once again we are

---

* In reality the endpoints are 45.5 and 54.5, because Dr. R. will reject the hypothesis if there are actually 45 or 55 heads. However, instead of laboring this point we would rather simply focus on the underlying ideas. Hence we shall use 5 rather than 4.5 as the difference in this and similar situations.

going to suppose that the coin experiment is repeated on a vast number of days, this time with the biased coin. But if the coin is not fair, then it may be biased to any degree. For instance, its probability of turning up heads on a flip might be 1/5 or 6/10 or 9/10 or any fraction. We have seen that the statement that the coin is fair was the specification of a particular value for P. However, the statement that the coin is biased is not a specification of a value for P, because a coin may be biased in any of an infinitude of ways. As we shall see, it typically occurs that a hypothesis may be true in only one way and false in many.

We want to evaluate both Dr. A's and Dr. R's rules to determine how successfully each man will manage to reject the hypothesis, now that it is proper to do so. However, the assumption that the coin is biased is insufficient to generate a picture of what would happen, so we cannot proceed yet. We must first specify a particular bias for the coin so that we can then use Theorem 11.1 to see what would happen.

Suppose for instance that the coin's probability of falling heads is 6/10. The experiment with this biased coin is carried out each day and the number of heads obtained each time is considered the outcome for the experiment for the day. According to Theorem 11.1 the distribution of outcomes has a mean of 60 $[NP=(100)(.6)=60]$. It has a standard deviation of about 4.9 $[\sqrt{NP(1-P)}=\sqrt{100(.6)(.4)}=4.9]$.

Fig. 11.6 shows the long-run distribution of outcomes with the particular biased coin.

In considering a coin with a specific bias, we can now determine how successful Drs. A and R would be in rejecting the fair-coin hypothesis. To begin with Dr. A, his decision rule was to accept the hypothesis so long as the number of heads obtained is between 10 and 90. One can see in Fig. 11.6 or compute using Theorem 11.1 that even a biased coin would, in virtually every experiment, produce a number of heads somewhere in Dr. A's acceptance region. In other words, when the coin's probability of falling heads is 6/10, Dr. A would virtually always mistakenly accept his hypothesis and thus make a type-two error.

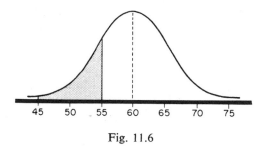

Fig. 11.6

As for Dr. R, his decision is to accept the hypothesis only upon obtaining a finding somewhere between 45 and 55 heads. He will mistakenly accept his hypothesis, making a type-two error, on those occasions when the biased coin produces a number of heads falling in this range. The shaded area in Fig. 11.6 shows the proportion of times that the biased coin will show a finding leading Dr. R to make a type-two error. Computing z values for each of the numbers 45 and 55 and using Appendix III, we find that the shaded area contains about 15 per cent of the total area. In other words, on those occasions when a coin with a 6/10 bias is used, the probability is about 15/100 that Dr. R will mistakenly accept his hypothesis or commit a type-two error.

There is still one final angle. When contending with a coin of the bias described, the probability is about 85/100 that Dr. R will properly reject his hypothesis. This fact is important. With the same coin Dr. A had virtually no chance of properly rejecting the hypothesis, whereas Dr. R has a probability of 85/100 of doing so. The statistician says that *under the alternative described* (that is, a coin with a 6/10 bias) Dr. A has virtually no power to detect that the hypothesis is wrong, whereas Dr. R has considerable power to do so. Dr. R's test gives him a probability of 85/100 of properly rejecting the hypothesis in the case of the particular bias described. To put it another way, Dr. R's test is said to have a *power of 85/100 against the alternative described.* The power of a test against a particular alternative is defined as the probability that the proper decision to reject will be made when the particular alternative occurs.

To summarize, the test used by Dr. A would make his rejection of the hypothesis extremely rare. When the hypothesis is true, Dr. A will scarcely ever mistakenly reject it or commit a type-one error, but Dr. R will commit a type-one error 32/100 of the time. When the hypothesis is false and the coin happens to have a 6/10 bias, Dr. A will virtually always make a type-two error, which means that he will accept the hypothesis. His test gives him much less power of detection against the particular alternative than the test of Dr. R. The test of Dr. R leaves him a probability of 15/100 of making a type-two error and a power of 85/100 when the coin used has the particular bias.

We must remember that this discussion is dependent upon our decision to consider what occurs with a coin of a particular bias. We have been considering only what would happen in the long run with a coin that in actuality has a probability of 6/10 of falling heads. Had we considered what would happen with some other kind of biased coin, we would have been led to consider a completely different distribution of outcomes, and for either Dr. A or Dr. R the chance of mistakenly accepting the hypothesis would have been different.

As one might expect, a coin with a more drastic bias would be even less likely to produce an outcome in the acceptance region of either man. For instance, let us consider an extremely biased coin, one that has a probability of 95/100 of falling heads. The distribution of outcomes of a vast number of experiments with such a coin would look like Fig. 11.7.

Both Dr. R and Dr. A are now virtually certain to reject the hypothesis. It should be seen that the probability of either man's making a type-two error is dependent upon the particular alternative—that is, the actual bias that

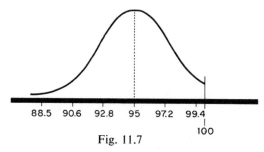

88.5   90.6   92.8   95   97.2   99.4
100

Fig. 11.7

happens to be the case.   When the reality is particularly divergent from what is hypothesized, both men are more apt to reject the hypothesis than when the reality only slightly belies the hypothesis.   Another way to put it is to say that when the coin was only slightly biased both men were more likely to be fooled into accepting the hypothesis.   Now that we are assuming that the hypothesis is a ludicrous misstatement, both men are more likely to reject it.   It will be true for every decision rule we discuss that when the hypothesis is only slightly wrong the chance of making a type-two error is relatively great, but when it is very wrong the chance of making a type-two error will be relatively smaller.   This property is favorable since it is usually less harmful to over-look a small inaccuracy than a large one.

In practice, the experimenter who wishes to test a hypothesis is, of course, in the blind position of deciding to accept or reject it without ever knowing whether he is right.   He would naturally like to accept only what is true and reject only what is false.   The trouble is that whatever decision rule he chooses leaves him open to risks.   He must specify some acceptance region with endpoints, and as soon as he does so he runs the risk that even if his hypoth-esis is true he will come upon some stray but random finding that is not in his acceptance region.   Furthermore, even if his hypothesis is wrong there exists some chance that his sample finding will be in his acceptance region.

The question of how the experimenter is to determine the width of his acceptance region now arises.   To use a narrow acceptance region means to run a high risk of making a type-one error.   Increasing the width of the acceptance region decreases the chance of making a type-one error but increases the risk of making a type-two error.   The point is that we cannot reduce one kind of risk without at the same time increasing the other.

As one might guess, the decision about how wide to make the acceptance region depends upon the readiness of the experimenter to make one kind of error as opposed to the other.   In the case of Dr. R and Dr. A, the two men differed essentially in terms of the risks they were willing to tolerate.   Dr. A seems to have been motivated by a dread of condemning a fair coin.   His expansive acceptance region suggests that he was relatively unconcerned with failing to detect a bad one.   Dr. R seems to have picked his narrow acceptance region in dread of being fooled into accepting a bad coin.   He seems scarcely to have cared that many good coins would be condemned by him in the long run.

153

In general, when one chooses an acceptance region he is motivated by his fear of mistakenly rejecting a true hypothesis and to some degree also by his fear of mistakenly accepting his hypothesis when it is false. The width of the acceptance region actually chosen represents a kind of compromise position.

This chapter has described what it is to test a hypothesis and has pointed up the risks always involved in reaching a decision. The next chapter will show specifically how one considers his own willingness to tolerate risk in actually testing a hypothesis.

Unlike Dr. A and Dr. R, one makes an explicit decision about tolerance of risk and then follows a formalized procedure for actually testing a hypothesis. We have discussed the considerations underlying this formal procedure; next we shall discuss the hypothesis-testing procedure itself.

PROBLEM SET A

11.1 Suppose a researcher hypothesizes that a dichotomous variable has $P = .5$. For samples of size $N = 400$, find the mean and standard deviation of the distribution of a vast number of outcomes.

11.2 For a hypothesis that $P = 1/4$, for samples of size $N = 64$, find the interval of outcomes that would permit the hypothesis to be accepted if the decision rule is to accept the hypothesis with a z score between -1.0 and $+1.0$.

11.3 For a hypothesis that $P = .5$, for samples of size $N = 100$, calculate the probability of a type-one error if the decision rule is to accept an outcome with a value that is in the interval from 5 units below the mean to 5 units above the mean.

11.4 For a hypothesis that $P = .5$, for samples of size $N = 256$, find the interval of outcomes that allows acceptance of the hypothesis if the decision rule is to accept the hypothesis with a z score between -0.5 and $+1.5$.

11.5 For a hypothesis that $P = .67$, for samples of size $N = 81$, suppose that, in fact, $P = .60$. If the decision rule is to accept an outcome with a value that is in the interval from 5 units below the mean to 5 units above the mean, calculate the probability of a type-two error and the power of the test.

11.6 Suppose a researcher hypothesizes that a dichotomous variable has $P = 5/8$. For samples of size $N = 256$, find the mean and standard deviation of the distribution of a vast number of outcomes.

11.7 For a hypothesis that $P = .75$, for samples of size $N = 400$, find the interval of outcomes that would permit the hypothesis to be accepted if the decision rule is to accept the hypothesis with a z score between -2.0 and $+2.0$.

11.8 For a hypothesis that $P = .83$, for samples of size $N = 180$, calculate the probability of a type-one error if the decision rule is to accept an outcome with a value that is in the interval from 2 units below the mean to 2 units above the mean.

11.9 For a hypothesis that $P = 1/3$, for samples of size $N = 455$, find the interval of outcomes that allows acceptance of the hypothesis if the decision rule is to accept the hypothesis with a z score between -2.0 and $+0.5$.

11.10 For a hypothesis that $P = .5$, for samples of size $N = 100$, suppose that, in fact, $P = .6$. If the decision rule is to accept an outcome with a value that is in the interval from 2 units below the mean to 2 units above the mean, calculate the probability of a type-two error and the power of the test.

PROBLEM SET B

11.11 Suppose a researcher hypothesizes that a dichotomous variable has $P = .25$. For samples of size $N = 64$, find the mean and standard deviation of the distribution of a vast number of outcomes.

11.12 For a hypothesis that $P = .5$, for samples of size $N = 100$, find the interval of outcomes that would permit the hypothesis to be accepted if the decision rule is to accept the hypothesis with a z score between -1.0 and $+1.0$.

11.13 For a hypothesis that $P = 2/3$, for samples of size $N = 81$, calculate the probability of a type-one error if the decision rule is to accept an outcome with a value that is in the interval from 5 units below the mean to 5 units above the mean.

11.14 For a hypothesis that $P = .4$, for samples of size $N = 250$, find the interval of outcomes that allows acceptance of the hypothesis if the decision rule is to accept the hypothesis with a z score between -0.5 and $+1.5$.

11.15 For a hypothesis that $P = 5/8$, for samples of size $N = 256$, suppose that, in fact, $P = .6$. If the decision rule is to accept an outcome with a value that is in the interval from 5 units below the mean to 5 units above the mean, calculate the probability of a type-two error and the power of the test.

11.16 Suppose a researcher hypothesizes that a dichotomous variable has $P = .75$. For samples of size $N = 400$, find the mean and standard deviation of the distribution of a vast number of outcomes.

11.17 For a hypothesis that $P = .83$, for samples of size $N = 180$, find the interval of outcomes that would permit the hypothesis to be accepted if the decision rule is to accept the hypothesis with a z score between -2.0 and $+2.0$.

11.18 For a hypothesis that $P = .67$, for samples of size $N = 300$, calculate the probability of a type-one error if the decision rule is to accept an outcome with a value that is in the interval from 2 units below the mean to 2 units above the mean.

11.19 For a hypothesis that $P = .75$, for samples of size $N = 432$, find the interval of outcomes that allows acceptance of the hypothesis if the decision rule is to accept the hypothesis with a z score between -2.0 and $+0.5$.

11.20 For a hypothesis that $P = .25$, for samples of size $N = 64$, suppose that, in fact, $P = .3$. If the decision rule is to accept an outcome with a value that is in the interval from 2 units below the mean to 2 units above the mean, calculate the probability of a type-two error and the power of the test.

twelve

the hypothesis-testing procedure

12.1 Introduction. There is a standard procedure for testing a statistical hypothesis, and it is used in psychology, education, the social sciences, and many other fields. In this chapter we shall discuss that procedure and illustrate its use in connection with several problems. We shall make practical use of the material in the previous chapter, including the discussion of a hypothesis and the relationships between the width of the acceptance region and the various risks of error.

One major factor must be discussed at the outset—the fact that evidence that is consistent with a hypothesis can almost never be taken as conclusive grounds for accepting it, whereas evidence that is inconsistent with a hypothesis does provide grounds for rejecting it. Of course errors can always be made either way, but the point is important and needs elaboration. The reason for not necessarily accepting consistent evidence is that a finding that is consistent with a hypothesis would be consistent with other hypotheses too, and thus does not necessarily demonstrate the truth of the given hypothesis as opposed to these other alternatives. For instance, a finding of 51 heads out of 100 flips is consistent with the hypothesis that a coin is fair, but this finding would of course also be consistent with the assumption that the coin slightly favors heads. Therefore the finding that is consistent with the hypothesis does not demonstrate its truth. Even a finding of 50 heads would scarcely imply that the coin did not have some minute bias.

On the other hand, a finding of 80 heads out of 100 flips could virtually be taken to contradict the hypothesis that the coin is fair. Since a fair coin would have less than one chance in ten thousand of showing as many as 80 heads, such a finding might properly lead us to conclude with only a minute risk of error that the coin was not fair.

In the same way, suppose we hypothesized that the members of a particular group had an average intelligence quotient of 100. A finding that the members of some sample taken from the group had a mean IQ of 102 might be consistent with the hypothesis. But the finding might also have been consistent with the hypothesis that the mean IQ was 101 or 99, and it certainly would have been consistent with the hypothesis that the mean IQ of the larger group was 102. Thus the finding would in no way demonstrate the truth of the hypothesis of 100 as opposed to these other possibilities. But now suppose the members of the sample had been found to have a mean IQ of 135. Assuming the sample size was large enough, we might show that, were the original hypothesis true, such a sample would virtually never be found. By inference, we might then properly construe the finding as contradictory to the hypothesis, and the risk of error in doing so might be infinitesimal.

What has been said amounts to the fact that the usual finding that is consistent with a hypothesis would have been consistent with other hypotheses too. Thus such a finding cannot be taken as grounds for accepting the particular hypothesis over its neighboring alternatives. However, we may always come upon a finding that would be so hard to reconcile with the hypothesis as to put it in considerable doubt. A hypothesis is like a precise story told by a defendant at a trial. He can never prove this precise account of what he did. Virtually any evidence he presents leaves open the possibility that he acted in a way somewhat different from that described. Yet a prosecutor can disprove his precise account by showing a single inconsistency.

12.2 The Null Hypothesis. It follows that when we fail to reject a hypothesis we are forced to conclude that it may be true. On the other hand, when we can reject it we conclude that it definitely is false.

This last statement is very important. When testing a hypothesis, we can reach a definite conclusion only if we can reject it. We therefore set up our experiment with the aim of *disproving* the hypothesis we test. That is, we state as our hypothesis an *alternative* to what we believe. By this approach, if we are in fact able to reject the hypothesis (to consider it disproven), we have then established our own belief.

For example, if we wish to show that men are taller than women, we test the hypothesis that there is *no difference* in their heights. We then hope to reject this hypothesis. To prove that members of one political party are different in some way from another, we test the hypothesis that there is *no difference* between the parties on whatever we are testing. Once again, our hope is that we can reject this "no difference" hypothesis and thereby establish our own thesis as the alternative.

Definition 12.1 *The hypothesis stating that there is no bias or that there is no distinction between groups is called the null hypothesis.*

We shall see many instances in which some form of a null hypothesis is tested, with the aim of rejecting it so as to "prove" its alternative.

Not every statement leads to a testable hypothesis. As we have seen, the statement that a coin is biased is not specific enough to be translated into a specific hypothesis, and this is often the case. It follows from what has been said that the only way an experimenter can proceed is when he can translate the alternative to what he is trying to prove into a specific hypothesis. Only then can he go ahead and test it, and hopefully reject it and so prove his theory.

The first major step for an experimenter then is to put into statistical form the hypothesis he hopes to *reject* in order to consider his theory proven. He is now up to the point of carrying out the regular hypothesis-testing procedure.

12.3 When to Reject a Hypothesis; the Significance Level. To focus on the logic of his procedure, let us suppose that our experimenter has already set up some hypothesis that he now hopes to reject in order to support his theory. (In our coin illustration let us say that he has decided to test the hypothesis that the coin is fair with the hope of rejecting it.*) Now he goes on to collect his sample and to calculate some value of interest from it (like the number of heads obtained in one hundred flips).

His next step is to ask himself a crucial question: "*If my null hypothesis is true, what was the probability that I would draw a sample with a value as deviant from the expectancy as the one I obtained?*" (Suppose the coin is really fair and would fall heads half the time on the average. How likely is it that there would have been a number of heads in my sample as deviant from the expectancy of 50 as the number I actually found?) Note the italicized question carefully. It is the key to virtually everything that is to follow.

The next step is to see whether, assuming the hypothesis is true, the obtained sample finding would have been likely or not. Exactly what is meant by this statement and the computations involved in connection with several problems will be discussed later. To go on with the logic, the probability of getting a sample value like the one obtained from the hypothesized population may be relatively high or low, and this fact obviously determines whether the ultimate decision is to accept or reject the hypothesis.

In general terms, suppose that after calculation it turned out that, if the hypothesis is true, the probability of having gotten a sample value like the one actually obtained would have been high. In such a case the obtained sample value would have to be regarded as representative of the hypothesized population (as would be the case if 52 heads were found). That is, a value

---

* The hypothesis that a coin is biased is not specific enough to be tested, as was mentioned. It is often the case that a hypothesis is not specific enough to generate a single distribution so that it can be tested. In other words, not every hypothesis is a testable hypothesis.

like the one obtained would have had too high a probability of turning up to allow us to reject the hypothesis.

However, suppose instead that the obtained sample value is so remote from the hypothesized expectancy that one like it would have had an extremely small probability of arising. (For instance, such would be the case if 93 heads had turned up.) To put it another way, suppose the sample value obtained is so deviant that calculations show that one like it would scarcely ever be found in a sample drawn at random, if the null hypothesis were true. Then the decision would be to reject the hypothesis. Rather than conclude that an incredibly unusual event has occurred, we would consider it more likely that the hypothesis was not true in the first place.

The process then is to assume the hypothesis is true temporarily; and, as will be seen, the next step is to calculate what the probability would have been of getting a value as deviant from the expectancy as the one actually obtained. The key question is how small should the probability of getting the sample value be (assuming for the time being that the hypothesis is true) before the experimenter decides to reject the hypothesis. Once again we get back to the issue of the stakes of the game, for the answer completely depends upon the experimenter's willingness to make a type-one error as opposed to a type-two error. The experimenter usually demands that the probability fraction be quite small, since the decision to reject the hypothesis, when it is made, is in effect a decision to believe a new theory. Naturally it is important to stop incorrect theories from being accepted. For the experimenter it is important to minimize the number of times that coincidental factors fool him into rejecting a hypothesis when it is true. Accordingly, before rejecting a hypothesis he demands that a sample value be found that would have had an extremely small probability of arising if in actuality the hypothesis is true.

In some scientific fields, the practice is to reject a hypothesis only if the sample value obtained is one rare enough to come from the hypothesized population 5/100 of the time or less. In other fields, the practice is not to reject a hypothesis unless the sample value obtained would be rare enough to have less than a 1/100 probability of coming from the hypothesized population. One thing is sure: an experimenter must demand that it be very unlikely that his sample would come from the hypothesized population before using the sample value as evidence that the hypothesis is false.

The probability fraction that the experimenter does pick is called the *significance level* of the experiment. His choice of significance level must be made before he actually collects his data so that nothing that he comes upon during the experiment will affect him in his choice of it. Having chosen his significance level, the experimenter calculates some outcome of interest from his sample. He then assumes for the sake of argument that his hypothesis is true, and under this assumption he determines whether a finding like his would be more or less probable than the significance level.

If, under the hypothesis, a finding like his would be more probable than the significance level, then he fails to reject the hypothesis, reasoning that his finding would not be sufficiently inconsistent with the hypothesis to warrant

159

his rejecting it. (For instance, suppose he chose the significance level of 5/100 and subsequently obtained 52 heads. If he calculates that a deviation of as many as two heads from the expectancy would occur more than 5/100 of the time by chance, then he cannot reject the hypothesis.)

Suppose instead he calculates that, under the hypothesis, a finding as deviant as his would be so rare as to occur with smaller probability than the significance level. Now he can reject the hypothesis. For instance, suppose that using the significance level of 5/100, he subsequently obtained a finding of 96 heads; and suppose he calculates that a deviation of this many heads from the expectancy would occur with a smaller probability than 5/100. The logic that leads the experimenter to reject the hypothesis in such a case would be this: "Assuming the hypothesis, a finding as deviant as mine would have been so improbable that I need not continue believing in the hypothesis any longer." When this happens, the experimenter considers his theory proven, and thus an experimenter always hopes to obtain a sample value that would have had *less* chance of arising under the hypothesis than the significance level.

Once the experimenter has chosen his significance level, he has in effect made a definite rule for when he will fail to reject and when he will reject a hypothesis. The significance level is a statement of when he will consider a finding to show a departure from the expectancy with small enough probability to induce him to reject the hypothesis.

Now suppose an experimenter has picked his significance level, which for the sake of discussion we shall say is 5/100. He assumes that his hypothesis is true, and according to protocol he gathers a random sample and computes some value of interest. We shall consider for a moment what is apt to happen when the hypothesis is actually true (though of course the experimenter does not know this fact). Under the condition that the hypothesis is true, the probability is 5/100 of obtaining a sample value so deviant from the expectancy as to induce the experimenter to reject his hypothesis mistakenly. *It follows that the significance level, which the experimenter chooses, tells precisely his probability of making a type-one error when the hypothesis is actually true.*

As another illustration, suppose that the experimenter picks the significance level of 1/100 at the start of his experiment. Now when he rejects a hypothesis he can say, "Well, if the hypothesis is true, a finding like mine was one with less than one chance in one hundred of arising; rather than believe that so great an unlikelihood occurred, I reject the hypothesis." But this means that, when the hypothesis is actually true, by definition there will turn up one time in one hundred (in the long run) the rarity that will fool the experimenter into mistakenly rejecting his hypothesis. By picking a significance level of 1/100, the experimenter is determining that if the hypothesis is actually true, there is one chance in one hundred that he will get a finding that will induce him to commit a type-one error.

In sum, as the experimenter picks his significance level, he is fixing the risk of a type-one error. The smaller his significance level, the less chance there is of his making a type-one error. However, the smaller his significance level, the harder it is for him to reject his hypothesis and the more likely it is for him to make a type-two error if his hypothesis is false.

The choice of a significance level should be seen as the choice of a decision rule. For our problems there would be no analogous way to fix the risk of making a type-two error as a way of determining a decision rule, because the risk of making a type-two error varies with the alternative, as we have seen. Having understood this much, we are now ready to discuss the formal procedure for testing a hypothesis, which entails fixing the significance level for an experiment and then testing our hypothesis according to it.

The steps of the hypothesis-testing procedure are here presented in formal order. It is important not only to go over them but to comprehend the total approach, since we shall make references to this approach repeatedly.

1. The experimenter must state the hypothesis that he is about to test, so that by rejecting it he will clearly be able to consider his theory confirmed. (In our example he may wish to show that a coin is biased, so he prepares to test the hypothesis that it is fair.)

2. He must arbitrarily pick a significance level based on the considerations already discussed. (For instance, he may decide to test his hypothesis at the 5/100 significance level, *usually written as the .05 significance level.*)

3. He must gather his random sample of observations and compute the sample value of interest to him.

4. He must assume the hypothesis is true and determine under this assumption how small the probability would have been of getting a finding as deviant from the expectancy as the one obtained.

5(a). If, assuming the hypothesis is true, his calculations tell him that the probability of having gotten a finding as deviant as his would have been greater than the significance level, he cannot reject the hypothesis. Some experimenters say that they "accept" their hypotheses when their findings do not result in their rejecting them. However, there are frequent objections to the use of the word "accept" in this context. To say that we accept a hypothesis suggests that we are accepting it as opposed to all others, whereas in most experiments the data allow acceptance of not one but many hypotheses. Moreover, findings are sometimes not deviant enough to allow rejection of the hypothesis but are deviant enough to incur doubts in an experimenter's mind about the validity of the hypothesis. Especially where these doubts lead to new experimentation, the statement that the hypothesis is accepted does not seem as accurate as the statement that on the basis of the findings one could not reject the hypothesis.

5(b). If, assuming the hypothesis is true, his calculations tell him that the probability of having gotten a finding as deviant as his would have been less than the significance level, then he rejects the hypothesis.

To conclude our coin example, suppose the finding is that of 55 heads. The hypothesis that the coin is fair is being tested at the .05 significance level.

We are up to Step 4. The experimenter temporarily assumes that the hypothesis is true. He supposes that in actuality the coin is fair, and he now conceives of the sample he drew as one of a vast number of random samples of 100 flips each. This is the standard logic. The experimenter assumes that, under the hypothesis, his particular experiment was one of a vast number of identical experiments that were carried out, and pictures the distribution of sample values that would be obtained.

The distribution of outcomes of a vast number of experiments, were they to be carried out with a fair coin, would be normal. This distribution, which is depicted in Fig. 12.1, would have a mean of 50 heads and a standard deviation of 5 heads (Theorem 11.1). The place of the obtained value of 55 heads has been noted by an X.

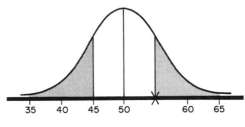

Fig. 12.1

Assuming that the hypothesis is true, the deviation from the exact expectancy was 5 heads.* The question is: Assuming the hypothesis to be true, what was the probability that the number of heads in a random sample of 100 flips would be as many as 5 from the expectancy? For the distribution of outcomes depicted in Fig. 12.1, the value 55 heads has a distance score of 5 from the expectancy. To get the z score of the value 55 heads, we divide its distance score by the standard deviation of the distribution, which is also 5. Thus the z score of the finding of 55 heads is 1.0.

To make things more straightforward for other similar problems, we shall use Theorem 11.1 to write a single formula to explain what was done. In general, the formula for the z score of an outcome when the sample consists of N observations of a dichotomous variable is as follows:

(12.1) $$z = \frac{\text{distance score}}{\text{standard deviation}} = \frac{O - NP}{\sqrt{(N)(P)(1-P)}}$$

where O is the obtained outcome (or number of times the event occurred in the sample), N is the sample size and P is the probability of the event's occurring.

Note that NP is the value of the hypothetical expectancy, which is the theoretical guess indicated by the hypothesis. One naturally anticipates finding some departure from this guess even when the hypothesis is true, but it is the degree of the departure as measured in z score units which becomes crucial.

*See the footnote on page 150.

In our case, $O = 55$, $N = 100$, and $P = .5$.   Thus

$$z = \frac{O - NP}{\sqrt{NP(1 - P)}} = \frac{(55) - (100)(.5)}{\sqrt{(100)(.5)(.5)}} = 1.0$$

To verbalize the formula, because of our null hypothesis we were actually assuming that $P = .5$, and under this assumption we found that our outcome would have had a z score of 1.0 in the distribution of outcomes if a vast number of identical experiments had been carried out.

Remember the experimenter's question: Assuming the hypothesis is true, what would have been the probability of obtaining an outcome that deviated from the expectancy by as many as 5 heads?   The question when translated becomes: What was the probability of randomly obtaining an outcome with a z score value of as much as one unit from zero?   Reference to Appendix III indicates that as many as approximately 32 per cent of the terms in the normal distribution have z scores at least one unit from the expectancy. That is, all those terms in the shaded regions of Fig. 12.1, or about 32 per cent of the terms, are more deviant than the outcome of 55.

Thus the outcome of 55 heads is such that, even were the hypothesis true, an outcome this deviant or more would have turned up by chance about 32 per cent of the time.   Since the probability of so deviant an outcome is higher than the significance level of .05 which we set, we cannot reject the hypothesis that the coin is fair.   It may be said that the finding would not have been unlikely enough to motivate us into rejecting the hypothesis.

Now suppose instead that the coin had turned up 65 heads.   This time the deviation from the expectancy is 15 heads and we see at a glance that a fair coin would be much less likely to produce such a finding. Fig. 12.1 indicates that, assuming the hypothesis is true, the obtained value would have been three standard deviations from the hypothetical expectancy.

Using Formula (12.1), we see more formally that the z score of the outcome of 65 heads is $z = 3$.

$$z = \frac{O - NP}{\sqrt{(N)(P)(1 - P)}} = \frac{65 - (100)(.5)}{\sqrt{(100)(.5)(.5)}} = 3.0$$

Appendix III tells us that the probability of having gotten so deviant a z score would have been about three in a thousand.   That is, the terms beyond the shaded tails of Fig. 12.1 comprise about 3/1,000 of all the terms in the distribution.   Thus a finding like ours was extreme enough to have had a probability of about one in a thousand of occurring.   Since this probability is less than the significance level, the decision must be this time to reject the hypothesis that the coin is fair and to regard as confirmed the theory that the coin is biased.

The purpose thus far has been to present and illustrate the hypothesis-testing method and to discuss its logic.   The basic steps and their rationales should be clear.   First there must be an exact statement of the hypothesis

163

to be tested. Then the experimenter chooses his significance level and calculates the outcome of interest (and sometimes other data) from his sample. Then, assuming that the hypothesis is true, he determines how small a probability there would have been of obtaining a finding as deviant from the hypothetical expectancy as was his particular finding. Where he decides that a finding as deviant as his would have been less probable than the significance level, he rejects the hypothesis. Unless his finding is this deviant, he cannot reject the null hypothesis and consider his original theory confirmed.

The basic methodology and logic we have been discussing is the same when applied to a whole variety of problems. Hypotheses concerning means, medians, standard deviations, and numerous other kinds of hypotheses are tested according to the identical procedure, though naturally the computational approach varies with the problem at hand.

12.4 Three Hypothesis-Testing Problems. We shall now look at three problems and their solutions. The first one will involve a dichotomous variable and thus be similar to the one already worked out. The other two will be hypotheses concerning the unknown mean of a population. It will be seen how the procedure already described is applied in each case.

Suppose a researcher believes that the voters in a particular state are not going to follow their usual trend in some particular year. Ordinarily 2/3 of them are Republicans and 1/3 of them are Democrats, but this time the belief is that some recent event in the state is going to have an unpredictable but strong impact on the voters' preference, That is, the researcher's theory is that the event will be meaningful, though he would not like to venture a guess as to which party will profit from it.

The researcher sets up the hypothesis that the usual ratio of voting preference pertains and plans to test this hypothesis at the .05 significance level with the hope of being able to reject it. Actually he is hoping to show merely that the usual 2/3 to 1/3 ratio does not obtain. The researcher polls 450 voters in all and it turns out that 225 of them are Republicans.

Under the hypothesis, $P=2/3$ and $N=450$. $O=225$. Note that the value $NP=300$. This value is the expectancy under the hypothesis. In other words, were the hypothesis true that 2/3 of the voters are Republican, then the expectancy for the sample is that 2/3 of its members would vote Republican.

Assuming that the hypothesis is true,

$$z = \frac{O-NP}{\sqrt{(N)(P)(1-P)}} = \frac{225-300}{\sqrt{450(\frac{2}{3})(\frac{1}{3})}} = \frac{-75}{10} = -7.5$$

The z score value of -7.5 has already been found. The logic of what it represents entails assuming for the sake of argument that the hypothesis that exactly 2/3 of the voters are Republicans was true. One must assume that a vast number of random samples of 450 voters each were gathered and that the distribution of Republicans per sample was created. Our particular sample of 450 cases included only 225 Republicans, and if it were one of this vast number of independent samples it would have had a z score of −7.5. Fig. 12.2 shows what the distribution of Republicans per sample would have looked like under the hypothesis.

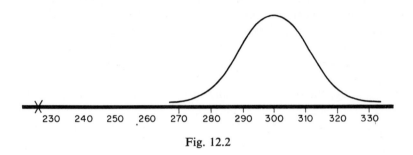

230 240 250 260 270 280 290 300 310 320 330

Fig. 12.2

It follows that, if the hypothesis was true, the researcher obtained a sample with a number of Republicans that was seven and a half standard deviations from the expectancy. The proportion of times such a remote outcome would occur is given by the proportion of times that a randomly obtained z value would be either less than −7.5 or greater than 7.5. The probability of getting a value so deviant from the expectancy is less than one in one hundred thousand, as may be verified by reference to Appendix III. Thus our researcher's decision must be to reject the hypothesis that the ratio of Republicans to Democrats is actually 2/3 to 1/3. Although of course he has not shown that the particular suspected event was a crucial factor, he may consider himself to have shown that the voting ratio is not what it usually is in the state. It should be clear that whenever a z score is obtained, it is actually what the z score of the obtained finding would be were the hypothesis true. The logic is that, where the magnitude of the z score would represent too improbable an event, the decision is made to reject the hypothesis altogether.

The procedure is identical when we test a hypothesis about a population mean, except that now we use our sample mean as the evidence instead of simply the number of times an event of interest occurred. We are now going to consider the simplest case of testing a hypothesis about a population mean—namely, when we know the standard deviation of the population.

As before, we begin by stating a hypothesis and setting a significance level. We then draw a random sample from the population, and this time the mean of our sample is measured for its discrepancy from the hypothetical expectancy. As before, the hypothesis is assumed to be true and the process is to envision that a multitude of equal-sized random samples have been drawn from the

165

population, and that the obtained sample is merely a member of this multitude. The *means* of these different samples would form a distribution, and, since we are assuming that the hypothesis is true, we can determine what this distribution would look like using Theorem 9.2.

Once we have constructed the distribution, the next step is to determine what the *z* score of the obtained sample mean would have been. From this *z* score, we can determine whether the obtained sample mean would have been commonplace, or whether the discrepancy noted would have had a small probability of occurrence. Once again, if the discrepancy is such that it would have occurred with larger probability than the significance level, then the decision is to accept the hypothesis. If the discrepancy would have occurred with smaller probability than the significance level, then the decision is to reject the hypothesis.

To illustrate, the question recently arose of whether boys of age eight who want to be firemen differ in IQ from the average. It was believed that as a group their IQ's did differ and that this could be shown using the Stanford Binet IQ Test. A study was planned in a city where the mean IQ of boys of age eight was 100 and the standard deviation of their IQ scores was 20 points. These facts had emerged from a city-wide study done previously.

To demonstrate that boys who wish to be firemen differ in IQ, it was decided to test the hypothesis that they are an exact cross-sectional representation so far as IQ is concerned. The significance level of .05 was picked and naturally the hope was to be able to reject the hypothesis. Sixty-four boys whose primary concern was to be firemen as determined from a questionnaire were selected randomly from the larger population. It turned out that the mean of their IQ scores was 108.

The null hypothesis that these boys were in no way special with respect to IQ was assumed to be true. According to this hypothesis, their 64 IQ scores could simply be viewed as a random sample of IQ scores chosen from the larger population of scores. The assumption that the hypothesis was true implied that the particular sample of 64 IQ scores could be viewed as merely one of similar randomly drawn samples of 64 IQ scores each. Theorems 9.3 and 9.4 tell us that, under the hypothesis, the distribution of means of these samples should itself have a mean of 100 and a standard deviation of 2.5 points. This distribution is illustrated in Fig. 12.3. The place of the obtained sample mean of 108 is indicated by an X.

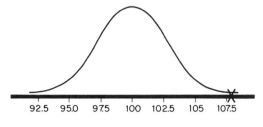

| 92.5 | 95.0 | 97.5 | 100 | 102.5 | 105 | 107.5 |

Fig. 12.3

Fig. 12.3 shows that, under the assumption that the null hypothesis is true, we have come upon a sample mean that is eight points above the balance point in a distribution that is normal and that has a standard deviation of 2.5 points. In other words, where we assume the null hypothesis is true, we must conclude that we have come upon a sample mean that is 3.2 standard deviations above its expectancy.

Appendix III tells us that, in a normal distribution, the probability is less than one in a thousand of getting a value by chance with a z score more than three units on either side of the balance point. Thus, assuming that the null hypothesis is true, the obtained sample value was so deviant from the expectancy that its probability of occurrence was less than one in a thousand. Upon obtaining such a finding, our decision would naturally be to reject the hypothesis, and to construe that boys who wish to be firemen differ as a group in IQ from others their age.

It will be useful to go back over what we have done and to use a formula for the z score of a sample mean in the distribution of sample means. This formula will enable us to do similar problems without spelling out the steps in so much detail. The z score of a sample mean may be determined as follows:

$$(12.2) \qquad z = \frac{\text{distance score for the mean}}{\text{standard deviation of means}} = \frac{\bar{X} - \mu}{\dfrac{\sigma}{\sqrt{N}}}$$

Here $\bar{X}$ stands for the obtained value of the sample mean (which in our problem was 108). μ stands for the mean of the distribution of sample means (we were working under the assumption that $\mu = 100$). σ stands for the standard deviation of the individuals' scores (which in our case was known to be 20) for the entire population.

To substitute the appropriate values in Formula (12.2):

$$z = \frac{\bar{X} - \mu}{\dfrac{\sigma}{\sqrt{N}}} = \frac{108 - 100}{\dfrac{20}{\sqrt{64}}} = 3.2$$

It was after using the z score value of 3.2 and making reference to Appendix III that the decision was made to reject the null hypothesis at the .05 level of significance.

We shall now look at one more problem. It was decided to test the hypothesis that eighth grade students in a particular school are a cross-sectional representation of students of their age in ability to spell. Actually these students had been taught spelling in a different way, which some contended was better and others contended was worse. At any rate, it was decided to test the null hypothesis at the .01 significance level. A city-wide spelling test was chosen for use, partly because a vast number of boys like those to be used were known to have achieved an average of 72 in this test.

It was also known that the standard deviation of their scores was 12 points. There were 36 boys in the eighth grade whose spelling scores were used as the sample for testing the hypothesis. The average spelling mark of these 36 boys turned out to be 74.

The assumption was made that the hypothesis was true so that the 36 spelling marks might be temporarily viewed as comprising a random sample from the defined population of marks. That is, the obtained sample mean of 74 could then be viewed as merely one value in a theoretical distribution of means of samples of 36 cases each drawn from the larger population. According to Theorem 9.2, the distribution of such sample means would be normal. Theorems 9.3 and 9.4 tell us that its mean would be 72 and its standard deviation would be $12/\sqrt{36}=2$. It would follow that the obtained sample mean of 74 was exactly one z score unit above the theoretical expectancy. Such a discrepancy from the exact expectancy would not have a smaller probability than the significance level, and so the decision was made not to reject the hypothesis. It was concluded that students in the particular school had not shown evidence of being different in spelling ability from comparable students elsewhere.

Formula (12.2) for obtaining the z score of the obtained sample mean, assuming the hypothesis to have been true, might have been used without discussion:

$$z = \frac{\bar{X}-\mu}{\dfrac{\sigma}{\sqrt{N}}} = \frac{74-72}{\dfrac{12}{\sqrt{36}}} = 1.0$$

Then using Appendix III, it might have been shown immediately that the obtained sample mean was not sufficiently different from the hypothetical expectancy to indicate the rejection of the hypothesis.

12.5 Some New Language. A sharp distinction must be made between what is called a population value and one obtained from a sample. For instance, the mean IQ of boys in a given community might be 100, whereas the mean of a sample of IQ's of boys taken from this community might be 93 or 110 or 135. A population value—for instance, the population mean of 100 in this case—is called a *parameter*. In contrast, a value obtained from a sample is called a *statistic*. The standard deviation of scores in a population would also be a parameter.

Note that any statement of a hypothesis is one about a parameter. To test a hypothesis, we gather a random sample and compute from it some statistic. We then use the statistic to facilitate our reasoning about whether to accept or reject the hypothesis. We have already used the phrase "to test a hypothesis at the .05 significance level." By this phrase, it should have been clear that we meant that we had chosen to test the hypothesis using the significance level of 5/100.   In the future, we shall regularly refer to testing a hypothesis at a given significance level, by which it shall be meant merely that the given significance level was selected.

Incidentally, experimenters who make regular references to the normal distribution when testing hypotheses get to know important relationships between z score distances and probabilities.   For instance, they know that the probability of randomly coming upon a term from the normal distribution with a z score further than 1.96 units from the mean is 5/100.   Thus when they test a hypothesis at the .05 level of significance, they automatically know not to reject it when z is between -1.96 and $+1.96$.   If z is less than -1.96 or greater than $+1.96$ they reject the hypothesis.

12.6   One-Tail Tests.   So far our purpose has been to reject the hypothesis that a parameter—such as a population mean—was some hypothesized value.   Our procedure was to test the hypothesis that the mean had the particular value and to reject it upon finding a sample value sufficiently above or sufficiently below the parameter value.   For instance, the theory was stated that boys of age eight who want to be firemen differ from the larger group in IQ.   The hypothesis that these boys have on the average an IQ of 100 was tested, and the hope was to be able to reject this hypothesis upon finding a sample of would-be firemen whose average IQ was either significantly higher or significantly lower than 100.

Now consider the situation in which a researcher wishes to show that members of some particular group differ in some specified way from those in a larger population.   That is, we not only speculate that our sample value will differ from the parameter, but we also speculate about the direction in which it will differ.   As an example, suppose we speculate that boys who wish to be firemen tend as a group to have higher IQ's than members of the large group.   Accordingly, we predict that the average IQ in the sample of would-be firemen will have a mean higher than 100.   Now we are no longer concerned with merely finding a discrepancy, but with predicting what statisticians call the "direction" of the discrepancy as well.   To find a sample mean even ten standard deviations lower than the parameter value will mean that we are wrong.

To proceed in this case, we begin by stating our null hypothesis—that the mean IQ of would-be firemen is identical to that of the larger population. We assume it is known that the distribution of IQ scores of the larger population has a mean of 100 and a standard deviation of 20 points.   Let us say once more that we decide to test the hypothesis at the .05 significance level and that we obtain the IQ's of 64 boys who wish to be firemen.   Once more we look at the sample mean, and under the hypothesis we assume that this mean is that of a random sample drawn from the larger population of IQ scores. In other words, we assume that the fact that we chose would-be firemen does not constitute a selective factor, so that we can construe our sample mean as one in the distribution of sample means that would be obtained if a vast   169 number of equal-sized samples had been drawn from the larger population. If the hypothesis is true, the distribution of sample means that would turn up should look like Fig. 12.3.

The question now is, which sample means, if obtained, should lead us to reject the hypothesis and consider our theory confirmed? To answer this question, remember that we have already set the .05 significance level for use in testing the hypothesis. This means we have decided that, when the hypothesis is true, we shall expose ourselves to the risk of making a type-one error exactly 5/100 of the time. We must pick some interval such that when the hypothesis is true only 5/100 of the sample means which would be obtained in the long run would fall into that interval.

Our theory is that the mean of would-be firemen is more than 100. Our hypothesis to be tested is that this mean is 100. Thus we certainly do not want to reject the hypothesis when we obtain any mean which is less than 100. We want to pick the interval such that when our theory is correct, we have the best chance to confirm it. Therefore we choose the interval containing the largest 5/100 of the sample means that would turn up.

According to Appendix III, the largest 5/100 of the terms in a normal distribution are the terms that are at least 1.64 standard deviations above the mean. In the distribution of sample means in Fig. 12.4, 1.64 standard deviations amount to 4.10 points. [(1.64)(2.5)=4.10] Thus the largest 5/100 of sample means that would be obtained if the hypothesis is true are those with values more than 4.10 points above the hypothesized population mean. In other words, the largest 5/100 of the sample means that would be obtained are those with values greater than 104.10 points. These are represented by the shaded region in Fig. 12.4.

Our decision rule must be to reject the null hypothesis and to consider our theory confirmed when a sample mean larger than 104.1 turns up. Otherwise the decision is to accept the hypothesis. We have found the particular .05 level decision rule that makes the most sense in view of our theory.

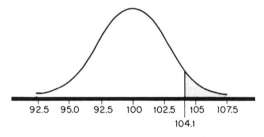

Fig. 12.4

The test of significance described is called a *one-tail test* because the rejection region is completely in one tail. Suppose we were trying to prove that the mean IQ of would-be firemen is less than 115. Then we would have wanted to reject the hypothesis that the mean is 115 only when we obtained a sample mean sufficiently deviant on the other side. We would have used analogous reasoning and the rejection region would have been the shaded area in Fig. 12.5. We have again assumed a standard deviation of 20 for the population and a sample size of 64. The reader should verify that the cut-off point is 110.9 for the .05 level.

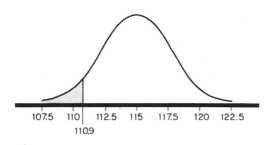

Fig. 12.5

When we were merely trying to prove that a mean was different from some specified value, we used a *two-tail test*. We compared our sample mean with the hypothetical expectancy and we construed an extreme deviation from the expectancy in either direction as evidence contradicting the hypothesis. But when we are trying to prove that the parameter differs in *some specified direction* from a hypothetical value, we use a one-tail test. The rejection region still is determined by the size of the significance level, but the logic of the problem tells us to put the entire rejection region in one tail. If we are trying to prove that the mean is larger than some value, we put the entire rejection region in the upper tail. If we are trying to prove that the population mean is smaller than some value, we put the entire rejection region in the lower tail.

The researcher whose theory predicts *how* the obtained sample mean differs from the expectancy profits by using a one-tail test. He gives himself a better chance to prove his theory using a one-tail test than by using a two-tail test. For instance, in our first illustration, if we were using a two-tail test our rejection region would be represented by the shaded area of Fig. 12.6.

Values between 104.1 and 104.9 would not have been sufficiently deviant from the expectancy to enable us to prove our theory using a two-tail test. But to obtain a value in this interval would allow rejection of the hypothesis using a one-tail test.

In essence in the case of the two-tail test, the theory merely predicts that a sample mean will depart from the expectancy. We consider the theory proven if the mean does in fact depart by a particular amount from the expectancy. In the case of the one-tail test, the theory predicts that the sample mean will

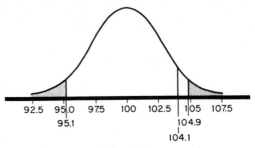

Fig. 12.6

be on a particular side of the expectancy.   The fact that this prediction comes true is itself partial confirmation of this theory.   In other words, it constitutes some evidence for the theory, and thus the magnitude of the departure needed to convince us fully of the theory is less than in the case of the two-tail test, since the prediction of direction has already been borne out.   The one-tail test precedure is used to test hypotheses about other kinds of parameters. For instance, the procedure is sometimes used to prove that a population variance is larger than some specified number. . We shall use another application of a one-tail test in later examples.

PROBLEM SET A*

12.1 Suppose that the probability of an event is $P = .5$ for a vast number of samples, each consisting of $N = 400$ observations of a dichotomous variable. If a researcher selects a random sample of $N = 400$ and obtains an outcome of $O = 175$, test the hypothesis that this outcome is the result of a random sample from the population, using a two-tail test at the .05 level of significance.

12.2 Suppose that the probability of an event is $P = .5$ for a vast number of samples, each consisting of $N = 50$ observations of a dichotomous variable. If a researcher selects a random sample of $N = 50$ and obtains an outcome of $O = 27$, test the hypothesis that this outcome is the result of a random sample from the population, using a two-tail test at the .01 level of significance.

12.3 Suppose that the probability of an event is $P = .25$ for a vast number of samples, each consisting of $N = 100$ observations of a dichotomous variable. If a researcher selects a random sample of $N = 100$ and obtains an outcome of $O = 25$, test the hypothesis that this outcome is the result of a random sample from the population, using a one-tail test at the .05 level of significance.

*In all one-tail tests in this and subsequent problem sets, assume that the obtained value (of O or $\overline{X}$) differs from the expectancy in the direction necessary to permit rejection of the hypothesis.

12.4 The distribution of a population of scores on a spelling test is normal, with a mean of 72 and a standard deviation of 8. Test the hypothesis that a sample of 100 having a mean of 76 is a random sample from this population. Use a two-tail test at the .05 significance level.

12.5 The time lengths of all films produced by a French film company are found to be normally distributed with a mean length of 92 minutes and a standard deviation of 23 minutes. Test the hypothesis that a sample of 30 having a mean of 100 is a random sample from this population. Use a two-tail test at the .01 significance level.

12.6 Suppose that the weights of the population of U.S. males are normally distributed with a mean of 180 and a standard deviation of 12 pounds. Test the hypothesis that a sample of 100 having a mean of 166 is a random sample from this population. Use a one-tail test at the .05 significance level.

12.7 The distribution of a population of scores obtained by high school seniors on a language aptitude test has a mean of 80 and a standard deviation of 10. Test the hypothesis that a sample of 400 having a mean of 73 is a random sample from this population. Use a two-tail test at the .05 significance level.

12.8 It is known that the mean systolic blood pressure for young men in the age range from 21 to 30 is 122 and the standard deviation is 9. Use a two-tail test at the .01 significance level to test the hypothesis that a random sample of 100 cases was obtained from this population if the mean of the sample is (a) 127, (b) 120, (c) 135.

12.9 It is known that 53 per cent of the pupils in a very large school system are males. Use a two-tail test at the .05 significance level to test the hypothesis that a random sample of 450 pupils was obtained from this population if the number of males it contains is (a) 180, (b) 210, (c) 225.

12.10 A manufacturer of light bulbs claims a mean life of 1,000 hours for his product, with a standard deviation of 50 hours. (a) Test the hypothesis $\mu = 1,000$ hours if a random sample of 25 bulbs has a mean life of 875 hours. Use a two-tail test at the .05 level of significance. (b) What is the largest value for the sample mean, under the conditions above, for the hypothesis $\mu = 1,000$ to be rejected? (c) Suppose the largest value of the sample mean calculated in (b) were obtained with the random-sample size increased by 100; should the hypothesis $\mu = 1,000$ still be rejected?

12.11 A bag contains circular disks, some of them black and some white. A test of the hypothesis that the number of black disks is equal to the number of white disks is performed by withdrawing 49 disks, one by one, and replacing each disk after it is drawn. The hypothesis will be accepted if between 21 and 28 black disks are drawn—otherwise the hypothesis is rejected. Find the probability of rejecting the hypothesis when it is actually correct.

12.12 (a) If you require the probability of rejecting the hypothesis, when it is actually correct, to be at most 0.01, what decision rule would you accept in Problem 12.11? (b) If the .05 level of significance were used, what would be the decision rule? (c) How would you test the hypothesis that there are more white than black disks in Problem 12.11? What null hypothesis would you use? What decision rule would you use at .05 level of significance?

PROBLEM SET B

12.13 Suppose that the probability of an event is $P = .5$ for a vast number of samples, each consisting of $N = 64$ observations of a dichotomous variable. If a researcher selects a random sample of $N = 64$ and obtains an outcome of $O = 35$, test the hypothesis that this outcome is the result of a random sample from the population, using a two-tail test at the .05 level of significance.

12.14 Suppose that the probability of an event is $P = .5$ for a vast number of samples, each consisting of $N = 80$ observations of a dichotomous variable. If a researcher selects a random sample of $N = 80$ and obtains an outcome of $O = 39$, test the hypothesis that this outcome is the result of a random sample from the population, using a two-tail test at the .01 level of significance.

12.15 Suppose that the probability of an event is $P = .5$ for a vast number of samples, each consisting of $N = 30$ observations of a dichotomous variable. If a researcher selects a random sample of $N = 30$ and obtains an outcome of $O = 24$, test the hypothesis that this outcome is the result of a random sample from the population, using a one-tail test at the .05 level of significance.

12.16 The physical-education performance scores of a population of ninth-grade boys are normally distributed with a mean of 82 and a standard deviation of 7. Test the hypothesis that a sample of 49 having a mean of 85 is a random sample from this population. Use a two-tail test at the .05 significance level.

12.17 The population of scores on a mechanical aptitude test are normally distributed with a mean of 79 and a standard deviation of 10. Test the hypothesis that a sample of 50 having a mean of 61 is a random sample from this population. Use a two-tail test at the .01 significance level.

12.18 A musical aptitude test was given to all third-grade classes in New York City. The mean score on this test was 75 and the standard deviation was 15 points. Test the hypothesis that a sample of 36 having a mean of 77 is a random sample from this population. Use a one-tail test at the .05 significance level.

12.19 Scores on a dental aptitude test are normally distributed with a mean of 65 and a standard deviation of 15. Test the hypothesis that a sample of 250 having a mean of 70 is a random sample from this population. Use a two-tail test at the .05 significance level.

12.20 Fifty-three per cent of a population of high school students are females. Use a one-tail test at the .05 significance level to test the hypothesis that a random sample of 450 students was obtained from this population if the number of females it contains is (a) 168, (b) 220, (c) 300.

12.21 A string manufacturer claims that his product has a mean breaking strength of 60 pounds, with a standard deviation of 3.5 pounds. Test the hypothesis that $\mu = 60$ pounds if a random sample of 49 pieces of this string is found, under test, to have a mean breaking strength of 55 pounds. Use a test at the .01 level of significance.

12.22 A patent-medicine manufacturer claims that his product is 90 per cent effective in relieving eczema for a period of 10 hours. Of 225 eczema victims sampled, 190 obtain such relief from the patent medicine. What would you conclude about the manufacturer's claim? Take $P = .9$.

12.23 In a statistics course at a certain college, over a long period of years the per cent of A's was 8.75. In one semester there were 35 A's awarded in a combined statistics course of 280 students. Test the significance of this grading at the .05 and .01 levels of significance.

12.24 A manufacturer of steel cable formerly claimed a mean breaking strength of 55,000 pounds, with a standard deviation of 500 pounds. Because of an improved manufacturing process, the manufacturer now claims an increased breaking-strength value. A sample of 50 cables is tested and the mean breaking strength is found to be 55,250 pounds. Does this support the manufacturer's claim at the .01 level of significance?

thirteen

---

estimation

13.1 Introduction. Suppose it is planned, as part of an experiment, to test the hypothesis that the mean IQ of eighth grade, private-school children in a large city is 110. This time, however, the standard deviation of the IQ scores of the members of the population is not known. Therefore, when the mean of the obtained sample is found, it is not possible to use the ordinary procedure for testing the hypothesis, as discussed in Chapter 12. To use that procedure it was necessary to know the value of the population's standard deviation.

The situation described is almost invariably the one that actually arises. A researcher who tests a hypothesis has no more knowledge than that contained in the data he has gathered. In particular, he has no definite knowledge of the standard deviation of scores in the distribution from which his sample came. What he must first do in order to test a hypothesis about an unknown population mean is to make some kind of enlightened guess about the value of the population's standard deviation.

The problem of using data in a sample to make an enlightened guess of a population value arises in diverse contexts. The guess that one makes is called an *estimate*. In this chapter we shall discuss various considerations that arise when one makes an estimate, and it will soon be seen that there is more to the issue of estimation than meets the eye.

13.2 Unbiasedness of an Estimate. To begin with, we shall consider the simplest case—namely, the case when our purpose is to use the data comprising a sample in order to estimate the mean of the population from which it was drawn. To be concrete, let us suppose that we are researchers whose task it is to estimate the mean IQ of the private-school children already described. We obtain a sample of five IQ scores of children randomly selected from this population of interest. These scores turn out to be 90, 100, 105, 145, and 100. Our task is to derive from this set of scores the best possible estimate of the mean of the population from which they came. The totality of our knowledge consists of these five scores. The question is, what shall we do to arrive at the best possible estimate? In other words, what measure shall we use as our *estimator* of the mean IQ of the private-school population? At this point, note that the word *estimator* denotes the kind of measure used for the purpose of estimation. An *estimate* is the numerical value of an estimator derived from a particular sample.

The answer, as one might guess, is that the proper procedure would be for us to compute the mean of the sample of five scores. We would then use this obtained sample mean, whatever it is, as our estimator of the population mean. The mean of the five scores turns out to be 108 [$\bar{X} = 108$]. Note that this particular estimate of 108 is merely a guess made on the basis of the sample obtained. To have obtained a different sample would almost certainly have meant to have arrived at a different guess.

The next question is more difficult: Why was it proper for us to use the mean of our sample as the estimator of the population mean? What made it better to use this estimate than, say, the midpoint between the lowest and highest values in the sample, which would be the value $117\frac{1}{2}$? To answer this question leads us to consider the more general issue of how to evaluate a good estimator. Actually there are many criteria for a good estimator, but here we shall consider only the most important one—namely, that of *unbiasedness*.

The way to evaluate an estimator is to evaluate the estimates that it would send forth from a multitude of different random samples drawn from the same population. (Of course in an actual experiment only one such sample is available.) Let us suppose this time that we have obtained five different IQ samples of five cases each. The purpose is to estimate the mean of the population of IQ scores from which these samples came, which in practice would be unknown to us.

In Table 13.1 the values of the five different random samples are listed, and the means of these samples are indicated.

TABLE 13.1 SAMPLES

1	2	3	4	5
90	120	105	150	110
100	110	120	113	95
105	100	95	107	115
145	115	130	90	95
100	90	115	115	110
$\bar{X}$ 108	107	113	115	105

Suppose now that we had access to the data comprising only the first sample. Using the sample mean as our estimator, the estimate we would obtain would be 108. Similarly, the estimate we would obtain from the second sample would be 107; from the third, 113, and so on. The mean of the five estimates listed in Table 13.1 is 109.6.

Consider now what would happen if we added more and more estimates from random samples of five cases each. As we added more estimates, the mean of these estimates would tend to get closer to the value of the unknown population mean. The reason is that, as more and more random samples are drawn from a population, the mean of their means approaches the mean of the population itself. For instance, by the time we had collected 5,000 samples, the mean of their means would be unlikely to differ from the population mean by as much as half a point.*

To sum up, suppose in general that a population mean is to be estimated and that the estimator to be used is the mean of a random sample drawn from the population. The estimate derived from one particular random sample may be relatively accurate or not. But were more and more estimates to be obtained from different samples, the mean of these estimates would approach the population mean. The sample mean, because it has this property, is said to be an unbiased estimator of the population mean. More generally, *an estimator of a parameter is called unbiased if the mean of estimates obtained from independent samples drawn from the population would approach the population value, were more and more samples to be included.* It is not implied that each new sample estimate must bring the mean closer to the unknown population value. The implication is only that, as new sample estimates are brought in, the trend is for the mean of these estimates to get closer to the population value.

Now we shall turn to the problem of computing from a sample an unbiased estimate of the variance of the population from which the sample was drawn. We shall contrast the case in which the mean is known with that in which it is not known. For convenience, we shall use for illustrative purposes the same data as before. That is, we are selecting our terms from a population of IQ scores of eighth grade private-school children. In theory we could start listing the scores as if we were able to look at each one in the vast population. We shall say that the first score we come upon is 90, the next is 100, the next is 105, and so on. The first fifteen scores we might come upon are listed in Column (1) of Table 13.2.

To get the actual population variance, we would have to subtract the true population mean from each score. Let us now pretend that we know this true population mean and that it is 110 [$\mu = 110$]. The subtraction was done

---

* To see why this is so, we may view the mean of the 5,000 sample means as an average of one gigantic sample of 25,000 cases. Even if the standard deviation of the IQ scores were as big as 20 points, there is little chance that the mean of a sample of 25,000 cases would differ from the population mean by as much as half a point. The reader should be able to verify this fact by using Theorem 9.4.

for the fifteen scores in Table 13.2. The distance scores appear in Column (2). Were we actually to have obtained distance scores for all the terms in the population, their sum would be zero since they are distances from the mean (Theorem 2.1). However, our concern is with the squared distance scores in Column (3) of Table 13.2. It is the mean of all the squared distance scores that is the variance. It is this that we would like to estimate.

TABLE 13.2   VARIANCE USING POPULATION MEAN

	(1)	(2) $X-\mu^*$	(3) $(X-\mu)^2$	
1	90	-20	400	
2	100	-10	100	
3	105	-5	25	
4	145	35	1225	
5	100	-10	100	1850
6	120	10	100	
7	110	0	0	
8	100	-10	100	
9	115	5	25	
10	90	-20	400	625
11	105	-5	25	
12	120	10	100	
13	95	-15	225	
14	130	20	400	
15	115	5	25	775

Suppose we had access to only the first five scores, and our purpose was to estimate the population variance. If we knew the population mean was 110, we might accordingly subtract 110 from each of these scores. We would obtain the five squared distance scores in Column (3), and their mean would give us an unbiased estimate of the variance. After all, we would simply be performing on a small sample what should be performed on the entire population in order to get the population variance. The sum of the first five squared distance scores is $400+100+25+1,225+100=1,850$. Our estimate of the population variance then would be $1,850/5=370$.

However, the usual situation is quite different. We usually have access to a sample of scores from a population but do not have knowledge of the value of the population mean. Let us imagine ourselves in this position, once more with access to only the first five scores. The most obvious thing that may come to mind may be to estimate the population mean first, by determining the sample mean. Then using our estimate of the population mean in place of the real one, it appears that we might be able to proceed as before. Table 13.3 shows what would happen if we were to proceed in this way.

---

* μ stands for the population mean.

To begin with, when we compute the value of the mean from our sample of five cases, it turns out to be 108. In Table 13.3, Column (2), we have subtracted this value from each of the terms in the sample, and in Column (3) we have squared the distance scores thus obtained.

The sum of the squared deviations from our sample mean turned out to be smaller than the sum of squared deviations of the same five terms from the population mean when we used it. Using the population mean, the sum of squared deviations was 1,850. Here it is 1,830. Thus our estimate of the variance using the sample mean is smaller than the estimate we just made using the population mean.

TABLE 13.3   VARIANCE USING SAMPLE MEAN

(1) X	(2) $X - \bar{X}$	(3) $(X - \bar{X})^2$
90	-18	324
100	-8	64
105	-3	9
145	37	1369
100	-8	64
		1830

$$\frac{\sum (X - \bar{X})^2}{N} = \frac{1830}{5} = 366$$

What happened here was not coincidence. It is always the case that a sum of squared deviations from a sample mean will be as small or smaller than what it would have been had we used the population mean. The reason is given by Theorem 2.2. For any set of terms, the sum of squared differences of the terms from their (own) mean is less than the sum of squared differences from any other point. We have been considering terms that comprise a sample, and their mean is the sample mean. With regard to these five terms, the population mean is an alien point. Thus it turned out necessarily that the mean of squared deviations of these terms from their sample mean was as small as or smaller than the mean of their squared deviations from the population mean.

Now look at the second sample of five terms listed in Table 13.2. There the mean of the sample is 107. Once again the sum of squared deviations from this sample mean turns out to be less than what it is when the population mean is used as a reference point. The reader should verify that this sum of squared deviations is less than 625.

What has been said boils down to the fact that one cannot make an unbiased estimate of a population variance from a sample by simply getting the mean of the squared deviations from the sample mean. Such an estimate would not be unbiased; or, more specifically, in the long run the mean of such estimates would be too small. We must now consider an important concept in statistics—namely, that of *degrees of freedom*. Once this concept becomes clear, we can go on to consider how to derive from a sample an unbiased estimate of the variance of the population from which it came. In fact, the concept of degrees of freedom is quite important and we shall refer to it in a later context.

13.3 The Concept of Degrees of Freedom. A meaningful distinction must be made between the number of independent facts that have been ascertained and the total number of facts ascertained in the data gathering. We shall look at this distinction in a general way first and then we shall consider it specifically later, because it is highly relevant to the problem of estimation.

In the simplest case, suppose we are told that John wrote an essay on which he got a score of 87. We are told also that he lost 13 points on the essay and, furthermore, that the top possible mark on the essay was 100.

It is clear that from any two of the given facts we might have deduced the third. Thus we have actually been given only two independent facts, though we have been given three facts altogether. A fact is said to be independent of one or more other facts when it is new in the sense of not having been implied by them. A fact is said to be dependent on one or more others when it might have been deduced from them.

It would not be advantageous for our purposes to single out two particular facts from the three concerning John's performance and to say that they are independent and that the third one is dependent upon them. It is sufficient to say that only two independent facts are contained in the set, and that the facts that comprise the total set are dependent. We use the phrase *degrees of freedom* to indicate the number of pieces of information that are free of each other in the sense that they cannot be deduced from each other. Thus two degrees of freedom comprise our information concerning John's performance on the essay question.

Note that we can communicate the total information concerning John's performance by presenting two facts. These might be John's score of 87 and the total possible score of 100. Or they might be John's score of 87 and the number of points he lost, which was 13. Or finally, we might have presented the number of points John lost, 13, and the total possible score on the test. The minimum number of independent facts that convey the total information is always the same no matter which facts we choose. Therefore we can always demonstrate the number of degrees involved by presenting independent facts that account for the totality of information. The number of facts entailed comprises the number of degrees of freedom involved. Incidentally, in the case described, the total possible test score would undoubtedly be known to be 100, so that for all practical purposes there would be one degree of freedom of interest. That would be the degree of freedom that provides information concerning John's performance.

The statistician must often make a careful count of the number of degrees of freedom that go into his making of an estimate. Actually we are discussing elementary techniques in this book, and there are simple rules for figuring up the number of degrees of freedom involved in making an estimate whenever the need to do so arises. As might be guessed, one can put relatively little stock in an estimate when it is based on relatively few degrees of freedom. As the number of degrees of freedom increases, the dependability of the estimate increases.

The main concern is with making the distinction between the number of degrees of freedom involved and the total number of facts reported, which may be considerably larger. A serious error is apt to occur when an estimate based on relatively few degrees of freedom is mistakenly thought to have been based on many more. The error is in the direction of overrating the soundness of the estimate.

The following medieval folk tale, treated by several German poets, may help the reader to distinguish between degrees of freedom and observations in general. The city states had not yet welded themselves together as the Prussian empire. The enlightened monarch of Wierstrasse, a mythical northern state, had died leaving his twin sons, Hans and Fritz, as heirs to his throne. The king had left written instructions with the council to choose as his successor the son with the superior talent of prophecy.

In accordance with the will of the king, a contest before the courtiers was arranged so that each boy could demonstrate his skill. Hans was the first to step before the crowd at the market place. He handed the prime minister a golden coin with an engraving of his deceased father on one side (heads) and one of the royal gardens on the other (tails). The minister leaned over a wooden table and then tossed the coin high in the air.

Hans shouted, "The picture of my father will face the sky when the coin has fallen." He was correct. A second time the minister threw the coin in the air and Hans shouted, "The picture of my father will be on the underside of the coin this time." Again he was correct, for the coin turned up tails and the picture of his father was on the underside as he had predicted.

The coin was thrown for a third and final time. Hans shouted, "Now the coin will have the tail side facing the sky." Again he was correct. The crowd applauded. Hans had made three correct predictions and his performance was over.

His brother Fritz now asked for the golden coin and handed it to the minister. The minister threw the coin in the air and Fritz shouted: "I make two predictions. First, I predict that the picture of my father will face the sky. Also I predict that the underside of the coin will have imprinted on it the picture of the royal gardens." Both predictions were correct. The heads side did face the sky and the tails did rest against the table on which it fell. Fritz smiled.

The minister threw the coin in the air a second time. "Again I make two predictions," Fritz shouted. "I predict that the tails side of the coin will face the sky and that the heads side of the coin will be the underside and rest against the table." Again both predictions were correct. Fritz now turned to the assemblage, for his performance was over.

He said, "I have made four correct predictions, whereas my brother has made only three. I have proven myself superior as a visionary. Humbly and gratefully do I accept the crown of Wierstrasse."

The courtiers were perplexed and began talking amongst themselves. The murmur grew and then one of them shouted, "Long live Fritz, King of Wierstrasse." Soon the others picked up the chorus. "Long live Fritz, King of Wierstrasse," they shouted three times, when Clarence, the King's philosopher stepped to the platform and showed his palms to the assemblage asking for silence.

"You have just witnessed a contest," he began. "You have witnessed three correct predictions made by Hans and four correct predictions made by his brother Fritz. Consequently, you esteem Fritz a better prophet than Hans." He paused. Then he said, "Ladies and gentlemen of the court of Wierstrasse—you have been duped!"

Loud talking began in the throng but Clarence brought silence as he began to speak again. "Let us examine the evidence closely. The evidence in favor of Hans consists of three correct independent observations in the form of predictions made by him." He emphasized the word *independent*. "I say that the observations are independent since knowledge of the outcomes of any two of them would give us no information about the outcome of the third. Our purpose is to estimate the ability of Hans as a prophet. The estimate we make is based upon three independent observations, or degrees of freedom, as they are called.

"As for Fritz, our four observations in the form of his correct predictions are not independent. Certain of these observations necessitated others. Fritz made two predictions at each flip of the coin. He predicted what its top side would be and he predicted what its underside would be. At each flip, we made two observations. The two observations made at each flip are dependent in the sense that either observation necessitated the other. Knowledge of the outcome of either one of these predictions would tell us the outcome of the other. The two observations made at each flip are based only upon one degree of freedom, and therefore our complete estimate of Fritz's skill is based upon two degrees of freedom.

"The soundness of an estimate depends upon the number of degrees of freedom used in making it, rather than the number of observations counted blindly. Our estimate of Hans' prophetic skill is based upon three degrees of freedom. Our estimate of Fritz' prophetic skill is based upon only two degrees of freedom."

The legend is that Hans made an excellent king and that his reign was a happy one. The legend emphasizes the importance of counting the number of degrees of freedom that go into an estimate. The number of degrees of freedom relates to the probability that the estimate is accurate, or, as we might say loosely, the dependability of the estimate.

13.4 Estimating the Variance. In our example of the population of IQ scores of eighth grade, private-school children, we have discussed two situations in which we might make an estimate of a population variance. The first, introduced only for the sake of discussion, occurs when we have the data of the sample and we know the population mean. We shall refer to it as Situation A. The second occurs when we have the data of the sample but have no outside information. We shall refer to the context when this occurs as Situation B.

Remember that degrees of freedom refer to the number of independent observations used to make an estimate, for we are now going to see what happens when we estimate the variance in Situations A and B. Let us return to the population of IQ scores of the eighth grade, private-school children, which we are pretending has a mean of 110. In particular, let us again contrast the problem of estimating the population variance, knowing the value of the mean and not knowing it.

To begin with, suppose our sample consists of only one score. We shall say that the IQ of the first child randomly selected happens to be 90. We shall compare what we would do to estimate the population variance in each situation.

Situation A: Since we know that the population mean is 110, we compute the difference between our single score of 90 and this reference point. The difference is -20 points. The fact of this difference is one piece of information on which to base our estimate. Our procedure would be to square the difference and divide by one to obtain the estimate of the population variance. Note that we are making an estimate based on one degree of freedom:

$$X - \mu \qquad (X - \mu)^2$$
$$(-20) \qquad\quad 400$$

Our estimate of the variance is

$$\frac{400}{1} = 400$$

Situation B: Now let us see what happens when we proceed in ignorance of the population mean. The mean of our sample of one term naturally has the same value as the term itself. Thus one term gives us no information about the variability of the terms in the population. We cannot estimate the population variance, since one term is not enough to give us one difference from a reference point.

Suppose that we now add a second IQ score to our sample, which we shall say happens to be 100.

Situation A: This second score gives us a new independent difference from the population mean. The first score led to a difference score of -20 points and now this second score leads to a difference score of -10. The first difference score would not permit us to infer the other. Note that these two independent differences are what go into the variance estimate. To put it in more technical language, our estimate of the variance is based on two degrees of freedom.

Situation B: The two scores give us a new location for the sample mean. Where there are two scores, their mean is the point midway between them. Since the two scores are 90 and 100, their mean is 95. Thus the first difference score with reference to the sample mean is -5 and the second difference score is 5. Remember that the differences of a set of terms from their own mean must add up to zero (Theorem 2.1). Therefore the two differences from the sample mean are mutually dependent. Were we to know either one, we could figure out the other. The two terms thus provide us with only one independent piece of information concerning the population variance. Even though we have a sample of two terms, our estimate is based upon only one degree of freedom.

Now to be general, let us suppose that our sample consists of some larger unspecified number of terms. We shall say it consists of N terms.

Situation A: Each IQ score in our sample gives us an independent difference score. That is, were we to know the value of all but one of these differences, we could still not possibly compute the value of the last one. Thus, when there are N terms in the sample, our estimate of the population variance becomes based on N independent difference scores. Technically, were we to know the value of the population mean, our estimate of the population variance would be based on N degrees of freedom.

Situation B: This time we have chosen the reference point ourselves. Now were we to know all the differences from the reference point except one, we could deduce the value of the last difference. Because the reference point is the sample mean, the sum of all the differences from this reference point must be just large enough to make the sum of all the differences total zero. For instance, if all but one of the differences total -4, then the last difference must be $+4$ in order to make the sum of all the differences exactly zero. In essence, there are N terms in our sample, which give us N differences. However, we have only $N-1$ degrees of freedom included in our N differences. Technically, there are $N-1$ degrees of freedom that go into our estimate of the population variance.

It may be shown mathematically that to make an unbiased estimate of a population variance from a sample, the proper procedure is to take the sum of the squared differences from the sample mean and to divide it by the number of degrees of freedom involved in making the estimate. In other words, the denominator must not be the number of terms in the sample in Situation B, but it must be one less than that number. For instance, if there are 30 terms in a sample, the procedure must be to compute the sum of the squared differences of these terms from the sample mean. This sum must then be divided by 29 in order to get an unbiased estimate of the population variance.

More generally, we can put what has been said in the form of a theorem. 185

Theorem 13.1 *Suppose a random sample of N terms is drawn from a population that has an unknown mean and variance. Then the sum of squared deviations of the N terms from the sample mean divided by $N-1$ provides an unbiased estimate of the unknown population variance. The square root of this value is an estimate of the population standard deviation.*

The usual letter used to designate a sample estimate of a population variance is s^2. That used to designate a sample estimate of population standard deviation is s. In other words, the value of s^2 provides an unbiased estimate of σ^2 for the population from which the sample was drawn. The value of s is our best estimate of σ for the population from which the sample was drawn.

The formulas for s^2 and s are as follows:

(13.1)
$$s^2 = \frac{\sum (X - \bar{X})^2}{N-1}$$

$$s = \sqrt{\frac{\sum (X - \bar{X})^2}{N-1}}$$

Here the random sample consists of N terms. Its mean is $\bar{X}$. The values of X are those of the terms of the sample.

Remember that it is only correct to use the denominator $N-1$ when the purpose is to obtain a value that is an estimate. That is, when the data are being construed as a sample drawn from some larger population that is of interest, it is proper to use $N-1$ when estimating the variance of that larger population. When the data themselves are thought of as comprising some total population, then the value computed is σ^2 and not s^2. Under this condition, the denominator must be N. It has become commonplace in the field to refer to the value of s^2 as a *sample variance*, and to refer to the value s as a *sample standard deviation*. It is especially important for the reader to have the distinction clear because this practice is so prevalent.

Finally, to complete a kind of cycle, let us get back to our initial problem. We are researchers whose task it is to estimate the variance of the scores of eighth grade, private-school children in a vast community. We obtain a sample of five IQ scores of children from this community. These scores are 90, 100, 105, 145, and 100. We quickly compute that the mean of this sample, $\bar{X}$, is 108. Now our purpose is to estimate the variance of the scores in the larger population. Following Formula (13.1), we find the difference between each score and the sample mean, square them, and then add up the squares. We then divide this sum of squares by 4 to get what is loosely called a sample variance. This value, which turns out to be 457.5, becomes our unbiased estimate of the population variance. Its square root, 21.4, is our estimate of the standard deviation of the larger population of scores of eighth grade, private-school children. The computations appear in Table 13.4.

TABLE 13.4 ESTIMATE OF
STANDARD DEVIATION

90	-18	324
100	-8	64
105	-3	9
145	37	1369
100	-8	64
		1830

$$\bar{X} = 108, \ N = 5, \ N - 1 = 4$$

$$s^2 = \frac{\Sigma (X - \bar{X})^2}{N-1} = \frac{1830}{4} = 457.5$$

$$s = \sqrt{457.5} = 21.4$$

13.5 Computing Formulas for s^2 and s. The formulas for s and s^2 (13.1) are not the simplest ones to work with when the mean is not a whole number as is usually the case. In earlier chapters we have used what were called computing formulas in addition to the literal formulas, as for instance when we discussed σ^2 and σ. The computing formulas, which make the actual arithmetic work much easier where one wishes to obtain the values of s^2 and s, are as follows:

$$(13.2) \qquad s^2 = \frac{\Sigma X^2 - \frac{(\Sigma X)^2}{N}}{N-1} \quad \text{or} \quad \frac{N \Sigma X^2 - (\Sigma X)^2}{N(N-1)}$$

$$s = \sqrt{\frac{\Sigma X^2 - \frac{(\Sigma X)^2}{N}}{N-1}} \quad \text{or} \quad \sqrt{\frac{N \Sigma X^2 - (\Sigma X)^2}{N(N-1)}}$$

13.6 Interval Estimation. So far we have considered how to derive from a random sample the single best guess of the mean and variance of the population from which the sample was drawn. However, as mentioned, even the best guess we make is virtually certain to deviate at least somewhat from the exact parameter value. What is not only desirable but necessary in practical situations is to make some determination of how close the guess is likely to be. That is, even the best guess may be dependable or it may be made as a shot in the dark, and, along with the guess itself, what is needed is some communication of the likelihood of its accuracy.

The language used in this section has been the subject of much controversy among statisticians, and the word "probability" is used here in a sense that is not acceptable to some of them. Therefore especially close attention is required. Once the conceptualizaton is clear, the use of the word "probability" may be seen as appropriate and need not be a source of confusion.

To begin with, when we make a best guess about a parameter value—like the mean of a population, for instance—we cannot talk about the probability that this guess is correct. For example, suppose that on the basis of a sample we make the guess that the mean height of adult American males is 5 feet 9 inches. It is virtually certain that this guess would prove to be at least somewhat off, were we actually able to measure the heights of all adult American males and average them. Therefore we cannot talk about the probability that such a guess is accurate, if for no other reason than that this guess is virtually certain to be at least somewhat off.

Instead of specifying a single point, the practice we are now going to consider is that of specifying an interval and giving the probability that it includes some unknown parameter between its endpoints. We call this kind of estimate an *interval estimate*. The practice is to specify some desired probability fraction (like 95/100 or 99/100) and then to find two points such that the probability that the unknown parameter is between them is the specified fraction. A typical statement would be that the probability is 95/100 that the average height of American adult males is between 5 feet 8 inches and 5 feet 10 inches.

Many statisticians argue that to use the word "probability" even here is incorrect on the grounds that a parameter value is fixed and thus one cannot talk about the probability that it is contained in a given interval. For them, the notion of probability is only applicable when one is talking about where a random-sample value will appear. Actually, those writers have begun with a definition of probability that does make it inappropriate for them to use it in the present context.* We have begun with a definition of probability that would not have been serviceable in all contexts, but that does make it quite easy for us to proceed here.

Before proceeding, let us look more closely at what we mean when we say that the probability is 95/100 that the height of the average American male is between 5 feet 8 inches and 5 feet 10 inches. To begin with, remember how we defined probability. We defined what we called an outcome of an event. We then said that the probability of the outcome occurring when the event takes place is defined as the proportion of times that the outcome would occur were the event to be repeated over the long run. In other words, the probability of a particular coin falling heads on a single flip is the proportion of times that the coin would fall heads were it to be flipped over and over again in the long run.

Our question now is, what do we mean when we say that the probability is 95/100 that the height of the average adult male is between 5 feet 8 inches and 5 feet 10 inches? In particular, what is the outcome of interest, and what is the event which we conceive of as being repeated over and over again? Note that we have gathered a single random sample and have set up an

---

* The interested reader may look further into the topic of logical probability and its use with set theory. Our definition of probability is often referred to as the *empirical* definition or the *relative frequency* definition.

interval, and we are predicting that this interval contains the unknown parameter value. We must conceive of ourselves going through the same procedure of gathering a random sample of heights from the same population and each day setting up a particular interval that we hope contains the unknown population mean. On Tuesday, following the same procedure, we might set up a different interval that also happens to include the height of the average American adult male. On Wednesday, our sample might be such that the interval we set up does not include this unknown parameter value, and so on. When we say that the probability is 95/100 that the interval we have set up today does contain the unknown parameter value, what we mean is the following: We are using a procedure which, if we repeated it each day, would in the long run lead us to establish intervals 95 per cent of which would include the unknown parameter value.

An analogy should make clear exactly what a probability fraction means when it is given in connection with an interval estimate. When we make an interval estimate, we can conceive of ourselves as operating like a bomber pilot hurtling over an enemy city. The pilot, who is out to destroy a munitions factory, has only one bomb, which he drops through the clouds. The range of the blast of the bomb is, let us say, ten miles. Thus the pilot knows that if the bomb strikes within ten miles of the factory, it will destroy it. Otherwise the blast will not extend to the factory and will thus leave it unharmed.

The bomb actually falls on a particular spot that corresponds to the bomber's estimate of the location of the factory. The blast of the bomb has a range of ten miles and this range corresponds to an interval estimate of a parameter. By now you may have guessed that the exact location of the factory is the unknown parameter. A bomb that catches the factory in its blast is like an interval estimate, which includes the actual value of the parameter between its endpoints.

Now remember the definition of probability. The probability of the outcome of an experiment is the proportion of times that the outcome would occur were the experiment to be repeated indefinitely. Thus, to determine the probability that the bomb blast destroys the factory, we must conceive of the identical mission being carried out over and over again.

Suppose we could show that, during 95 out of every 100 missions in the long run, the pilot would destroy the factory with the blast of his bomb. Then it would be correct to say that the probability is 95/100 that the pilot will destroy the factory with his single bomb blast. Analogously, our purpose is to set up an interval such that were we to repeat our procedure identically, 95/100 of the interval estimates we would make in the long run would include the unknown parameter value. Referring to this fact we shall say that the probability is 95/100 that our interval estimate includes the parameter.

13.7 Interval Estimation of the Population Mean. We are now ready to consider specifically the procedure for making an interval estimate of a 189 population mean. The procedure involves making a best guess of the population mean as before. This time, however, we are going to construct an interval around our best guess such that the probability that the population mean is inside of the interval is some specified fraction. For instance, instead of guessing that the mean IQ of a population is 117, we might now end up saying that the probability is 95/100 that the population mean is between 113 and 121.

It is of course desirable to have an interval estimate as narrow as possible. The narrowness of an interval indicates the specificity of the estimate. For example, compare the following two interval-estimate statements:

1. The probability is 95/100 that the average weight of people with disease K is between 100 and 180 pounds.

2. The probability is 95/100 that the average weight of people with disease K is between 139 and 141 pounds.

The same probability fraction, 95/100 is attached to each statement. However, the narrowness of the interval in the second statement pinpoints the stipulation and makes it much more meaningful.

Note also that the higher the probability fraction attached to a statement, the stronger is the interval-estimate statement. For example, compare the following two statements:

1. The probability is 90/100 that the mean life span of fox terriers is between 10 and 14 years.

2. The probability is 99/100 that the mean life span of fox terriers is between 10 and 14 years.

We might say that the second statement is stronger than the first in the probability sense. Consequently a researcher puts more stock in the second statement than in the first.

We are now ready to attack the problem of making an interval estimate of a population mean. To think in terms of a specific problem, suppose we are interested in setting up an interval estimate of the mean IQ of American medical students. Specifically, we wish the endpoints of our interval to have the probability of 95/100 of including between them the mean IQ of these students. We shall suppose in this chapter that we know that the standard deviation of the population of medical students' IQ's is 16 points. Later on we can deal with the more complex case where the standard deviation of the population is unknown. Let us say that we gather a random sample of the IQ's of medical students and that it turns out that the mean IQ of these scores is 118.

As mentioned, we are going to conceive of a multitude of researchers, each of whom follows the same procedure as we do. That is, each of these researchers must be conceived of as gathering a random sample of 64 cases from the same population, computing its mean, and setting up an interval estimate as we are about to do. Our purpose now is to discuss a procedure which, if followed, would lead 95/100 of all the researchers to set up intervals that actually include the unknown population mean.

To proceed, we must reason in a somewhat general way. The population of IQ scores of medical students, from which our sample emanated, has a mean that is of course unknown. Its standard deviation is 16 points. Now conceive of each researcher, including our research team, as computing the mean of his obtained sample of 64 cases. According to Theorem 9.2, the distribution of obtained sample means is normal. Its mean is unknown, but its standard deviation is 2 points.

$$\sigma_M = \frac{\sigma}{\sqrt{N}} = \frac{16}{\sqrt{64}} = 2$$

σ_M is called the "standard error of the mean" or "the standard deviation of means." In the distribution of sample means each term is the mean of a sample of 64 IQ scores. More specifically, each term in it is the best guess of the mean IQ of medical students that a particular researcher would make, having drawn his particular sample. One sometimes refers to such a best guess as a *point estimate* to distinguish it from the interval estimate, which we are now considering.

Since this distribution of sample means is normal and since it has a standard deviation of 2 points, it follows (according to Appendix III) that roughly 68/100 of the sample means included will be less than one standard deviation away from the unknown population mean. This means that roughly 68/100 of the obtained sample means will be within two points of the population mean.

According to Appendix III, in a normal distribution approximately 95/100 of the terms are less than 1.96 standard deviations from the mean. Therefore in Fig. 13.1 roughly 95/100 of the sample means are within 1.96 standard deviations of the unknown population mean. *More specifically, about* 95/100 *of the point estimates that would be made by researchers would be within* 3.92 *points of the unknown population mean.* This is true because 1.96 standard deviations amount to 3.92 points.

Now let us look at what must be the point of view of any researcher who has computed the mean of this particular sample. He may say, "Of the vast number of my fellow researchers, 95/100 have computed sample means that are within 3.92 points of the unknown population mean. I may be among this 95/100 or I may be among the 5/100 of researchers whose sample means are more than 3.92 points from the population mean. Since 95/100 of researchers who operated like myself are in the first category, the probability is 95/100 that I am too. In other words, the probability is 95/100 that my particular sample mean is within 3.92 points of the unknown population mean."*

---

* To satisfy the mathematician, he might say, "Before I gathered my sample the probability was 95/100 that I would obtain a mean within 3.92 points of the population mean. I have no information on the subject. Therefore the probability is still 95/100 that my sample mean is within 3.92 points of the population mean."

Suppose in particular that as a research team we have computed the mean of our sample to be 118 as specified. We are able to say that the probability is 95/100 that our particular sample mean is within 3.92 points of the population mean (which is the average IQ score of medical students). This means that the probability is about 95/100 that the population mean is within 3.92 points of 118. To put it another way, the probability is about 95/100 that the mean IQ of medical students is between 114.08 and 121.92. We have constructed the interval containing all values that are within 3.92 points of our obtained sample mean. This is called a 95 *per cent confidence interval* for the mean.

Note that were we to repeat our procedure each day we would get a different sample mean with each new sample. Therefore, each day the center of our interval estimate would be different. We can only say that, on about 95 out of every 100 days in the long run, the interval that we managed to construct would include the value of the unknown mean IQ of medical students.

We might have wanted to construct an interval such that the probability was 99/100 that it included the value of the unknown population mean. In this case, our initial logic would have told us that about 99/100 of the terms in a normal distribution are within 2.58 standard deviations of its mean. Therefore, the probability is about 99/100 that our obtained sample mean is within 2.58 standard deviations of the unknown population mean. And it would follow that the probability is about 99/100 that our obtained sample mean of 118 is within 5.16 points of the unknown population mean. Finally we would have said that the probability is about 99/100 that the unknown mean IQ of medical students is between 112.8 and 123.2. This is called a 99 *per cent confidence interval* for the mean.

To sum up, these are the steps we followed in this process: We built our interval estimate around our obtained sample mean. We computed the standard deviation of the mean (which is what the standard deviation of the distribution of a myriad of such means would be). Where our probability fraction was 95/100, we arrived at the z score value of 1.96. The lower end of our interval turned out to be 1.96 standard deviations of the mean below our obtained sample mean, and the upper end was 1.96 standard deviations above our obtained sample mean. In terms of a formula, the probability is about 95/100 that the following statement is true:

$$(13.3) \qquad \left[\bar{X} - (1.96) \frac{\sigma}{\sqrt{N}} \right] < \mu < \left[\bar{X} + (1.96) \frac{\sigma}{\sqrt{N}} \right]$$

Formula (13.3) gives the 95 per cent confidence interval for the mean. Similarly, the probability is about 99/100 that the following statement is true:

(13.4) $$\left[\bar{X}-(2.58)\frac{\sigma}{\sqrt{N}}\right] < \mu < \left[\bar{X}+(2.58)\frac{\sigma}{\sqrt{N}}\right]$$

Formula (13.4) gives the 99 per cent confidence interval for the mean.

In Formulas (13.3) and (13.4) μ stands for the unknown population mean. $\bar{X}$ stands for the obtained sample mean, σ stands for the standard deviation of the original terms, N stands for the sample size, and $<$ stands for "less than." Incidentally, $\frac{\sigma}{\sqrt{N}}$ stands for the standard deviation of the mean.

In the problem that we just discussed, $\bar{X} = 118$, $\sigma = 16$, and $N = 64$. Therefore, using Formula (13.3), our interval turned out to be

$$\left[118-(1.96)\frac{16}{\sqrt{64}}\right] < \mu < \left[118+(1.96)\frac{16}{\sqrt{64}}\right]$$

$$114.1 < \mu < 121.9$$

Using Formula (13.4), our interval was

$$\left[118-(2.58)\left(\frac{16}{\sqrt{64}}\right)\right] < \mu < \left[118+(2.58)\left(\frac{16}{\sqrt{64}}\right)\right]$$

$$112.8 < \mu < 123.2$$

Note that the 95 per cent confidence interval in Formula (13.3) will always be smaller than the 99 per cent confidence interval in Formula (13.4), as a result of the fact that the value 1.96 is less than 2.58. This makes sense because, where less probability of being correct is demanded, one can issue a statement that is relatively precise. As one demands a higher probability that the interval constructed will actually contain the unknown parameter value, it becomes necessary to construct an accordingly wider interval.

PROBLEM SET A

13.1 In a competitive event, a sample of 25 college students receives scores of 102, 85, 109, 83, 112, 105, 98, 115, 91, 117, 88, 95, 116, 105, 130, 80, 115, 84, 95, 121, 85, 89, 94, 111, and 106. Determine (a) the unbiased estimate of the mean, (b) the degrees of freedom for estimating population variance, (c) the unbiased estimate of the population variance, and (d) the estimate, s, of the population standard deviation.

13.2 The weights (in pounds) of a sample of ten people are 156, 162, 170, 177, 180, 181, 183, 196, 205, and 209. Determine (a) the unbiased estimate of the population mean, (b) the degrees of freedom for estimating population variance, (c) the unbiased estimate of the population variance, and (d) the estimate, s, of the population standard deviation.

13.3 The numbers of dental cavities in a sample of eighth-grade students are 3, 11, 5, 12, 9, 8, 16, 13, 12, 11, 6, 19, 16, 11, 15, 12, and 0. Determine (a) the unbiased estimate of the population mean, (b) the degrees of freedom for estimating population variance, (c) the unbiased estimate of the population variance, and (d) the estimate, s, of the population standard deviation.

13.4 The table below shows the grouped frequency distribution of spelling quiz scores for a sample of third-grade students. Determine (a) the unbiased estimate of the population mean, (b) the degrees of freedom for estimating population variance, (c) the unbiased estimate of the population variance, and (d) the estimate, s, of the population standard deviation.

Interval	Frequency
10	7
9	8
8	7
7	5
6	3
5	1

13.5 The distribution of the ages of a sample of children in a school district is given in the table below. Determine (a) the unbiased estimate of the population mean, (b) the degrees of freedom for estimating population variance, (c) the unbiased estimate of the population variance, and (d) the estimate, s, of the population standard deviation.

Interval	Frequency
12	10
11	23
10	44
9	46
8	42
7	48
6	37

13.6 The mean of a sample of 100 scores on a mechanical aptitude test is 81. Assume the population standard deviation to be 8. Find the 95 per cent confidence interval for the mean of the population.

13.7 The mean height of a sample of 144 American soldiers is 5 feet 7 inches. Assume the population standard deviation is $3\frac{1}{2}$ inches. Find the 99 per cent confidence interval for the mean of the population.

13.8 The mean weight of a sample of 100 U.S. adult males is 165 pounds. Assume the population standard deviation to be 15 pounds. Find the 95 per cent confidence interval for the mean of the population.

13.9 The mean of a sample of 400 dental students on a dental aptitude test is 79. Assume the population standard deviation to be 9. Find the 99 per cent confidence interval for the mean of the population.

13.10 A sample of 36 hybrid corn plants has a mean height of 47 inches, and the population standard deviation is 2.1 inches. Find the 95 per cent and 99 per cent confidence intervals for the mean of the population.

13.11 The mean monthly income of students at a certain college is to be estimated from a random sample. If the mean income is to be estimated within $20.00 with a probability of 95 per cent, find the smallest size sample needed for this estimate. Assume that the standard deviation is $100.

13.12 A drink-vending machine dispenses 64 cups with a mean filled value of 7.6 ounces. The population standard deviation is .5 ounces. Find the 95 per cent and 99 per cent confidence intervals for the dispensed amount of drink from this machine.

PROBLEM SET B

13.13 The heights of a sample of 11 players on a college basketball team are 5 feet 9 inches, 5 feet 10 inches, 5 feet 11 inches, 6 feet, 6 feet, 6 feet, 6 feet 1 inch, 6 feet 2 inches, 6 feet 3 inches, 6 feet 2 inches, and 6 feet 6 inches. Determine (a) the unbiased estimate of the mean, (b) the degrees of freedom for estimating population variance, (c) the unbiased estimate of the population variance, and (d) the estimate, s, of the population standard deviation.

13.14 On a social studies test, a sample of 25 high school juniors receives scores of 107, 90, 114, 88, 117, 110, 103, 120, 96, 122, 93, 100, 121, 110, 135, 85, 120, 89, 100, 126, 90, 94, 99, 116, and 111. Determine (a) the unbiased estimate of the population mean, (b) the degrees of freedom for estimating population variance, (c) the unbiased estimate of the population variance, and (d) the estimate, s, of the population standard deviation.

13.15 Below is a grouped frequency distribution of arithmetic achievement scores for a sample of fourth-grade students. Determine (a) the unbiased estimate of the population mean, (b) the degrees of freedom for estimating population variance, (c) the unbiased estimate of the population variance, and (d) the estimate, s, of the population standard deviation.

Interval	Frequency
96–100	3
91–95	6
86–90	9
81–85	14
76–80	8
71–75	5
66–70	4
61–65	1

13.16 Below is the grouped frequency distribution of chemistry midterm scores for a sample of eleventh-grade students. Determine (a) the unbiased estimate of the population mean, (b) the degree of freedom for estimating population variance, (c) the unbiased estimate of the population variance and (d) the estimate, s, of the population standard deviation.

Interval	Frequency
70–75	2
64–69	5
58–63	8
52–57	7
46–51	7
40–45	4
34–39	2

13.17 Below is the grouped frequency distribution of aptitude scores for a sample of high school seniors. Determine (a) the unbiased estimate of the population mean, (b) the degrees of freedom for estimating population variance, (c) the unbiased estimate of the population variance, and (d) the estimate, s, of the population standard deviation.

Interval	Frequency
70–73	2
67–69	8
64–66	32
61–63	26
58–60	10
55–57	2

13.18 The mean of a sample of 49 high school graduating seniors on an aptitude test is 85. Assume the population standard deviation to be 7. Find the 95 per cent confidence interval for the mean of the population.

13.19 The mean length of a sample of 64 films is 90 minutes. Assume the population standard deviation to be 23 minutes. Find the 99 per cent confidence interval for the mean of the population.

13.20 The mean of a sample of 36 pupils on a musical aptitude test is 77. Assume the population standard deviation to be 15. Find the 95 per cent confidence interval for the mean of the population.

13.21 The mean of a sample of 250 individuals on a language aptitude test is 75. Assume the population standard deviation to be 10. Find the 99 per cent confidence interval for the mean of the population.

13.22 A sample of 60 pills is tested and found to contain 35 milligrams of active ingredient, and the population standard deviation is 4.2 milligrams. Find intervals within which the population mean is expected to lie with a probability of (a) 95 per cent, (b) 98 per cent, (c) 99 per cent.

13.23 The mean aptitude score for 36 job applicants is 7.8, and the population standard deviation is .9. What are the 95 per cent and 99 per cent confidence intervals for the mean of the population of all job applicants?

13.24 For a certain brand of cigarette, a random sample of 49 cigarettes is tested and found to contain an average of 12.0 milligrams nicotine. The population standard deviation is 1.5 milligrams. Find a 98 per cent confidence interval for the mean nicotine content of this brand of cigarettes.

the *t* distribution
and *t* tests

14.1 Introduction. In this chapter we are going to discuss procedures that embody much of what has already been said.   However, up to now we have presupposed that the experimenter testing his hypothesis about a population mean had direct knowledge of the standard deviation of the population from which his sample was drawn.   This is seldom the case. Nearly always the situation is such that a researcher has as his only information the data comprising the sample he has drawn.   On the basis of the data comprising his sample and virtually no other information, he must contrive to test whatever hypothesis he states concerning a population mean or some other parameter.

We shall now consider a method of testing a hypothesis concerning an unknown population mean when the experimenter has no information other than that contained in his sample.   *If N is small, the one new assumption that must be made is that the terms in the population from which the sample is drawn are normally distributed, or at least there must not be a drastic departure from normality in their distribution.   If N is moderately large, this assumption is virtually always unnecessary.*   Thus, the hypothesis tests to be discussed have very wide applicability.   Given the assumption, it is possible for an experimenter to use his sample mean to test a hypothesis about a population mean, even when he does not have actual knowledge of the standard deviation of the population from which he drew his sample.

The best way to make the logic and procedure clear is to think in terms of a specific experiment, and we shall begin with one that is arbitrarily simple. Suppose an anthropologist wants to demonstrate that the mean height of natives on a particular island is smaller than 5 feet 2 inches. He decides to test the hypothesis that the mean height of these natives is exactly 5 feet 2 inches. He might have set up a confidence interval, as described in the last chapter. He might then have asked whether the value 5 feet 2 inches was inside that interval. Instead, let us say, he chooses to test the hypothesis that $\mu = 5$ feet 2 inches. He chooses the .01 significance level and hopes to be able to reject the hypothesis by getting a sample mean that is too small.

We shall assume that the distribution of heights of natives on the island is normal or approximately so. The anthropologist randomly selects five natives and determines their heights (though of course in practice he would not be content with so small a sample). The heights of these five natives turn out to be 4 feet 6 inches; 4 feet 8 inches; 4 feet 9 inches; 5 feet; and 5 feet 3 inches. The anthropologist now wishes to base upon his sample a statistical test that will tell him whether to accept or reject his hypothesis.

To set the stage, let us go back for a moment and suppose he knew the exact value of the population standard deviation. For instance, suppose he knew that the standard deviation of natives' heights was 5 inches. He could then test the hypothesis in the manner described in Chapter 12. He would conceive of his random sample as one of a vast number gathered by similar anthropologists. Under the hypothesis, the distribution of the means of such samples would have a mean of 5 feet 2 inches. Knowing the value of the standard deviation of the natives' heights to be 5 inches, he could compute the standard deviation of the theoretical distribution of sample means. That is, given that $\sigma = 5$, he would compute

$$\sigma_M = \frac{\sigma}{\sqrt{N}} = \frac{5}{\sqrt{5}} = 2.24 \text{ inches}$$

In other words, assuming the hypothesis was true, he would ascertain that his obtained sample mean would have been 4 inches below the expectancy in a distribution of sample means with a standard deviation of 2.24 inches. Assuming that the hypothesis was true, by dividing he would see that the sample mean he had obtained would have had a z value of -1.79. We may sum up what his calculations would have been:

$$\text{(14.1)} \qquad z = \frac{\overline{X} - \mu}{\sigma_M} = \frac{4'10'' - 5'2''}{2.24''} = \frac{-4''}{2.24''} = -1.79$$

On the basis of the obtained z score, he would have decided whether to accept or reject the hypothesis.

But we shall not pursue his calculations further. The point is that now our anthropologist does *not* know the actual value of the standard deviation of the natives' heights. He cannot proceed in the same way. This time he is going to test his hypothesis in ignorance of σ. What he must do first of all is to make an enlightened guess of the population standard deviation using the data of his sample. He must compute the value of s, which he knows will be an estimate of σ though not identical with it.   The value of s that he would compute, using Formula (13.1), is 3.54.   Note that this value of s is an estimate of the actual value of σ.   Next our anthropologist must make an estimate of the value of the standard error of the mean.   We shall call this estimate s_M and remember that it is an estimate of σ_M.   As one might guess, our anthropologist makes use of Formula (14.2), which is analogous to our Formula (9.1) for σ_M.   The essential distinction is that here we are using s, an estimate of σ, whereas in (9.1) it was assumed the experimenter knew σ. Here we write:

$$s_M = \frac{s}{\sqrt{N}}$$

In the case described, $s=3.54$ and $N=5$.   Therefore

$$s_M = \frac{s}{\sqrt{N}} = \frac{3.54}{\sqrt{5}} = 1.58$$

If our anthropologist could be certain that his estimate, s_M, was perfectly accurate and was the actual value of σ_M, he could proceed smoothly.   He could compute the z value of his obtained sample mean using Formula (14.1) with s_M in place of σ_M.   That is, he could determine this value by solving for the value of $\frac{\bar{X}-\mu}{s_M}$ instead of having to solve for the value of $\frac{\bar{X}-\mu}{\sigma_M}$.

The trouble is that the value s_M is only an enlightened estimate of σ_M and is not identical with it.   Therefore, the best our anthropologist could do would be to compute what he might view as an estimate of the z score value of his sample mean.   We shall call the value he computes a t value and write

(14.3) $$t = \frac{\bar{X}-\mu}{s_M}$$

In particular, our anthropologist would compute the t value of -2.53 for his sample mean.

$$t = \frac{\bar{X}-\mu}{s_M} = \frac{4'\ 10''-5'\ 2''}{1.58''} = -2.53$$

Note that in the case described, the t value obtained was numerically greater than the z value would have been.   The reason is that the estimate, s, was smaller than the hypothetical value we used for σ.   Had the value of s, computed from the sample, been larger than the value of σ, then the t value computed would have been smaller than the actual z value.   The point is that the value of t in any particular case depends upon not only the obtained sample mean but also the value of s that happens to be computed from the particular sample.

14.2   The *t* Distribution.   Let us assume that the mean height of natives on the island is actually 5 feet 2 inches, so that the null hypothesis is actually true.   Under this assumption, let us suppose further that our anthropologist is one of a vast number of anthropologists, each of whom gathers a random sample of the heights of five natives. We shall assume that each anthropologist now computes the mean and s value of his sample, and then the t value of his particular sample mean. In other words, assuming that the mean height on the island is actually 5 feet 2 inches, there would come into existence a distribution of t values obtained by the different anthropologists. In each case, the particular anthropologist would be obtaining his t value as a result of using Formula (14.3). This distribution of t values, which would turn up in the long run, is crucial to know. With reference to it, a particular anthropologist like the one described can determine whether his particular t value would be commonplace or rare.

The mathematician has addressed himself to the problem and has calculated what is called the t distribution. This is the distribution of t values that would be obtained from independent random samples of equal size drawn from a normal population. For now we shall stick to the case where each sample consists of five cases. The graph of the distribution of t values appears in Fig. 14.1. Note that the distribution of t values, based on samples of five cases each, is similar in appearance though not identical to the normal distribution. The proportion of cases in various sectors are indicated. Observe that the t distribution depicted here resembles the normal distribution. For different degrees of freedom, the t distribution looks different, and thus the t distribution is a whole family of curves. The more the degrees of freedom determining the particular shape, the more closely the t distribution resembles the normal distribution. Fig. 14.2 shows the t distributions for 5, 10, and 20 degrees of freedom.

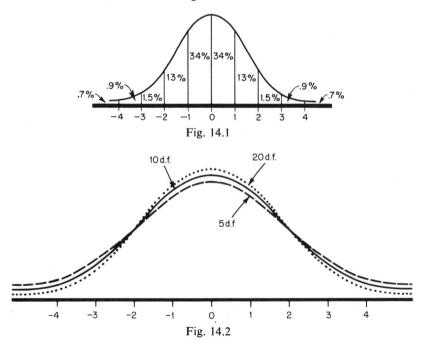

Fig. 14.1

Fig. 14.2

The researcher who has obtained a *t* value for the mean of a sample of five cases may make reference to the graph in Fig. 14.1. Remember that one assumes that a hypothesis is true before computing a *t* value. Our anthropologist was prepared to reject his hypothesis only if he obtained a *t* value so small as to have a probability of less than 1/100 of appearing by chance. According to Fig. 14.1, this means that only if his obtained *t* value was less than -3.75 would he reject his hypothesis. Actually his obtained *t* value was -2.53 which means that it was not sufficiently small to indicate rejecting his hypothesis. Accordingly, his decision must be not to reject his null hypothesis that the mean height of natives on the island is 5 feet 2 inches.

Remember that the number of degrees of freedom used to compute *s*, which was the estimate of σ, is one less than the sample size. In the case described, the anthropologist based his estimate on four degrees of freedom because his sample consisted of five cases. This number of degrees of freedom used to compute *s* also becomes described as the number of degrees of freedom used to compute *t*. In other words, we say that the anthropologist based his *t* value on four degrees of freedom. Were it the case that his sample consisted of 50 terms, then it would be the case that 49 degrees of freedom went into his estimate of *t*.

The number of degrees of freedom determines the shape of the *t* distribution. Or to put it another way, for any given number of degrees of freedom there is a particular *t* distribution with its unique set of critical values for significance. Therefore the worker whose sample consists of five cases must think of his obtained *t* value as a member of the *t* distribution based on four degrees of freedom and not any other. This fact becomes crucial when the time comes for him to determine whether his obtained *t* value is significant (that is, whether it indicates rejecting the hypothesis).

Appendix IV makes it possible for the experimenter to determine what proportion of *t* values would be smaller than the one he obtained, if in fact his null hypothesis is true. The entries at the top are proportions of cases with *t* values less than those indicated. Each row relates to a *t* distribution with a unique number of degrees of freedom. For instance, our anthropologist would have to go to the fourth row because his *t* value must be considered a member of a *t* distribution based on four degrees of freedom. The first entry in this fourth row, which is in the column headed .005, happens to be -4.60. This means that under the hypothesis only 5/1,000 of the *t* values from random samples would be less than -4.60. Similarly, the value in the second column is -3.75, meaning that if the hypothesis is true 1/100 of the obtained *t* values over the long run would be less than -3.75. Note that it was this fourth row in Appendix IV that was used in plotting Fig. 14.1.

Note also that the fifth column, headed .500, has the value zero in every row. This means that regardless of the number of degrees of freedom in the particular *t* distribution, exactly half of the terms (or sample means in our case) have *t* values less than zero. As one might guess, there is perfect symmetry to the *t* distribution.

201

One more fact of interest is that, as the samples yielding their means become larger, the t distribution approaches the normal distribution. When the size of the samples is large, the values of s computed from the different samples become relatively close to σ as a group. As a result, the t values computed from the different samples become virtually identical to what the z values would have been. Thus in the bottom row of Appendix IV, where the number of degrees of freedom for the distribution is 33, the cut-off points for the t distribution are virtually the same as those for the normal distribution. As a matter of fact, it is generally thought adequate to use the normal distribution table so long as one's sample is made up of at least 30 cases.

To return finally to our anthropologist, his sample mean was computed to have a t value of -2.53. The critical question for him to ask is whether the probability of having gotten a sample mean as small as his was less than $1/100$. According to Appendix IV, the t value of -3.75 is larger than exactly $1/100$ of the t values in the distribution. Thus our anthropologist's obtained t value of -2.53 must have been higher than more than $1/100$ of the terms. (To put it another way, Fig. 14.1 shows that the sample mean with a t value of -2.53 is too large to get into the .01 rejection tail.) The anthropologist has not found a sample mean small enough and must accept his original hypothesis that the mean height of natives on the island is exactly 5 feet 2 inches.

This time, suppose a researcher knows that the average score obtained by twelve-year-old boys taking a mechanical aptitude test is 72. He is willing to assume that scores on this test are roughly normally distributed. He is interested in showing that boys who apply for the boy scouts at age twelve are higher as a group than average in mechanical ability. Our experimenter plans to obtain the mechanical aptitude scores of 20 would-be boy scouts. He wishes to test the hypothesis that their scores comprise a random sample taken from the larger population. He plans to test this hypothesis at the .05 significance level with the hope of rejecting it by obtaining a sample mean that is significantly high.

The mean of his sample of scores, $\overline{X}$, turns out to be 86. The value of s, which he computes from his sample, is 8. Now, still supposing that the hypothesis is true, he computes the t value of his obtained sample mean. Remember that μ, which is the hypothetical population mean, is 72. $N=20$. Our experimenter uses Formula (14.3) and computes the t value of his obtained sample mean:

$$ t = \frac{\overline{X} - \mu}{\dfrac{s}{\sqrt{N}}} = \frac{86 - 72}{\dfrac{8}{\sqrt{20}}} = 7.83 $$

This time our experimenter based his s value on 19 degrees of freedom since there were 20 scores in his sample. He must make reference to the nineteenth row in Appendix IV. In this row he must look up the t value under the column headed .95. The value he finds is 1.73. The interpretation is that under the hypothesis the probability of obtaining a sample mean with a *t* value less than 1.73 is .95, and the probability of obtaining one with a *t* value greater than this number is .05. Since our experimenter has found a sample mean with a *t* value of 7.83, he may conclude that under the hypothesis it is high enough to have had less than 5 chances in a hundred of appearing. Thus it constitutes a rarity sufficient to allow him to reject the hypothesis. In fact, the value in the nineteenth row and under the column headed .995 is only 2.86. This implies that the obtained *t* value was so high that under the hypothesis it would have turned up by chance less than five in a thousand times in the long run.

Our experimenter may reject the hypothesis as a consequence of his finding. He may conclude that his sample of boy scouts' scores is not merely a random sample taken from the population of mechanical aptitude tests. Having rejected the hypothesis, he may conclude that there is some relation between having mechanical aptitude and choosing to be a boy scout (at least for the particular age group studied). Note that we used a one-tail test in each of the foregoing examples, since in each case we were concerned with a theory that involved difference in a specific direction from the hypothetical value. Otherwise we could have used a two-tail test, proceeding in the same way as when we used such a test with *z* values.

The test of significance, which has been described in this section, is often called a *t* test. The critical ratio given in Formula (14.3) is often called a *t* ratio. As one may imagine, there is much more practical use made of the *t* distribution than even the normal distribution when one tests a hypothesis. Even though one must conceive of his sample mean as a member of a distribution that is normal, the fact that one can usually only estimate the standard deviation of this distribution causes the crucial statistic computed to be a *t* value and not a *z* value.

We have discussed the *one-sample* case. A single sample mean is obtained to test a hypothesis about a population mean. We shall now consider one more hypothesis test for which we make reference to the *t* distribution. This test is also known as a *t* test and it is more complex than the one we have considered. We shall use virtually all of our previous logic in considering this test, which will be the last one of its kind to be presented in this chapter.

14.3 The *t* Test for the Difference between Means. The context that we shall now consider is the one that arises most often for the researcher. He has no information concerning the mean or standard deviation of a population. He comes to a field with no validated facts, but only with the desire to demonstrate some phenomenon that he thinks he has either observed or inferred. For instance, a teacher may have designed a method that he believes is more efficient than a usual one. He has no validated facts concerning the old method, much less the new one, but he would like to demonstrate that his new method is superior. Or a drug company may wish to compare two drugs for efficiency, and because both drugs are new there may be no information concerning either.

As the reader may have noticed, several chapters back we started with the simplest kind of hypothesis test, for which we assumed considerable previous knowledge on the part of the researcher. We have subsequently considered tests where he has had successively less information, and these tests have been successively more subtle to grasp. At the same time we have been moving toward considering more usual contexts. Now we are considering the one in which the researcher has no prior knowledge about any parameter and yet wishes to prove a theory. The test that we are about to consider requires that the experimenter deal with two sample means and compare them.

This time it will be simplest to begin with a theoretical discussion. Suppose that there is a normally distributed population of terms to which we have access. We gather a random sample of, let us say, 15 terms and we compute its mean and variance. We shall call this our first sample, since we are going to gather another independent one. Accordingly we shall use the symbol N_1 to stand for the number of cases in this sample, so that here $N_1 = 15$. We shall use the symbol $\bar{X}_1$ to stand for the mean of the first sample and the symbol $s_1{}^2$, to stand for its variance.

We do exactly the same for a second sample which, let us say, consists of 20 cases. We designate the number of cases in this sample by N_2, so that here $N_2 = 20$. We designate its mean by $\bar{X}_2$ and its variance by $s_2{}^2$.

We now compute the difference between the means of the two independent samples. Specifically we shall compute $(\bar{X}_1 - \bar{X}_2)$. Since both samples came from the same population, the expectancy is that the value of $(\bar{X}_1 - \bar{X}_2)$ will be zero. (This is true because the mean of each sample has an expectancy of being the same as the population mean.) However, in practice where the samples are random we can obviously anticipate that the mean of one sample will be higher than that of another, though of course we cannot say which will be higher. Thus in practice the value of $(\bar{X}_1 - \bar{X}_2)$ will almost certainly be some value either less than zero or more than zero.

If it turns out that $\bar{X}_1$ is smaller than $\bar{X}_2$, then $(\bar{X}_1 - \bar{X}_2)$ will be a negative number. If it turns out that $\bar{X}_1$ is larger than $\bar{X}_2$, then $(\bar{X}_1 - \bar{X}_2)$ will be a positive number. Now suppose that each one of a vast number of researchers was to do as we have done. That is, each researcher goes to our same normally distributed population and draws two random samples, the first comprising 15 terms and the second comprising 20 terms. Each individual researcher computes the value of $(\bar{X}_1 - \bar{X}_2)$ from his particular pair of samples. For instance, suppose for one pair the values $\bar{X}_1 = 50$ and $\bar{X}_2 = 60$; then $(\bar{X}_1 - \bar{X}_2) = -10$. For the next researcher the value of $(\bar{X}_1 - \bar{X}_2)$ might be 2. For the next one it might be $-.5$, and so on.

Table 14.1 shows what might happen for the first few researchers from among the vast multitude. For each researcher there is an obtained value of $\bar{X}_1$ and one of $\bar{X}_2$, and also a difference score, $(\bar{X}_1 - \bar{X}_2)$. To simplify we are going to use the letter d in place of $(\bar{X}_1 - \bar{X}_2)$ from now on.

It is the column of d values, each of which was obtained by a different researcher, that shall be of concern to us from now on. When we conceive of these differences as terms in a distribution, we find that in the long run the mean of these differences is zero and that they are normally distributed.

TABLE 14.1   SAMPLE DIFFERENCES OF MEANS

	$\bar{X}_1$	$\bar{X}_2$	$d=(\bar{X}_1 - \bar{X}_2)$
First researcher	50	60	-10
Second researcher	48	46	2
Third researcher	41	41.5	-0.5

The standard deviation of these d values may be designated by the symbol σ_d. σ_d is sometimes called the "standard error of the difference."

(14.4)
$$\sigma_d = \sqrt{\frac{\sigma^2}{N_1}+\frac{\sigma^2}{N_2}} = \sigma\sqrt{\frac{1}{N_1}+\frac{1}{N_2}}$$

where σ is the standard deviation of the population from which each pair of samples was drawn.

But remember that we are talking from the position of a single researcher, in particular the first one, who has no knowledge of the value of σ.   The sum total of his information is contained in the two samples he has drawn.   Suppose this researcher starts to think of his obtained d value (of -10) as a member of the distribution of d values similarly obtained.   He knows that the mean of this distribution is zero and that it is normal, but he can only estimate the standard deviation of this distribution.   That is, he must use the information he has gathered to estimate σ_d.   He computes the value of s_d, which is the symbol we shall use for his estimate of σ_d.   As might be suspected (from our previous use of s in place of σ), it can be shown that

$$s_d = s\sqrt{\frac{1}{N_1}+\frac{1}{N_2}}$$

Now, however, s^2 is a variance based on *both* of the samples the researcher has drawn rather than on just one sample as in Section 14.1.   It can be shown mathematically that

$$s^2 = \frac{(N_1-1)s_1^2+(N_2-1)s_2^2}{N_1+N_2-2}$$

(This may be described as a "weighted mean" of s_1^2 and s_2^2, each weighted by the number of degrees of freedom on which it is based.)   When the square root of this is taken to obtain s and it is then substituted in the expression for s_d, the result is the rather formidable-looking Formula 14.5.

(14.5)
$$s_d = \sqrt{\frac{(N_1-1)s_1^2+(N_2-1)s_2^2}{N_1+N_2-2}} \cdot \sqrt{\frac{1}{N_1}+\frac{1}{N_2}}$$

In other words, the first researcher may think of his obtained d value as a member of a normal distribution whose mean is zero and whose standard deviation he estimates by Formula (14.5). He cannot obtain the z score of his obtained difference but he can obtain a t value for this difference. To get this t value he must compute how far his obtained difference is from the balance point of this theoretical distribution of differences. But we have already seen that the balance point of this distribution of differences is zero. Therefore our researcher whose obtained d value was -10 has found a d value that is ten units to the left of the balance point in the theoretical distribution of differences.

To get the t value of his obtained difference, our researcher would first have to compute the value of s_d. That is, he would have to make his estimate of the standard deviation of the theoretical distribution of differences depicted in Fig. 14.3. Suppose he computed that $s_1{}^2 = 210$ and $s_2{}^2 = 220$.

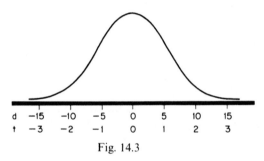

Fig. 14.3

He would then use Formula (14.5) and get

$$s_d = \sqrt{\frac{(N_1-1)s_1{}^2+(N_2-1)s_2{}^2}{N_1+N_2-2}} \cdot \sqrt{\frac{1}{N_1}+\frac{1}{N_2}} =$$

$$\sqrt{\frac{14(210)+(19)(220)}{33}} \cdot \sqrt{\frac{1}{15}+\frac{1}{20}} = \sqrt{\frac{2940+4180}{33}} \cdot \sqrt{\frac{7}{60}} =$$

$$(\sqrt{215.7})(\sqrt{.1167}) = 4.99 \text{ or } 5$$

Now to find the t value of his obtained difference of -10, he would have to divide the distance score by his estimate, s_d.

$$t = \frac{d-0}{s_d} = \frac{-10}{5} = -2$$

In the theoretical distribution of differences, our researcher has found a difference with a t value of -2. What he has done in effect is to construe his obtained difference, d, as the finding of interest. He has viewed this finding, d, as a member of a normal distribution with a mean of zero. The members of this distribution are other d values that in theory would be found by other researchers doing exactly the same thing. Our researcher has made an estimate of the standard deviation of the distribution of d values—that is, he has computed s_d. Finally, since he knows the distance score of his particular finding, he divides this distance score by the value s_d to get a t value for his obtained difference.

We can imagine each of our multitude of researchers doing the same thing. That is, we can imagine each researcher computing the value of d and also computing s_d. Thus each researcher divides d by s_d to get the t value for the obtained difference. Table 14.2 shows what the computations of a few such researchers might look like. The table has been left for the reader to complete.

It should be clear that each researcher has used the assumption that his particular pair of independent samples came from the same normally distributed population. Having had no further knowledge, he has computed a t value for the difference, d. That is, each researcher has obtained a t value for the difference $(\bar{X}_1 - \bar{X}_2)$ computed from his pair of samples.

TABLE 14.2 COMPUTATION OF t VALUES

	$\bar{X}_1$	$\bar{X}_2$	d	$s_1{}^2$	$s_2{}^2$	s_d	$t=\dfrac{d-0}{s_d}$
First researcher	50	60	-10	210	220	5	-2
Second researcher	48	46	2	200	230		
Third researcher	41	41.5	$-$ .5	170	190		

We must now specify the number of degrees of freedom that went into the t distribution just described. Note that each researcher computed a value of $s_1{}^2$ based on N_1-1 degrees of freedom and a value of $s_2{}^2$ based on N_2-1 degrees of freedom. Both estimates went into the estimate of s_d so that this latter statistic was based on $(N_1-1)+(N_2-1)$ degrees of freedom inasmuch as the two samples are independent. In other words, the t distribution described is based on N_1-1+N_2-1 degrees of freedom. To simplify, it is based on N_1+N_2-2 degrees of freedom. Specifically for the instance described, since $N_1=15$ and $N_2=20$, the distribution of differences—some of which were listed in column 7 of Table 14.2—is a t distribution based on 33 degrees of freedom. The reader may have noticed that the expression N_1+N_2-2 appears in Formula (14.5).

What has been said may be summarized in the form of a theorem, which shall soon be of special use to us.

Theorem 14.1 *Suppose a vast number of pairs of independent random samples of size N_1 and N_2 are drawn from the same normally distributed population. For each pair the value $\dfrac{\bar{X}_1 - \bar{X}_2}{s_d}$ is computed, where $\bar{X}_1$ is the mean of the first sample, $\bar{X}_2$ is the mean of the second, and s_d is as defined in Formula (14.5). Then the distribution of values thus obtained is the t distribution with N_1+N_2-2 degrees of freedom.*

Even if the original population is not normal, Theorem 14.1 holds provided that the samples are large enough.

14.4 Application of the t Test for the Difference between Means. We are now going to make use of Theorem 14.1 to describe the hypothesis-testing procedure used most often in the fields of education, psychology, and the social sciences. An illustration will be useful to present the technique. Suppose a French teacher has taught vocabulary by two methods, one of which involves using visual aids whereas the other does not. He decides to do an experiment to determine which method is more effective for teaching French vocabulary to his classes.

We shall say that this French teacher has two comparable classes, one of 15 students and the other of 20, and that he decides to use the visual aid method with one class and the other method with the second class. At the end of the semester, he tests both classes on vocabulary words learned during the year. Actually he has taught 300 new words in all, and he obtains for each student in each class a score indicating the total number of words the student has learned.

As a researcher, the instructor is essentially interested in learning which of the methods was more effective, and he assumes that he is typical of an instructor using either. We shall say that he used the visual aid method on Class A, which consisted of 15 children, and that he used the other method on Class B, composed of 20 children. It turns out that the mean number of words learned by Class A was 220, whereas the mean number learned by members of Class B was only 200. The question our researcher asks is whether the difference is meaningful. In other words, may he infer from the finding that the advantage shown by Class A means that the visual aids method was really better? Or was the difference merely one that might have been expected to occur by chance so often that he can draw no conclusion from it whatsoever?

The first crucial step for our researcher is to set up the *null* hypothesis that the two methods are identically effective. Now he may conceive of a vast population of children with whom he has used a uniform method of teaching French vocabulary. Under this hypothesis, Class A would be merely a random sample of 15 children whom he has taught and Class B would be an independent random sample consisting of 20 children. The obtained difference between the mean of Class A, which we shall call $\bar{X}_1$, and that of Class B, which we shall call $\bar{X}_2$, is considered by him merely a difference between the means of two random samples from the same population. According to the hypothesis, this difference, $(\bar{X}_1 - \bar{X}_2) = 220 - 200$, might as easily have been zero or have gone the other way.

The next step is for our researcher to conceive of himself as only one of a vast number of similar researchers who have performed the identical experiment. Each of these other researchers would also obtain a difference and a t value corresponding to it. That is, each researcher would, from his pair of samples, compute $(\bar{X}_1 - \bar{X}_2)$ and also s_d and then divide the former by the latter to get a t value. Our researcher must construe the particular t value that he obtains to be a member of this t distribution based on 33 degrees of freedom.

We may suppose for instance that he is testing his hypothesis at the .05 level of significance. The difference he has obtained between the mean vocabulary scores in the classes is 20, and we shall say that he computes the value of s_d to be 15. Therefore

$$t = \frac{\bar{X}_1 - \bar{X}_2}{s_d} = \frac{20}{15} = 1.33$$

The question he now asks is whether his obtained *t* value is extreme enough to have had a probability of less than 5/100 of occurring. The assumption that the hypothesis is true led him to determine his *t* value. Now he is to accept the hypothesis unless it turns out that his obtained *t* value happens to be too extreme. According to Appendix IV, where *t* is based on 33 degrees of freedom, the smallest $2\frac{1}{2}$ per cent of *t* values are those less than -2.03, and the largest $2\frac{1}{2}$ per cent are those greater than 2.03. Therefore the most extreme 5/100 of *t* values are those less than -2.03 and those greater than 2.03, taken together. The *t* distribution based on 33 degrees of freedom is shown in Fig. 14.4.

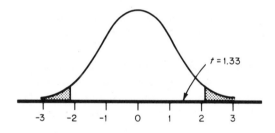

Fig. 14.4

The obtained *t* value of 1.33 is not in the rejection region of Fig. 14.4. Therefore the null hypothesis that the two classes represent random samples from the same population cannot be rejected. This is the hypothesis that there is a uniform effectiveness to the two teaching methods, so that the difference between average achievement in the two classes cannot be considered meaningful.

Suppose instead that the obtained *t* value had been 4.00. This obtained *t* value would be in the rejection region of Fig. 14.4. The hypothesis that Class A and Class B represent random samples from the same population of French vocabulary achievement scores would have to be rejected. Since the two classes may not be considered as samples from the same population, one cannot reject the alternative explanation that the actual teaching methods used in the classes must have made them different. The conclusion must be that the method used with Class A, the visual aid method, is different in effectiveness from the method used with Class B. Note that reference to the appropriate row in Appendix IV would have been sufficient to make the decision. It was not necessary to have drawn Fig. 14.4.

The procedure for doing a t test like that described is quite straight-forward in practice. The experimenter picks his significance level. Then he determines the values of N_1, N_2, $\bar{X}_1$, $\bar{X}_2$, $s_1{}^2$, and $s_2{}^2$. He then solves for s_d using Formula 14.5. Then he looks up his obtained t value using the degrees of freedom given by $N_1 + N_2 - 2$ in Appendix IV. If the obtained t value indicates that he cannot reject the null hypothesis, he concludes that whatever difference was found between the two group means was apt to have been the result of chance. If he is able to reject the hypothesis, he concludes that the two groups did not come from the same population in the first place. The conclusion then is that the difference between the two groups was the result of the different procedures used with them or to some other factor that distinguished them in the first place.

The point is that the experimenter knows which range of t values to expect when comparing means of two samples from the same population. He compares the means of the two groups treated differently by assuming that they were treated similarly and so are from the same population. If, using Appendix IV, he gets a t value not in the rejection region, he holds on to his assumption that the differential treatment did not have a differential effect. Otherwise he concludes that it did.

Perhaps the reader can already see the wide applicability of the t test for the difference between means. Whenever the effect of a method or treatment is to be studied, the t test for the comparison of group means is likely to be appropriate. Often the effect of a single process applied to human beings is evaluated by the researcher's comparing those on whom it has been applied with an equivalent group of subjects on whom it has not been applied. For instance, when the effect of a drug is studied, a frequent procedure is to administer it to one group and to administer a placebo to another group. Then a t test of the kind described is done and the means of the two groups are compared to determine whether the drug itself has had an effect. It should be noted that we have *not* discussed a situation in which the *same* group is used twice (see Sections 14.5 and 19.4).

It is worthwhile illustrating the procedure already described once more because of its applicability and importance. An experimenter wants to determine whether there is a difference in the musical aptitudes of eight-year-old boys and girls as measured by a particular test. He chooses two groups of subjects, one of 15 eight-year-old boys and the other of 15 eight-year-old girls, such that the members of these groups are roughly equivalent in intelligence. He gives to all of his subjects a standard musical aptitude test and adopts the null hypothesis that the two groups represent random samples from the same population with regard to aptitude. He plans to test this hypothesis at the .05 level of significance.

For convenience we shall refer to the boys as comprising the first group. Since there are 15 boys, $N_1 = 15$. The mean of the boys' scores will be designated $\bar{X}_1$ and the variance of their scores, $s_1{}^2$. Since there are 15 girls, $N_2 = 15$. The mean of the girls' scores will be designated $\bar{X}_2$ and the variance of the girls' scores, $s_2{}^2$.

We shall say that the 15 scores obtained by the boys were 35, 42, 50, 60, 68, 72, 74, 74, 76, 80, 81, 83, 85, 90, 93.

The 15 scores obtained by the girls were 40, 44, 48, 65, 70, 73, 73, 73, 79, 84, 86, 87, 89, 90, 91.

The reader should verify that $\bar{X}_1 = 70.87$, $\bar{X}_2 = 72.80$, $s_1^2 = 294$, and $s_2^2 = 287$. It follows that

$$s_d = \sqrt{\frac{(N_1-1)s_1^2 + (N_2-1)s_2^2}{N_1+N_2-2}} \cdot \sqrt{\frac{1}{N_1}+\frac{1}{N_2}} = 6.21$$

Therefore,

$$t = \frac{\bar{X}_1 - \bar{X}_2}{s_d} = \frac{-1.93}{6.21} = -.31$$

The obtained t value is small enough in magnitude to be one that would be expected if two samples of size 15 each were in fact randomly selected from the same population. In other words, the disparity between observed means is not large enough to indicate rejecting the hypothesis. On the basis of his finding, our experimenter must say that he is unable to conclude that eight-year-old boys and girls differ in musical aptitude. Note that our experimenter would have had to make reference to the t distribution based on 28 degrees of freedom, since $N_1 + N_2 - 2 = 28$. However, it would scarcely matter, because the obtained t value would not have been significant when related to any t distribution regardless of its number of degrees of freedom.

Only one other point might be mentioned in connection with the foregoing. We have used two-tail tests in our examples. A one-tail test could be used if the theory involved a prediction that a specific one of the two means was higher than the other. In this case, the test for the differences between means using a one-tail test would proceed in the same way as the one-tail tests presented earlier in the text.

14.5 The t Test for Matched Groups. Sometimes there is a natural pairing between scores in one group and those in another. This occurs, for instance, when subjects are each tested twice. The subject obtains two scores, and the *change* in the subjects' scores are of interest to us.

For instance, suppose 20 boys in a class are given a test of prejudice, are afterward shown a film calculated to reduce prejudice, and are then given the same (or an alternate form of the same) test. Of interest to us is whether the film changed their attitudes, as measured by the test.

We shall assume that the stability of the test is such that, unless there was some new learning, the scores would not change. This assumption would of course have to be demonstrated for any particular test, since practice sometimes produces change.

With natural pairings, we do not use the t test for means mentioned in the previous section because a more powerful test is available. The t test for the difference between means, mentioned in Section 14.4, is subject to a sort of error that we can eliminate when we use matched pairs of scores. 211

Chance differences between subjects will not add to our error likelihood this time, because our approach is to deal wholly with the *differences* between performances by the same subject under the two conditions. The approach is also used when the experimenter has matched subjects so that corresponding to each subject in one group there is a particular subject in the other group serving as his counterpart. For instance, the matching may be done on the basis of IQ or on any other control variable.

Table 14.3 shows the prejudice scores of 20 boys both before and after seeing a particular educational film. Column 4 lists the difference scores for each subject. On this particular test, the higher the score the more the prejudice. It is desired to show that the film reduced prejudice—that the post-film scores are lower than the pre-film scores.

This time we operate entirely on the column of differences to do our significance test. We shall choose the .05 level of significance, and note that we are doing a one-tail test, since our theory asserts that the scores after seeing the film will be lower than (and not merely "significantly different" from) the earlier scores.

TABLE 14.3   SCORES ON PREJUDICE TESTS

SUBJECT	PRE-FILM SCORE	POST-FILM SCORE	DIFFERENCE SCORE
1	74	70	-4
2	35	34	-1
3	46	47	1
4	79	55	-24
5	55	30	-25
6	67	60	-7
7	44	41	-3
8	23	20	-3
9	78	67	-11
10	66	55	-11
11	43	32	-11
12	21	19	-2
13	90	45	-45
14	44	55	$+11$
15	35	40	$+5$
16	69	60	-9
17	88	82	-6
18	34	32	-2
19	25	12	-13
20	76	30	-46

$$N = 20 \qquad\qquad \Sigma d^2 = 6230$$

$$\Sigma d = -206 \qquad\qquad N\Sigma d^2 = 124600$$

$$\bar{d} = \frac{\Sigma d}{N} = \frac{-206}{20} = -10.3$$

The null hypothesis we are testing is that the film had no effect. From this it would follow that the difference scores average out to zero in the long run. Our finding of $\bar{d} = -10.3$ is, under this hypothesis, a random departure from the expected value of $\bar{d}$, which is zero. Note that in this design N stands for the number of difference scores, which means that it is the number of matched pairs.

The question we are to answer by means of our significance test is whether our obtained $\bar{d}$ is significantly far away from zero. The t value used to test the null hypothesis, based on $N - 1$ degrees of freedom, is given by the following formula:

$$t = \frac{\Sigma d}{\sqrt{\dfrac{N\Sigma d^2 - (\Sigma d)^2}{N - 1}}}$$

$$t = \frac{-206}{\sqrt{\dfrac{124600 - (-206)^2}{19}}}$$

$$t = \frac{-206}{65.75}$$

$$t = -3.13$$

Here, $N - 1 = 20 - 1 = 19$. As the reader should verify, the obtained t value is significant at the .05 level of significance. We are able to reject the hypothesis that the film made no difference in the expression of prejudice. We may wish to conclude that the film reduced the degree of prejudice felt by the subjects or, at the very least, that it curtailed their expression of prejudice.

As long as the groups can be matched pairwise, the approach given in this section can be used.

PROBLEM SET A*

14.1 A sample of 16 graduating medical students taking a final exam receives $\bar{X} = 78$ and $s = 6$. Use a one-tail t test at the .05 significance level to test the hypothesis that this sample comes from a population having $\mu = 82$.

14.2 A sample of 20 political interview tapes yields lengths having $\bar{X} = 97$ and $s = 25$. Use a two-tail t test at the .05 significance level to test the hypothesis that this sample comes from a population having $\mu = 85$.

14.3 A sample of 25 language aptitude test scores has $\bar{X} = 75$ and $s = 7$. Use a one-tail t test at the .01 significance level to test the hypothesis that this sample comes from a population having $\mu = 78$.

*In all one-tail tests in this and subsequent problem sets, assume that the obtained value (of O or $\bar{X}$) differs from the expectancy in the direction necessary to permit rejection of the hypothesis.

14.4 A sample of 13 students taking a mathematics aptitude test receives $\overline{X} = 84$ and $s = 6$. Use a two-tail t test at the .01 significance level to test the hypothesis that this sample comes from a population having $\mu = 82$.

14.5 Two samples of psychology students taking a final exam have $N_1 = 16$, $\overline{X}_1 = 78$, $s_1 = 6$ and $N_2 = 16$, $\overline{X}_2 = 84$, $s_2 = 6$. Use a one-tail t test at the .05 significance level to test the null hypothesis that the two samples come from the same population.

14.6 Two samples of adult males, when weighed, have $N_1 = 16$, $\overline{X}_1 = 170$, $s_1 = 15$ and $N_2 = 10$, $\overline{X}_2 = 185$, $s_2 = 12$. Use a two-tail t test at the .05 significance level to test the null hypothesis that the two samples come from the same population.

14.7 Two samples of business students taking a business-aptitude test have $N_1 = 12$, $\overline{X}_1 = 75$, $s_1 = 6$ and $N_2 = 10$, $\overline{X}_2 = 87$, $s_2 = 7$. Use a one-tail t test at the .01 significance level to test the null hypothesis that the two samples come from the same population.

14.8 Two samples of young adult females were measured for systolic blood pressures. The samples have $N_1 = 6$, $\overline{X}_1 = 120$, $s_1 = 9$ and $N_2 = 5$, $\overline{X}_2 = 128$, $s_2 = 9$. Use a two-tail t test at the .01 significance level to test the null hypothesis that the two samples come from the same population.

14.9 A feeding experiment is conducted on two groups of guinea pigs using different diets. The results are shown in the following table.

	High-Fat Diet	Low-Fat Diet
Number of guinea pigs	15	8
Mean weight gain (grams)	150	98
Standard deviation	23.5	17.5

Does the experiment show a significant difference between gained weight for the two groups of guinea pigs? Use .01 as the level of significance.

14.10 Suppose a company experiments to measure the effect of coffee breaks on work productivity. Ten workers are selected for the experiment and the procedure is to measure their productivity on a day with a 15-minute coffee break and then to measure their productivity on a day with no coffee break. The plant industrial engineer establishes the following scores for the workers' productivity:

Worker	Coffee Break	No Coffee Break
A	45	47
B	53	50
C	42	40
D	38	38
E	51	48
F	36	39
G	39	39
H	46	42
I	37	33
J	35	34

Does a coffee break increase productivity as indicated by the study? Use .05 as the significance level.

14.11 A group of sixth-grade students were given Form A and Form B of an intelligence test. The following data were obtained.

Child	Form A	Form B
1	9	10
2	12	14
3	15	13
4	10	7
5	11	7
6	15	12
7	8	9
8	9	12
9	7	6
10	10	14

Assume the researcher predicted that the scores on one form of the test would not be higher than on the other form. Using .05 as the significance level, would you reject the null hypothesis that there was no difference between Form A and Form B?

14.12 A group of eight sophomores were given an English achievement test before and after receiving instruction. Their scores are presented below.

Student	Before	After
A	20	18
B	18	22
C	17	15
D	16	17
E	14	8
F	14	20
G	12	9
H	9	7

The English teacher predicted that the students would do better after instruction. Perform a t test to determine whether or not to reject the null hypothesis. Use .01 as the significance level.

PROBLEM SET B

14.13 A sample of nine high school graduates taking a vocational competence test receives $\bar{X} = 85$ and $s = 9$. Use a one-tail t test at the .05 significance level to test the hypothesis that this sample comes from a population having $\mu = 79$.

14.14 A sample of seven music students taking a musical aptitude test receives $\bar{X} = 77$ and $s = 9$. Use a two-tail t test at the .05 significance level to test the hypothesis that this sample comes from a population having $\mu = 85$.

215

14.15 A sample of 37 children taking a drawing aptitude test receives $\bar{X} = 70$ and $s = 8$. Use a one-tail t test at the .01 significance level to test the hypothesis that this sample comes from a population having $\mu = 77$.

14.16 A sample of 30 young adult females yields systolic blood pressures with $\bar{X} = 120$ and $s = 9$. Use a two-tail t test at the .01 significance level to test the hypothesis that this sample comes from a population having $\mu = 122$.

14.17 Two samples of musical tapes when measured for length, in minutes, have $N_1 = 10$, $\bar{X}_1 = 88$, $s_1 = 28$ and $N_2 = 10$, $\bar{X}_2 = 102$, $s_2 = 20$. Use a one-tail t test at the .05 significance level to test the null hypothesis that the two samples come from the same population.

14.18 Two samples of science students taking a science aptitude test have $N_1 = 20$, $\bar{X}_1 = 90$, $s_1 = 6$ and $N_2 = 15$, $\bar{X}_2 = 78$, $s_2 = 8$. Use a two-tail t test at the .05 significance level to test the null hypothesis that the two samples come from the same population.

14.19 Two samples of prisoners were given IQ tests. They have $N_1 = 50$, $\bar{X}_1 = 109$, $s_1 = 16$ and $N_2 = 40$, $\bar{X}_2 = 115$, $s_2 = 12$. Use a one-tail t test at the .01 significance level to test the null hypothesis that the two samples come from the same population.

14.20 Two samples of adult females had their bust measurements taken. They have $N_1 = 10$, $\bar{X}_1 = 30''$, $s_1 = 4''$ and $N_2 = 18$, $\bar{X}_2 = 36''$, $s_2 = 5''$. Use a two-tail t test at the .01 significance level to test the null hypothesis that the two samples come from the same population.

14.21 An improved manufacturing process is developed. The quality-control tests show the following scores:

Old process: 11.5, 12.5, 13.5, 12.5, 14.0
New process: 12.5, 13.0, 13.5, 14.0, 16.2, 15.0

Does the new process show improved quality at the .05 level?

14.22 An educator undertook an assessment of the relative effectiveness of the "look-say" and the phonics approaches to teaching reading in the first grade. He drew a sample of 24 first-graders and formed two groups—or 12 matched pairs—by matching individuals on initial reading ability. He taught one group phonically, the other by means of the "look-say" method. At the end of three months he tested pupils for reading achievement, with the following results:

Matched Pairs	Phonics	"Look-Say"
A	43	40
B	39	39
C	54	52
D	51	51
E	63	60
F	37	40
G	20	19
H	32	33

Reading Scores

I	64	60
J	57	58
K	31	29
L	41	43

If the educator uses the .05 level of significance, what should he conclude?

14.23 For one year, two groups of children were instructed in art by different methods. Each child receiving Method A was matched according to IQ score with a child receiving Method B. At the end of the year, each sample of children was given a creativity test. The scores are presented below.

Pairs of Children	Method A	Method B
1st	42	36
2nd	39	38
3rd	37	32
4th	37	31
5th	34	25
6th	32	28
7th	31	21
8th	27	20

Assume the researcher did not predict which method would be more effective. Perform a t test to determine whether or not to reject the null hypothesis. Use .05 as the significance level.

14.24 A typing instructor predicted that his students made more typing errors in the afternoons than they did in the mornings. He gave a group of students a test at both times and determined the errors to be as follows:

Student	Morning	Afternoon
A	2	4
B	2	4
C	1	3
D	4	3
E	4	5
F	1	2
G	3	4
H	3	3
I	2	2
J	3	1

Perform a t test to determine whether he should reject the null hypothesis. Use .05 as the significance level.

217

fifteen

independence
and the
chi square distribution

15.1 More Than One Variable. The preceding chapters have dealt with problems encountered in describing distributions of one variable and in making predictions based on samples involving one variable. Many practical problems, however, involve more than one variable. For example, consider such questions as "To what extent is it possible to predict college success from entrance examination scores?" and "Is it possible to predict scores on the Graduate Record Examination with any success from a knowledge of college grade point averages?" In each of these situations it is possible to identify two variables. In fact, much of science may be described as the discovering of new relationships between variables. Man's power to predict and control nature arises largely from his knowledge of existing relationships in which two or more variables are involved.

In the remainder of this book we shall be concerned primarily with problems involving two variables. The present chapter deals primarily with approaches to problems involving discrete variables in which the values of the variables are given as categories. Subsequent chapters consider methods more appropriate for variables that are continuous or that are viewed as continuous. As in our previous work, we shall consider both descriptive and inferential situations.

15.2 Independence. We shall begin by illustrating and defining the situation of unrelatedness or independence between two variables. When we have understood the meaning of independence, the meaning of a relationship and some of its implications will become evident. Note that we have

218

already considered one kind of independence, namely that between two or more observations. Now we are considering independence in another form —that is, the independence of variables.

As usual our starting step is to specify a population of interest. For simplicity, say that Population A consists of 12 people and that we have scored each person on two variables, gender and political-party preference as indicated in the last election. The variable, gender, may assume either of two values, male and female. We shall say that the variable, party preference, must assume either the value Republican or the value Democrat. Table 15.1 summarizes the findings for the twelve people.

TABLE 15.1 GENDER AND PARTY PREFERENCE— POPULATION A

	REPUBLICAN	DEMOCRAT
Male	6	2
Female	3	1

We now pick one of the two variables at will and use its different values to break down the population into subpopulations. Specifically, we shall call the variable, gender, our primary variable. Each value of this variable characterizes a subpopulation and we are about to look at these subpopulations one by one.

First, we shall consider only the cases in which our primary variable assumes the value *male*. This means that we shall look at only the first row of Table 15.1. This row shows the 8 cases having the value male. We see that when our primary variable assumes this value, the other variable, party preference, assumes the value Republican 75 per cent of the time and the value Democrat 25 per cent of the time. In other words, when we are considering only males, we find that 75 per cent are Republicans and that 25 per cent are Democrats. *The distribution of voting preferences among the males is such that 75 per cent are Republicans and 25 per cent are Democrats.* (We are going to use the word distribution repeatedly to refer to the way that terms distribute themselves without regard to how many there are in the subpopulation.)*

Now we shall let our primary variable assume the value *female*. This means that we shall look at only the second row of Table 15.1, which shows the four cases having the value female. We see that when our primary variable assumes this value, the other variable, party preference, assumes the value Republican 75 per cent of the time and the value Democrat 25 per cent of the time. But this is exactly the same situation that we had when the

---

* Thus a population of eight Republicans and two Democrats has the same distribution as a population of 80 Republicans and 20 Democrats. The percentages at the different values are the same.

primary variable assumed the value male. Thus at each value of the variable, gender, the distribution of political preferences is the same. We therefore say that gender and political party preference are unrelated (or *independent*) in Population A. This leads us to make the following definition:

Definition 15.1 *A. Two variables are called unrelated or independent in a population if at every value of one of them the distribution of the other is the same as at any other value.*

B. Two variables are said to be related when it is not true that at every value of one of them the distribution of the other is the same as at any other value.

15.3 Illustrations of Relatedness. We shall now look at two illustrations of relatedness involving the same variables as before. Suppose that once again we record the gender and political party preference of each of 12 people and they are as indicated in Table 15.2. Remember that this is a *population* that we are using for illustrative purposes.

TABLE 15.2 GENDER AND PARTY PREFERENCE—
POPULATION B

	REPUBLICAN	DEMOCRAT	TOTALS
Male	7	1	8
Female	1	3	4
Totals	8	4	12

As before we arbitrarily pick the variable, gender, as the primary variable. We focus on those cases for which that variable assumes the value male. Among the males, 7 out of 8 are Republicans. In other words, when the variable, gender, assumes the value male, the distribution is such that $87\frac{1}{2}$ per cent of the subpopulation are Republicans and $12\frac{1}{2}$ per cent are Democrats.

However, when the primary variable assumes the value female, the distribution is such that 75 per cent (3 out of 4) of the subpopulation are Democrats. Thus, at different values of the primary variable, the distribution of voting preferences differs. In Population B, the variables, gender and political preference, are related. Note that among the males the Republicans are the majority, whereas among the females the Democrats are the majority.

TABLE 15.3 GENDER AND PARTY PREFERENCE—
POPULATION C

	REPUBLICAN	DEMOCRAT	TOTALS
Male	1	7	8
Female	1	3	4
Totals	2	10	12

Finally, consider Population C (Table 15.3). In this population, when the variable, gender, assumes the value male, $87\frac{1}{2}$ per cent of the members are Democrats and $12\frac{1}{2}$ per cent are Republicans. When it assumes the value female, 75 per cent of the members are Democrats and 25 per cent are Republicans. Once again the variables, gender and party preference, are related. When we go from one value of the primary variable to another, the distribution of the other variable changes. Note that in Population C both subgroups are predominantly Democrats. Yet the majority is greater among the males than among the females.

The implication of relatedness for making predictions is important. Once again we shall start by considering the situations of unrelatedness. Suppose we are told that a person has been chosen at random from Population A. We are now asked to make a predictive statement concerning his party preference. Since three-quarters of the people in Population A are Republicans, we say that the probability is 3/4 that the person in question is Republican.

Next we are given the added information that the person is male. However, since three-quarters of the males are Republicans, we say again that the probability is 3/4 that the individual is a Republican. The knowledge that the person was female would have given rise to exactly the same probability statement. We have been given information concerning a variable unrelated to the variable being predicted. This information has not altered our predictions.

We shall now consider the effect of the same kind of information when the variables are related. This time suppose we are making a prediction about a person chosen from Population B and we know nothing about the person. Since 8/12 of the members of Population B are Republicans (Table 15.2), the statement we make is that the probability is 8/12 that this individual is Republican. As before, suppose we are told in addition that the person in question is a male. Since, among the males in Population B, 7/8 are Republicans, we now say that the probability is 7/8 that the person is a Republican. The information that the person is a female would have given rise to the statement that the probability is 3/4 that the person is a Democrat. The reason is that 3/4 of the females in this population are Democrats. The fact that gender is related to party preference in Population B implies that the gender of an individual, when it is known, influences our probability statement concerning his party preference. (Note the assumption that we are making predictions only from Population B.)

Our "blind" prediction for an individual from Population C would be that the probability is 10/12 that the person is a Democrat (Table 15.3). The information that the person is a male gives rise to the statement that the probability is 7/8 that the person is a Democrat. The information that the person is female gives rise to the statement that the probability is 3/4 that the person is a Democrat. In Population C, as in Population B, the gender of an individual when it is known influences our probability statement concerning his party preference.

More generally, we have seen that when two variables are independent, knowledge of the obtained value for one variable does not affect our prediction about the value for the same case for the other. When two variables are dependent and the relationship between them is known, knowledge of the obtained value on one variable does affect our prediction concerning the other.*

The important problem is nearly always to determine the existence and nature of the relationships between variables in some vast population, not in some small collection of cases. As usual, we consider our observations as comprising a sample and we end by making inferences about some unseen population. Remember that in the foregoing we have been discussing populations. (Small populations were used for simplicity of discussion.)

We are now ready to consider inferences from a sample. The existence of some relationship in an obtained sample in itself tells us little. The reason is that even if two variables, like gender and political party preference, are independent in some vast population, any particular sample drawn from this population is apt to show some relationship between the two variables merely by chance. That is, in any particular sample, it is likely that the proportion of Republicans among the males will not be exactly the same as the proportion of Republicans among the females. Therefore the important issue is usually whether or not the relationship in the sample is marked enough to suggest that a relationship is actually present in the vast population that yielded the sample.

In experimental work, the purpose is usually to make a general statement about what is called an *infinite* population. An infinite population is one that contains so many terms that we can use our knowledge of theoretical distributions (such as the normal distribution) to make inferences about it.

For instance, the purpose of an experiment might be to determine whether or not the variable, gender, is related to the variable, "reactivity to a particular drug." Almost certainly some kind of relationship between these two variables would appear in a sample. That is, it would be surprising if in a small sample the males and females reacted in exactly the same way to the drug. The issue, however, is whether or not the two variables are related in the vast population of people from which the sample was derived. In theory, we would even like to make a generalization to people not yet born. Thus we view our observations as comprising a random sample drawn from an infinite population.

15.4 Presentation of Findings in a Contingency Table. We are now ready to consider a method of inferring from a sample whether or not two variables are independent in the population that yielded the sample. The

---

* More specifically, dependence implies that there is at least one value of the known variable which, when it turns up, is capable of altering the prediction we would have made in complete ignorance. In most practical applications, every value of the obtained variable gives rise to a unique set of predictive statements about the other variable.

technique is best applied to categorical variables, although it is often applied to discrete variables that assume numerical values and to continuous variables.

To gain focus, we shall consider a specific problem.   Suppose that we are again investigating voting preference (Republican or Democrat) among men and women. Accordingly, we obtain a sample of 150 people, of whom 100 are men and 50 are women. Analysis of the data reveals that among the men there are 66 Democrats and 34 Republicans and that among the women there are 27 Democrats and 23 Republicans. The findings appear in Table 15.4, which is called a *contingency table*. A contingency table is defined as a two-way table in which the categories are discrete.

TABLE 15.4   CONTINGENCY TABLE FOR
VOTING PREFERENCE

	DEMOCRAT	REPUBLICAN	TOTALS
Male	66 (1)	34 (2)	100
Female	27 (3)	23 (4)	50
Totals	93	57	150

Note that the two cells in the first row pertain to men and that the two in the second row pertain to women.   The two cells in the first column pertain to Democrats and the two in the second column pertain to Republicans.

Each individual cell is in a particular row and a particular column.   The large number in each cell is the frequency for that cell.   Each cell has also been assigned a number in parenthesis for purposes of identification.   To determine the meaning of the frequency of any given cell, we must look at the title of its row and that of its column.   For instance, Cell (1) (in the first row and first column) pertains to men and to Democrats.   The interpretation of the number 66 is that there are 66 men who are Democrats in the sample. Similarly, Cell (4) is in the second row and second column.   Thus its frequency of 23 means that there are 23 women who are Republicans in the sample.

The numbers in the margins of Table 15.4 are called *marginal totals*. The marginal total of the first column is 93, telling us there are 93 Democrats in the whole sample.   The marginal total of the second column tells us that there are 57 Republicans in all.   The marginal total of the first row tells us that there are 100 men in the sample, and the marginal total of the second row tells us that there are 50 women.   The grand total of members of the sample, 150, appears in the lower right corner.

Let us digress to see whether or not the variables are related in the sample.   We shall arbitrarily pick one of them, gender, and see what happens at each of its values.   Among the men the proportion of Democrats is 66/100 (which is 33/50).   Among the women the proportion of Democrats

223

is 27/50. Thus the distribution of party preference is not identical at each value of the variable, gender. The variables, gender and party preference, are related *in our sample*.

15.5 Expected Frequencies. We shall soon decide by means of what is called the "chi square test" either to accept or to reject the hypothesis that the variables are independent *in the population*. This hypothesis of independence in the population gives rise to the expectancy that the variables will be independent in the sample also. The theoretical values that would constitute independence in the sample are called *expected frequencies*.

We must compute the expected frequency for each cell, because it is the number that would have appeared were the two variables absolutely independent in the sample. Once we have computed the expected frequencies for the cells, we may determine how far our obtained frequencies differ from them. The data provide us with marginal totals that in and of themselves tell us nothing concerning whether or not the variables are related. The row totals tell us merely that we have 100 men and 50 women. The column totals tell us merely that the sample contains 93 Democrats and 57 Republicans. Table 15.5 shows these marginal totals in *italics*. The marginal totals are now going to be used to determine the expected frequencies for each of the four cells.

TABLE 15.5   EXPECTED FREQUENCIES

	DEMOCRAT	REPUBLICAN	TOTALS
Male	62 (1)	38 (2)	*100*
Female	31 (3)	19 (4)	*50*
Totals	*93*	*57*	150

The proportion of Democrats in the sample is 93/150. Independence would mean that the proportion of Democrats among the men is *identical* to the proportion of Democrats among the women. That is, independence would imply that 93/150 of the women are Democrats and 93/150 of the men are Democrats. There are 100 men in the sample and 93/150 of 100 is 62 $\left(\frac{93}{150}(100)=62\right)$. Similarly, the expected frequency of Democrats among the 50 women is 31 $\left(\frac{93}{150}(50)=31\right)$.

By analogous reasoning, 57/150 of the members of the sample are Republicans. Independence implies that 57/150 of the 100 men are Republicans and that the same proportion of the 50 women are Republicans. Thus the expected frequency of the Republicans among the men is 38 and among the women is 19. The expected frequencies just computed have been entered in the cells of Table 15.5. Note that the expected frequencies in any particular row or column add up to the appropriate marginal total. (For instance, in the first row $62 + 38 = 100$.)

In practice, the procedure for finding the expected frequency of a cell is mechanical. Each cell has a corresponding row total (the total of its row) and a corresponding column total (the total of its column). For instance, Cell (3) has a corresponding column total of 93 (since it is in the first column) and a corresponding row total of 50 (since it is in the second row). In effect, the expected frequency of any cell has been found by multiplying its corresponding row total by its corresponding column total and dividing the product by the grand total for the sample. The grand total for our sample is 150. Thus in effect the expected frequency for Cell (3) has been found by multiplying 93 by 50 and dividing the product by 150. Note that we have computed each expected frequency by this procedure.

15.6 The Chi Square Test. We are now ready to test the hypothesis that the two discrete variables are independent *in the population that yielded our sample*. Specifically, we shall choose the .05 significance level. We are going to compute the chi square statistic, which indicates the discrepancy between our obtained frequencies (Table 15.4) and the expected frequencies (Table 15.5). The size of the chi square statistic will tell us whether to accept or reject the hypothesis.

Our procedure will be to compare our obtained chi square value with a value given in a chi square table. We shall accept or reject the hypothesis depending upon whether the value we obtain is smaller or larger than the appropriate value in the chi square table.

Table 15.6 shows the obtained frequencies and the *expected frequencies in italics* for each of the cells. The chi square statistic, to be computed, will be a measure of the discrepancies between these frequencies.

TABLE 15.6 OBTAINED AND EXPECTED FREQUENCIES

	DEMOCRAT	REPUBLICAN	TOTALS
Male	66 *62* (1)	34 *38* (2)	100
Female	27 *31* (3)	23 *19* (4)	50
Totals	93	57	150

Our first step is to list the obtained and the expected frequencies and to compute the difference between them for each cell. The magnitudes of these differences give us an impression of how far the variables depart from inde-

225

pendence in the sample. The measure that we compute will be based on the squares of these differences. (Were we simply to add the differences, we would get a total of zero.) The differences and squared differences are shown in Columns (4) and (5) of Table 15.7.

In our example, each of the squared differences happened to have the value of 16. A squared difference, such as 16, in and of itself tells us little. We must consider the squared difference of each cell in proportion to the expected frequency of the same cell. Consequently our next step is to divide the squared difference for each cell by the expected frequency for that cell. The quotients for the different cells are listed in Column (6) of the table (for example, $16/62 = .258$; $16/31 = .516$; and so on).

The sum of the quotients in Column (6) is called the *chi square* statistic, and it is our final measure of departure from independence in the sample.

TABLE 15.7   COMPUTATION OF CHI SQUARE

(1) CELL	(2) OBSERVED FREQUENCY O	(3) EXPECTED FREQUENCY E	(4) DIFFERENCE $O - E$	(5) SQUARE $(O - E)^2$	(6) $\dfrac{(O-E)^2}{E}$
1	66	62	4	16	.258
2	34	38	-4	16	.421
3	27	31	-4	16	.516
4	23	19	4	16	.842

$$\Sigma = 2.037 = \chi^2$$

The larger the chi square statistic, the stronger is the relatedness in the sample. We see in Table 15.7 that the value of the chi square statistic (denoted by χ^2) yielded by our sample is 2.037. Note that the procedure for computing the chi square value of 2.037 has been described, but it has not been justified here. The proof that it is a meaningful procedure is beyond the scope of this book. Stated symbolically we have

(15.1) $$\chi^2 = \sum \frac{(O-E)^2}{E}$$

where O is the obtained frequency and E is the expected frequency for the same cell.

Appendix V gives the values of chi square that would be needed for significance at various levels. However, the reader will note that this table is set up for various numbers of degrees of freedom. Before we can interpret our value of 2.037 correctly, we must determine the number of degrees of freedom that went into our particular chi square statistic. To this end, let us suppose once more that we have computed only the marginal totals from the data. These marginal totals, in *italics* in Table 15.8, provide no information concerning the possible relatedness of the variables.

TABLE 15.8   DEGREES OF FREEDOM IN
OBTAINED FREQUENCIES

	DEMOCRAT	REPUBLICAN	TOTALS
Male	66 (1)	34 (2)	100
Female	27 (3)	23 (4)	50
Totals	93	57	150

Suppose we give ourselves *one* piece of information concerning the relationship of the variables; that is, suppose we allow ourselves knowledge of the obtained frequency in one cell.   Specifically, we shall put the value *66* in the first cell.   This has been done in *italics* in the table.   Now the determination of the obtained frequency of one cell has simultaneously fixed the frequencies of each of the other cells.   The frequencies of the two cells in the first column must total 93, so the frequency of Cell (3) must be 27 *(93−66=27)*.   By further subtracting, it follows that the frequency of Cell (2) must be 34 and the frequency of Cell (4) must be 23.   The frequency of one cell and the marginal totals (the numbers in *italics*) determine the frequencies for the other three cells.

Thus the totality of information concerning the relationship in the sample is contained in a single cell frequency.   That is, given one frequency we are able to determine all the others by making proper subtractions from the marginal totals.   It follows that the chi square measure of relatedness that we just computed is based upon one degree of freedom.   A general method of enumerating degrees of freedom in chi square problems is discussed in the next section.

Our next step is to compare the chi square value derived from our sample with the appropriate value in the table in Appendix V.   The result of this comparison will determine whether we accept or reject the hypothesis that the variables are independent *in the population*.   If the chi square value that we have computed from our sample turns out to be smaller than the appropriate value in Appendix V, then our decision must be to accept the hypothesis. If the chi square value that we have computed turns out to be larger than the appropriate value in the table, then our decision will be to reject the hypothesis.

Remember that we are testing our hypothesis at the .05 significance level. We go back to Appendix V and use the row for one degree of freedom (which is the top row).   The appropriate value given in this row for the .05 significance level is 3.84, and we must compare our finding to this value.   Since our chi square value was 2.04 and it is smaller than the value in the table, the conclusion is that we must accept the hypothesis.   In essence, our obtained value of 2.04 was a measure of the discrepancy between the expected frequencies under the hypothesis and those frequencies that were obtained.   227

The fact that our obtained chi square value was smaller than the necessary value for significance means that we did not find enough of a discrepancy from independence to lead us to reject the hypothesis.

The chi square distribution is discussed in various elementary texts but we shall not examine its properties here. Actually, there exists a different chi square distribution for each number of degrees of freedom, as is shown by the different rows in Appendix V. We computed a chi square value based upon one degree of freedom, so we referred to a value given in the first row of the table. We also used the column of the table that gives values for the .05 significance level. The other columns give cut-off points for other levels of significance and give the experimenter more freedom in choosing a significance level.

We can observe in Appendix V that the value 3.84 is a cut-off point in the distribution of chi square with one degree of freedom. To see the implications of this cut-off point, consider a population in which two discrete variables are actually independent. Suppose that a vast number of random samples like ours are drawn from this population and that a chi square value is computed from each sample in the manner described. We are assuming that each chi square value is based on one degree of freedom. Under these conditions only five out of 100 samples would yield values of chi square as large or larger than 3.84.

We hypothesized that our data comprised a random sample drawn from a population characterized by independence. Our sample yielded a chi square value (based on one degree of freedom) smaller than 3.84. Thus the finding was not sufficiently improbable to lead us to reject the hypothesis that it came from a population characterized by independence. We were prepared to reject the hypothesis only upon obtaining a sample with a chi square value at least as large as 3.84. For instance, if our sample had yielded a chi square of 6 we would have deemed it so improbable under the hypothesis that we would have rejected the hypothesis.

As we shall see, the procedure and the logic are the same when the chi square test is based on any number of degrees of freedom. For each case the appropriate cut-off point is indicated in the chi square table for each of the various significance levels. Thus when we consider a chi square test using four degrees of freedom at the .05 level, we shall use 9.49 as the cut-off point. That is, we will accept the hypothesis that the variables are independent in the population if the obtained chi square value is less than 9.49. We will reject the hypothesis upon obtaining a chi square value as great or greater than 9.49.

15.7 Chi Square for More Complex Problems. The chi square test is the same when the variables of interest take on more than two possible values. For instance, suppose that we are concerned with the reaction of Negroes, whites, and Puerto Ricans to a proposed bill in New York State. We wish to test the hypothesis that each of these groups holds the same attitude toward the bill.   Members of the three groups are asked to indicate whether their reactions are favorable, indifferent, or unfavorable.   (Incidentally, the variable, reaction to the bill, might be viewed by some as continuous; but so long as we are assigning its values to distinct categories, we may use the chi square test.)

Suppose that data are collected for a sample of 100 Negroes, 200 whites, and 100 Puerto Ricans, and that the obtained values are as given in Table 15.9 without italics.

TABLE 15.9   A THREE-BY-THREE CONTINGENCY
TABLE

	FAVORABLE	INDIFFERENT	UNFAVORABLE	TOTALS
Negro	80 *52.5* (1)	10 *22.5* (2)	10 *25* (3)	100
White	55 *105* (4)	70 *45* (5)	75 *50* (6)	200
Puerto Rican	75 *52.5* (7)	10 *22.5* (8)	15 *25* (9)	100
Totals	210	90	100	400

The *expected frequency* for any cell is calculated exactly in the manner described in the previous sections although the variables are not dichotomous. For each cell, we multiply its corresponding row total by its corresponding column total and divide the product by the grand total for the sample.   For example, the expected frequency for Cell (5) in Table 15.9 is $(90)(200)/400 = 45$. The logic is that the proportion of indifferents in the whole sample is 90/400. Independence implies that the same proportion of indifferents is found among the 200 whites; therefore the expectancy is that 45 whites will be indifferent to the proposed bill.   The expected frequency for each of the cells, found in the same way, appears in *italics* in the table.

The computation of the chi square statistic for the values in Table 15.9 appears in Table 15.10.   The procedure is exactly the same as that used in the

TABLE 15.10   COMPUTATION OF CHI SQUARE
FOR A THREE-BY-THREE TABLE

(1)	(2)	(3)	(4)	(5)	(6)
CELL	O	E	$O - E$	$(O - E)^2$	$\dfrac{(O - E)^2}{E}$
1	80	52.5	27.5	756.25	14.40
2	10	22.5	-12.5	156.25	6.94
3	10	25	-15	225	9.00
4	55	105	-50	2500	23.81
5	70	45	25	625	13.89
6	75	50	25	625	12.50
7	75	52.5	22.5	506.25	9.64
8	10	22.5	-12.5	156.25	6.94
9	15	25	-10	100	4.00

$$\Sigma = 101.12 = \chi^2$$

229

previous section except that now we have to work with nine cells instead of four. The sum of Column (6) is the obtained value of chi square and it is 101.12. It is the value given by Formula (15.1).

Before we can use our obtained value of chi square and Appendix V to test our hypothesis, we must determine the number of degrees of freedom involved. We will also state a general rule concerning the number of degrees of freedom to be used in similar problems of any degree of complexity. Once again we begin with just the marginal totals that in and of themselves tell us nothing concerning the possible relatedness of the variables. Again we shall allow ourselves knowledge of the frequency in one cell, say Cell (1). This obtained frequency and the marginal totals have been written in *italics* in Table 15.11. But one piece of information is now insufficient for us to determine the frequencies in the other cells. If we give ourselves a second obtained cell frequency and then a third we still cannot fill out the chi square table. Suppose, for example, that we are given the frequencies for Cells (1), (2), and (4). These are italicized in Table 15.11. It is seen that these are sufficient to fix the frequencies in some of the other cells, namely Cells (3) and (7). However, the frequencies of the remaining cells, (5), (6), (8), and (9), are still free to vary.

TABLE 15.11   DEGREES OF FREEDOM FOR A
THREE-BY-THREE TABLE

	FAVORABLE	INDIFFERENT	UNFAVORABLE	TOTALS
Negro	*80* (1)	*10* (2)	10 (3)	*100*
White	55 (4)	⑦⓪ (5)	75 (6)	*200*
Puerto Rican	75 (7)	10 (8)	15 (9)	*100*
Totals	*210*	*90*	*100*	**400**

One more independent piece of information is all that we need, however. Once we know the frequency of Cell (5), we can determine all of the rest by subtraction from the marginal totals. This frequency has been encircled in Table 15.11. It is thus seen that four independent pieces of information have proved sufficient. The frequencies of the four cells in the upper left corner together with the marginal totals are sufficient to determine the frequencies of all of the cells in the third row and the third column.

In general, we can always infer the frequencies in the last row and in the last column by making subtractions from the marginal totals. *The number of degrees of freedom is equal to the number of cells remaining in the chi square table after the last row and the last column are eliminated.* Symbolically we have

(15.2) $$\text{d.f.} = (R - 1)(C - 1)$$

where there are R rows and C columns.

We now know that our obtained value of chi square, 101.12, is based upon four degrees of freedom $[(3-1)(3-1)=2\cdot2=4]$. The cut-off point given in Appendix V is 9.49 for the .05 level. Since our obtained chi square value is much larger than the appropriate value in the table, we reject the hypothesis that the sample came from a population in which the variables are independent. We conclude that attitude toward the proposed bill is related to ethnic group.

Incidentally, after a significant chi square value has been obtained, one may attempt to make more specific statements by considering the data in some of the cells and not others. In our own example, we might, for instance, compare the Negroes and whites. To do so, we necessarily disregard the final row of Table 15.9 and revise the marginal totals as in Table 15.12.

TABLE 15.12   A TWO-BY-THREE CONTINGENCY TABLE

	FAVORABLE	INDIFFERENT	UNFAVORABLE	TOTALS
Negro	80 *45* (1)	10 *27* (2)	10 *28* (3)	100
White	55 *90* (4)	70 *53* (5)	75 *57* (6)	200
Totals	135	80	85	300

As usual the *expected frequencies* have been added in *italics*. The reader should verify that these new expected frequencies are correct. They have been necessitated by the change in the marginal totals.

The reader should also verify that the chi square value for Table 15.12 is 74.2 as found by Formula (15.1). It is based upon two degrees of freedom as verified by Formula (15.2) and is significant at the .05 level. Thus we reject the hypothesis that the whites and Negroes are identical in attitude toward the proposed bill.

The chi square test is often used in connection both with discrete variables that take on numerical values and with continuous variables. Successive numerical intervals are set up before the data are gathered and these intervals serve as categories. For instance, we might wish to see whether IQ is related to scores on a music appreciation test, in which case the table for recording the data might look like Table 15.13.

When we put the values of a continuous variable into categories, we are admittedly losing information. For instance, by treating the data as in Table 15.13 we are not differentiating between individuals whose IQ's are between 101 and 115. However, the possible relationships become so numerous when one or both variables are continuous that we must either categorize the variables, as we have done here, or make certain assumptions about them in order to test for independence. One need make no assumptions about the distribution of continuous variables in order to use the chi square test. An alternative method that also requires no assumptions is discussed in Chapter 19.

TABLE 15.13 PATTERN FOR A FIVE-BY-FOUR CONTINGENCY TABLE

Music Appreciation Score

IQ	BELOW 26	26–50	51–75	76–100	TOTALS
71–85					
86–100					
101–115					
116–130					
131–145					

Totals

PROBLEM SET A

15.1 In an experiment involving two methods (A and B) of teaching Swahili, the numbers of pupils who made a gain or made no gain are as follows:

	Gain	No Gain
Method A	65	35
Method B	35	15

Complete a contingency table and test using (a) the .05 significance level of chi square and (b) the .01 significance level of chi square.

15.2 A researcher wishes to determine whether there is a difference between men and women in their choice of four brands of cigarettes. He takes a sample of each sex and obtains the following results.

	Brand A	Brand B	Brand C	Brand D
Men	60	50	40	50
Women	0	10	30	30

Apply a chi square test at the .01 significance level.

15.3 An instructor wants to determine whether a certain question on a standardized test distinguishes between the students who receive good and poor scores on the test as a whole. Of the 50 students in the upper quarter on the test, 38 answer the particular question correctly; of the 50 students in the lower quarter, only 23 answer it correctly. Test at the .05 significance level of chi square.

15.4 In an attempt to ascertain whether gentlemen prefer blondes, the following data are obtained on the number of dates per month of unmarried females, ages 18–25.

	Number of Dates per Month		
Hair Color	Under 9	9 to 13	Over 13
Blonde	20	45	15
Brunette	15	35	20
Redhead	5	20	10

Apply a chi square test at the .05 significance level.

15.5 The results of the use of two drugs in the treatment of a certain disease are as follows:

	Recovered	No Change	Died
Drug A	40	18	12
Drug B	50	8	7

Test for significance using chi square at the .01 level.

15.6 Three preparations are tested for effectiveness against dandruff with the following results:

	No Improvement	Good Improvement
A	11	20
B	17	23
C	16	19

Is there a difference in the effectiveness of the preparations? Apply a chi square test at the .05 significance level.

15.7 A drug manufacturer claims that a certain product, when combined with various treatments of patients, will cure a specific disease. The following results are obtained:

	Drug with Treatment			
Patient Response	A	B	C	D
Favorable	115	171	68	72
Unfavorable	45	29	32	28

Is there a significant difference between the results for each pair of drugs? Complete chi square tests at the .01 significance level for each pair of drugs.

15.8 A group of 100 individuals is sampled for brand preference on a certain item, with the following results:

	Male	Female
Brand X	62	22
Brand Y	9	7

Is there a difference between male and female preference? Apply a chi square test at the .05 significance level.

15.9 A physical education instructor wishes to determine preferences of boys and girls for three activities. He asks a sample of 40 boys and 30 girls to state their preferences. Here are the results of his survey.

233

	Basketball	Volleyball	Kickball
Boys	20	5	15
Girls	7	13	10

Apply a chi square test at the .05 significance level.

15.10 A music instructor wishes to determine preferences of boys and girls for classical music. He asks a sample of students "Do you enjoy classical music?" and obtains the following results:

	Yes	Undecided	No
Boys	46	10	30
Girls	20	18	50

Apply a chi square test at the .01 significance level.

PROBLEM SET B

15.11 Suppose that in a sample of cigarette smokers it is found that of 400 women 280 prefer Brand Y and 120 prefer Brand Z, while of 300 men 190 prefer Brand Y and 110 prefer Brand Z. Test using the .05 significance level of x^2.

15.12 In a survey to determine boys' and girls' preferences for three different brands of candy, the following results are obtained:

	Candy A	Candy B	Candy C
Boys	20	26	30
Girls	70	50	20

Apply a chi square test at the .01 significance level.

15.13 In a comparison of two methods of teaching a concept in statistics, an instructor finds the following results:

	No Improvement	Moderate Improvement	Good Improvement
Method A	15	25	50
Method B	15	15	45

Apply a chi square test at (a) the .05 significance level, (b) the .01 significance level.

15.14 A study is made to determine the possible relationship between age at first marriage and income level of the bride's family. The findings are as follows:

	Age of Bride		
Income Level	Under 18	18–21	Over 21
Low	45	25	15
Middle	35	60	25
High	10	28	24

Apply a chi square test at the .05 significance level.

15.15 A study is made to determine the possible relationship between religious affiliation and attitude toward a certain proposed piece of federal legislation. The data are as follows:

	For	Against	Indifferent
Catholic	60	40	20
Jewish	35	30	10
Protestant	50	65	25

Apply a chi square test at the .05 significance level.

15.16 The results of a test for germination of two brands of zinnia seeds are as follows:

	Germinated	Did Not Germinate
Brand B	80	45
Brand C	60	25

Test for significance using chi square at the .05 level.

15.17 A study is made of the relationship between a certain disease and the healing of wounds among certain hospital patients. The findings are as follows:

	Normal Healing	Prolonged Healing
Diseased	95	45
Not Diseased	70	30

Is there any evident relationship between the disease and healing time? Apply a chi square test at the .05 significance level.

15.18 The following results are obtained on a test for color blindness:

	Male	Female
Color Blind	25	9
Not Color Blind	270	196

Apply a chi square test at the .05 significance level.

15.19 A frequent TV commercial states that 9 out of 10 housewives prefer Brand X. Random sampling of 1,000 housewives reveals that 79 per cent prefer Brand X and 21 per cent prefer some other brand. Is the claim made on the TV commercial sustained? Apply a test at the .05 significance level.

15.20 A television network wishes to determine whether certain types of programs appeal to different nationality groups. An interviewer asks people in four nationality groups to indicate their preference among five types of television programs. Here are the results of his survey.

	French	German	Italian	Chinese
Program Type A	20	18	16	14
Program Type B	16	14	18	16
Program Type C	14	14	12	18
Program Type D	20	16	14	16
Program Type E	18	20	14	18

Apply a chi square test at the .05 significance level.

sixteen

regression and prediction

16.1 Introduction. The number of possible relationships between two continuous variables is infinite. Of course there may be no relationships at all, but, in the simplest case where one does exist, it may be that high scores on one variable tend to accompany high scores on the second. For instance, such is the case with the variables, age and vocabulary. The younger one is, the fewer words he is likely to know, and the older he gets the more he is likely to have learned. The relationship described is sometimes called a positive one.

A second kind of relationship is one in which successively higher values of one variable tend to accompany lower values on the other. For instance, such is the relationship between degree of education and crime rate. Those who are poorly educated are the most likely to commit crimes, and those who tend to have successively more education are to that extent less likely to commit crimes. This type of relationship is often called a *negative* or *inverse relationship*.

More subtle kinds of relationships also exist quite commonly. For instance, successive increases in one variable may first accompany increases in the other. Then further increases in the first variable may go along with decreases in the second. The relationship between age and physical strength is of this type. As age increases, so does physical strength, up to a point; but beyond this point further increases in age are accompanied by decreases in physical strength. Thus, at the low age levels, to know that one person is five years younger than another leads to a best guess that he is weaker; but to know the same thing for two people at advanced ages would be to make a best guess that the younger person is stronger.

236

Keep in mind as we proceed that we are talking merely about relationships between two variables and not necessarily causal relationships. For instance, one must not infer that a causal relationship exists between education and abstinence from crime merely from what was said. It may be that a third variable, like degree of economic security, is crucial in accounting for the relationship that was described. That is, the degree of one's economic security may account both for one's abstinence from crime and also for one's decision to increase his education. It is important to remember that relationships may be causal or they may be consequential relations that exist because of some other variable that was not of interest to us. Our concern shall be only with the relationships as they exist and with their implications, not with how they came into existence. The latter problem is for the theorist and the researcher himself.

16.2 Blind Prediction. One of the primary advantages of knowing about a relationship between two variables is that one can use the knowledge to facilitate making predictions. Specifically, when one has exact knowledge of an individual's score on one of two variables, then he can use knowledge of the relationship to increase the accuracy of a prediction of the individual's score on the other variable. For instance, suppose there is a known relationship between IQ and students' freshman averages in a particular state university. Then an admissions office worker can use his knowledge of an individual's IQ score to aid him in making a prediction of what the student's freshman average will be. The fact of a relationship between two continuous variables is often relevant to the whole issue of prediction, though sometimes the relationship is not strong enough to produce sizeable predictive advantages and at other times there is no practical concern with making predictions. At any rate, some comments must be made about the issue of prediction before going ahead, since we shall make reference to this issue throughout the discussion.

To begin with, by a prediction we mean a best guess of what a single value or score will be. We often have occasion to make predictions defined in this way. For a college admission office, the value to be predicted may be the average that a boy will get as an engineering student. For the student, the value predicted may be the income that a graduate from the same college will earn ten years after graduating. For an insurance company, the value to be predicted may be the number of years that a woman of 62 with high blood pressure will live. Or the value to be predicted may be the number of years that an industrial machine will run before breaking down.

Definition 16.1 *A prediction is a guess about the value of a term to be drawn from a specified population.*

To make any kind of meaningful prediction, we must have at least some knowledge about the population that is to yield the single case. We are going to assume throughout that we know at least the mean and variance of

237

this population.   For instance, if we are going to make a prediction about the height of an adult American male, we are going to assume that we know at least the mean and variance of the heights of adult American males.   We shall say, for the sake of discussion, that we know the mean height of American adult males to be five feet eight inches and that the variance of these heights is nine.

The simplest kind of prediction to consider is that of the value of a term when we know nothing more than the mean and variance of the population of which the term is a member.   Such a prediction shall be called a *blind prediction*.   The use of the blind prediction is worth considering so that we may make comparisons with it later on.   Suppose, for instance, that we are going to go through a telephone book and pick out the name of an adult American male in order to predict his height.   The name turns out to be Joe Brown, and now we must make our prediction.   We are making a blind prediction because it is to be based on no information other than that concerning the population from which the single case has been derived.

Before going any further, note that there is a world of difference between the topic of estimation that we have already discussed and that of prediction which is being introduced.   An estimate is a guess of a parameter value, like a population mean, and is typically based upon information derived from a sample.   When we make a prediction we know all we are going to know concerning the population, and the prediction itself concerns a single case drawn from the population.

Now we return to our problem of making our blind guess of the height of Joe Brown.   It turns out that the best we can do is to take the population mean of five feet eight inches and use this value as our guess.   In other words, our best blind guess is that five feet eight inches is the height of Joe Brown.   In general, the population mean is the best blind guess of the value of a single case drawn from a population.

The reason for using the population mean as the blind guess may be seen by considering the situation when one makes many blind guesses, one after the other.   For instance, suppose Joe Brown's actual height is five feet ten inches and one guesses that his height is five feet eight inches.   Then the error attached to this guess, defined as the actual value of the term minus the guess, is two inches.   (Five feet ten inches minus five feet eight inches equals two inches.)   We shall say that once more a name is randomly chosen from the telephone book and once more the guess of five feet eight inches is made.   This time the actual height of the person turns out to be five feet seven inches, so the error turns out to be negative one inch.   The process is repeated *ad infinitum*.

To begin with, note that, when the mean is used as a blind guess, the error each time is really the deviation of the term from the mean of its population.   The sum of these errors to be made in the long run is zero (Theorem 2.1).   Also the sum of the squares of these errors is really the sum of the squared deviations of the terms from the mean of their own population.   Suppose some value other than the population mean was repeated as a blind guess each time. Then for one thing the errors would not balance out to zero. Also the sum of the squares of the errors made in the long run would be larger than when the mean itself was used. The reason is that, for any population, the sum of the squared deviations about the mean is less than the sum of squared deviations about any other value (Theorem 2.2).

It would not help to vary one's guess from one time to another. Only when the population mean is used as the guess of the unknown case each time is it true that the errors average out to zero in the long run and that the squared deviations are a minimum. Note that when the errors as a group are large, whether positive or negative, then the squares of these errors are large. The squares of the errors as a group may be small or large, but when they are small they indicate that the predictions as a group are relatively accurate. When the squares of the errors tend to be large, they indicate that the guesses as a group tend to be inaccurate, because the errors themselves must be running large in order for their squares to be large.

The classic measure of how accurate the guesses are as a group is the mean of the squares of the errors. For instance, if the errors in inches are $-3, 2, 0, 1, 2$, then the squares of these five errors are $9, 4, 0, 1, 4$. The mean of these five squares is $18/5$. That is, the mean of the squared errors for the five guesses mentioned is $18/5$. But in the long run, when one guesses the mean each time, each squared error is a squared deviation from the mean. As a vast number of guesses are made, the mean of the squared errors becomes the population variance. In sum, when one makes blind guesses, the mean of his squared errors over the long run is the population variance, and thus the measure of his accuracy is the population variance. Later we shall see how a system *using other information* is apt to lead to more accurate predictions over the long run. A system is considered more accurate than the blind predictive method when the mean of the squared errors made according to this system over the long run would be less than the population variance.

It should be very clear how each particular error has been found—by subtracting the population mean from the actual value of the term of interest. Where the term was five feet ten inches, the error was two inches. Where the term was five feet seven inches, the error was negative one inch. Table 16.1 shows some terms and the errors accompanying them when the population mean was used as a guess.

TABLE 16.1 ERRORS OF GUESSES

GUESS	TERM	ERROR
5'8"	5'10"	+2
5'8"	5'7"	−1
5'8"	5'8"	0
5'8"	5'11"	3
5'8"	5'2"	−6
5'8"	6'1"	+5

Note that to find the error in each case the process was to take the original terms and subtract the population mean from them. In other words, from each term a constant was subtracted to find the corresponding error. This means that the variance of the original terms is identical to the variance of the errors made in predicting them. With reference to Table 16.1, the variance of the values in the third column is identical to the variance of those in the second column, a fact that the reader should verify. We may talk about either the population variance or the variance of the errors made when the mean is used as the predictive guess. To facilitate comparisons later on, we shall talk about the latter and call it the *error variance*. For instance, we are supposing that when the mean is used repeatedly to predict the heights of adult American males, the variance of the errors made is nine. We shall call this value, nine, the error variance. The error variance in the case described is the measure of the accuracy of predictions made when we have no information aside from knowing the mean of the variable of interest. It should be noted that the error variance is *always* the mean of the squared errors.

16.3 Predictor and Predicted Variables. Consider once more the variable, freshmen average in some large state university. We shall say that any student who graduates from an accredited high school in the particular state is allowed to enroll as a matriculated student in the state university. Furthermore, we shall say that freshmen averages, defined as averages obtained by students at the end of their first year, have a mean of 75 and a variance of 100. Note that we are not specifying anything about the distribution of these averages, so it may or may not be normal.

It follows that if a particular student were to come along, and we knew only that he was to begin at the university next September, our best guess would be that his average after one year would be 75. For each student in the same situation, our best procedure would be to make this prediction. Obviously we would make errors nearly every time, but in the long run at least they would balance out. The sum total of our errors in one direction would equal the sum total of our errors in the other. The variance of our errors would be 100.

We are going to consider now the simplest context in which we have information relevant to making predictions about individual students. Suppose we know the following facts to be true as a result of an extensive survey:

Those whose high school averages were below 80, considered as a group, received the poorest college freshman grades. We shall call them Group A. This collection of students got freshman grades that had a mean of only 70. Considered together, they got freshman grades that had a variance of 60.

Those whose high school grades were from 80 to 90 inclusive, considered as a group, got freshman grades that averaged 78. We shall say that these students compose Group B. Considered together, they got freshman grades with a variance of 60 also.

Those whose high school grades were over 90, considered as a group, got college freshman grades that averaged 85. We shall say that these brightest students compose Group C. Considered together, their grades as college freshmen also had a variance of 60.

We are going to assume that the facts given have been roughly true over a long time period with only trivial variations from one year to the next. Therefore, there is reason to believe that they will continue to hold, at least roughly, for next year, too. For instance, students who fall into Group A and who are about to begin at the state university next year may be expected to get freshman averages with a mean of about 70 and a variance of 60.

Where we are concerned with making predictions, we must consider the variable, high school average, as the predictor variable. A *predictor variable* is one that provides relevant information for predicting what scores will be on some other variable. The variable, college freshman average, is called our predicted variable. A *predicted variable* is one about which predictions are made. We have not yet discussed the exact method of making predictions, but it should be clear that exact knowledge of the relationship between the predictor variable and the predicted variable is necessary if we are to make the most accurate predictions possible.

As might be expected, when we do have an interest in making predictions, we must designate as our predictor variable the one that yields scores to which we may have access before we know anything about scores on the predicted variable. The distinction is usually chronological in that values of the predictor variable come into existence first. But there are exceptions, such as when we wish to infer from later events what happened before them. For instance, one might wish to make inferences about a person's childhood from his behavior, in which case scores relating to the individual's current behavior constitute values of the predictor variable and those relating to his childhood constitute values of the predicted variable.

Note that the grouping of students as described has been wholly according to their high school performances. For grouping to be meaningful, it must always be in terms of the predictor variable, as we shall see. Note also that those who compose any particular group in our example tend to perform more similarly as college freshmen than would a random sample of applicants. We know this from the fact that those composing any particular group, like Group A, get freshman grades with a variance of only 60, whereas applicants in general get grades with a variance of 100. This means, for instance, that members of Group A, considered exclusively, show less scatter in their freshman performances than do members of the larger population, considered altogether.

Note that in our illustration the students composing any one group on the predictor variable performed with the same variance of 60 on the predicted variable. When changes in category on the predictor variable do not affect the variance of the predicted variable, the discussion becomes greatly simplified, although the main points remain the same. This phenomenon of unchanging variance on the predicted variable is in fact often the case, and is called homogeneity of variance or *homoscedasticity*.

Now suppose that we are members of the board of admissions of the state university described and an applicant comes before us. For various purposes, such as advising him concerning his program for the following year, we would like to predict what his freshman average will be. If we know nothing of his high school average, our best guess—as we have seen—is that his freshman average will be 75, since this is the average grade achieved by freshmen considered as a group. Were we to make this same prediction for each applicant, our predictions would have an error variance of 100, which is the variance of freshmen's grades. In sum, our blind predictions that each student will receive a 75 average have an error variance of 100. But suppose instead that we have the information given concerning students' high school averages and their relationship to freshmen grades. We are going to consider how to make best use of this information, and we are also going to measure to what extent it is of advantage to us. To sum up what we know, it is that students with high school averages below 80 perform as freshmen with a mean of 70 and a variance of 60. Those with high school averages from 80 to 90 perform as college freshmen with a mean of 78 and a variance of 60, and finally those whose high school averages are over 90 perform as college freshmen with a mean of 85 and a variance of 60. For convenience we have described these students as composing Groups A, B, and C, respectively.

Our procedure must be first of all to determine for each student his high school average and to use this average to place him in one of the three groups. Then we will use the mean of his particular group as the prediction of what his freshman average will be. In other words, if he is a member of Group A, based on his high school average, then our prediction must be that his freshman average will be 70. If he is a member of Group C then our prediction must be that his college freshman average will be 85.

To illustrate: When the first high school graduate comes before us, we promptly look up his high school average and determine that it was 74. Clearly this applicant is a member of Group A. That is, he is a member of a group who perform as freshmen with a mean of 70, and therefore we use the value 70 as our prediction of what his freshman average will be. The next applicant, we shall say, has a high school average of 93 and thus is a member of Group C. He is in effect a member of a group who perform as college freshmen with an average of 85 and, therefore, we predict that his freshman average will be 85. One by one as the applicants come before us, our procedure in each case is first to place each applicant in his appropriate group and then to use the mean of that group as our prediction of what his freshman average will be.

Previously, when we used the mean of freshmen averages as our guess for each student, it turned out that the variance of our errors was 100. Now that we are using knowledge of the subject's high school performances to facilitate making predictions, it turns out that we are more accurate in the long run; or, to put it another way, the variance of our errors is less. Specifically, suppose that we have made a vast number of predictions using the new method described. To determine the variance of our errors, we must consider the applicants as falling into the three groups individually, though naturally they did not come in any specific order.

We made the prediction that all those who composed Group A would obtain a freshman average of 70. The variance of the errors made concerning members of this group was 60. We predicted that each member of Group B would get a 78, and here too the variance of the errors was 60. Our prediction for those in Group C was that each member would get an 85, and for this group also the variance of the errors made was 60. In other words, the variance of the errors made in each group was 60. In sum, the blind guesses led to an error variance of 100, whereas when high school averages were used the variance of the errors made was only 60. One might say that the information made it possible to reduce the variance of the errors to 3/5 of its original size.

We shall now consider one way of presenting findings graphically when each individual has received scores on each of two variables. We designate the predictor variable as the *X variable* and represent it along the horizontal axis, so that in our case the horizontal axis will be used to stand for high school averages, which we shall say may range from 65 to 100. We call the predicted variable the *Y variable* and shall represent it along the vertical axis. This variable, freshman averages, shall be said to have a possible range of from 50 to 100. (Note, however, that even when we are not concerned with the problem of prediction we may use the same mode of representation by arbitrarily choosing either variable as the *X* variable.)

In illustration, suppose we have obtained for a multitude of subjects their high school averages and also their college freshman averages. The first student attained a high school average of 70 and a college freshman average of 80. We shall use a minute square to represent his pair of scores in Fig. 16.1.

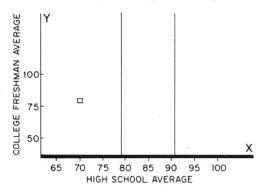

Fig. 16.1

With respect to the horizontal axis, representing high school averages, this square rests on the value 70. With respect to the vertical axis, representing college freshman grades, it is opposite the value of 80. Note also that it is in Group A, which includes all those high school averages that were between 65 and 80.

Let us say that the next student attained a high school average of 92 and a college freshman average of 78. And a third student attained a high school average of 80 and a freshman average of 60. The squares representing the scores of these two students appear in Fig. 16.2 along with the square already shown representing the scores of the first student.

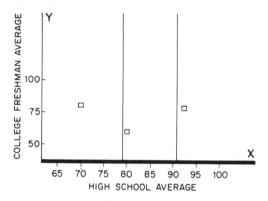

Fig. 16.2

It might turn out that a fourth student would have gotten the same pair of scores as did the first one, in which case the square representing his pair of scores should go right on top of the square used for the first student. In other words, there should actually be a height dimension indicating where more than one student got the same pair of scores. The squares representing the scores of some of the multitude of students might assemble themselves as in Fig. 16.3.

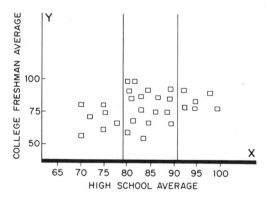

Fig. 16.3

There is only one trouble with the mode of representation using squares as in Fig. 16.3. Where two students perform similarly but not identically, the squares representing their scores must be at a minute distance from each other. In fact, no matter how small we were to make our squares, sooner or later we would find, since our variables are continuous, that we had insufficient room to place some pair of squares side by side. Thus from now on as we proceed, instead of squares we shall use points, because they have no actual dimension. Note that the location of each point indicates how a single student fared both in high school and in college. That is, each point, like the square that it is replacing, has a value on the horizontal axis and also on the vertical one. It remains true for Fig. 16.4 that, where one student performed identically with another, one point should be precisely on top of the other, and the illustration does not actually make a perfect representation of this phenomenon. But this cannot be helped, since no height dimension is possible here. In any event, for our purposes it will be satisfactory to use points, and where they cluster most thickly we can imagine that they would be most likely to coincide. Fig. 16.4 is called a *scattergram*, which is a graph showing the scatter of points, each of which indicates values on a pair of variables. It shows the scores of only a few of the students.

Our next step is to indicate the various means, or values, that we use as our predictions. To begin with, the mean of the freshmen averages taken all together is 75. In Fig. 16.5 this value is indicated by the horizontal double bar. Note that this double bar intersects the value 75 on the vertical (or Y) axis and moves horizontally across the whole graph.

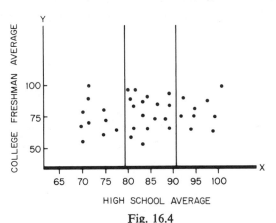

Fig. 16.4

The single horizontal line in each group is at the level to indicate the value of the mean performance of students in that group. For instance, considering Group A alone, the mean freshmen grade of the students composing this group was said to be 70. This mean is represented by the single horizontal line at the level 70, which appears only in Group A and not in the other groups. The mean of students in Group B so far as freshmen average is

245

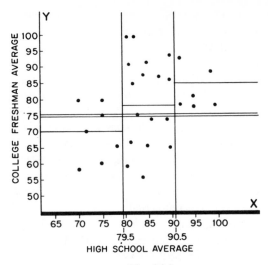

Fig. 16.5

concerned is 78, and this value is indicated by the bar at the appropriate level in Group B. Finally, members of Group C got freshmen averages that had a mean of 85, and this value is indicated by the bar at the appropriate level in Group C.

We are now ready to come back to the problem of prediction in the individual case. This time we shall make reference to Fig. 16.6. Consider first a student whose high school average is 85 and whose college average is 81. His pair of scores is indicated by the star in Fig. 16.6. We may consider what would have been the case had he merely walked into our admissions office, and without knowing anything about him we had attempted to predict what his freshman average was to be. As mentioned, our best guess would have been that his freshman average was to be 75. His actual freshman average of 81 would have been 6 points over our best guess. Our error is indicated by the broken line in Fig. 16.6, which is 6 units long.

On the other hand, had we known the facts relating high school averages to freshmen grades and had we known also that the particular student's

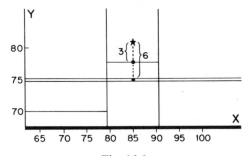

Fig. 16.6

high school average was 85, we would have proceeded otherwise.   First we would have properly considered him as a member of Group B, and then we would have predicted his college freshman average to be 78.   This time his actual average of 81 would have been only 3 points away from our predicted value.

In Fig. 16.6, the line connecting the actual value of 81 with our predicted value of 78 may be seen to be only 3 units long.   Here the actual freshman average of the student was such that there resulted an error of only 3 points when we made our enlightened prediction.

Next, Fig. 16.7 shows what would have occurred in two other cases.   The first one is that of a student whose high school average was 96 and whose college freshman average was 80.   His pair of scores is indicated by the star where the value on the horizontal axis is 96 and where the one on the vertical axis is 80.   Once more our blind prediction would have been 75, and this time our error, indicated by the broken line, would have been 5 units.   On the other hand, had we first thought of this student as a member of Group C, we would have predicted that his freshman average was to be 85.   Here too our error, indicated by the solid line, would have been 5 units, but in the other direction.

Finally, we may consider, also in Fig. 16.7, the case of a student whose high school average was 93 but whose college freshman average turned out to be only 74.   With no information we would have used the value of the double line, 75, as our prediction of his college freshman average, and as indicated in Fig. 16.7 our error would have been only one point.   On the other hand, the enlightened prediction we would have made, after properly locating the student in Group C, would have been the value at the level of the single line, and this value is 85.   In this case our prediction would have turned out 11 points too high.   In other words, this particular time our blind prediction would have been more accurate than the enlightened one.   As one might guess, one does better by making enlightened predictions, and this last case cited goes against the trend.

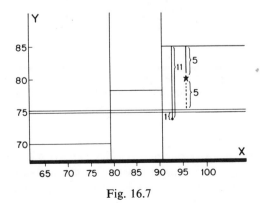

Fig. 16.7

247

From the illustration given, several facts should be clear; the first is that when we are without meaningful information, our prediction is that a student's freshman average is to be the value of the double line. Our error corresponds to the distance that the double line happens to be from the point representing the particular student's actual performance. When we have enough information to locate the student in one of the three groups, then our prediction is that his freshman average is to be the value of the single horizontal line representing the mean of that particular group. Our error in the particular case is the distance from that single line to the point representing the particular student's actual performance.

In sum, using the appropriate single bar for each subject in turn results in more accurate predictions in the long run than using the double bar (the population mean) repeatedly. Graphically, the best prediction of each student's freshman average is made by first placing the student in his proper group and then using the value of the single horizontal bar in that group as the prediction of his score. The single horizontal bar in any group comes closer to the points in that group than does the double bar. The three single bars, taken together, come closer to the total collection of points than does the double bar.

Now let us go back to a particular statement made earlier—that use of the population mean leads to an error variance of 100, whereas use of the group means leads to an error variance of only 60. Graphically, the first part of this statement means that using the double bar as the prediction throughout results in a mean squared error of 100. For example, in Fig. 16.8 where the double bar is used, the mean squared length of the error lines indicated is 100.

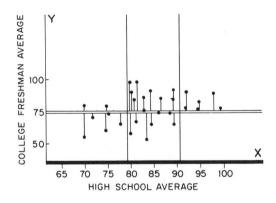

Fig. 16.8

The fact that using the group means as predictions led to an error variance of 60 means that using the single bars led to a mean squared error of 60. The error lines in Fig. 16.9 are such that their mean squared length is 60.

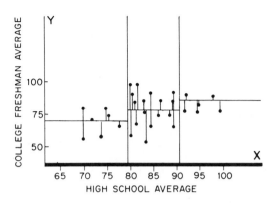

Fig. 16.9

16.4 Regression. The use of one variable to make predictions about another is called *regression*. More specifically, regression entails using known values of one variable to predict unknown values of the other. Here the known values actually used were the categories of the variable, high school average. Note also that using the mean of a group for predictive purposes meant that within any group the errors made averaged out to zero. Therefore, not only did the blind predictions lead to a mean error of zero, but also the regression predictions led to a mean error of zero. Note that in Fig. 16.5 as we move from left to right toward higher values of the variable, high school average, the mass of points taken together tends to rise. The fact that successive group means rise implies, intuitively speaking, that these lines are "going along" with the mass of points. Our information relating the two variables might have been more specific. Suppose for instance we had known the facts in Table 16.2.

TABLE 16.2   GROUPED DATA

GROUP	HIGH SCHOOL AVERAGE	COLLEGE MARKS
A	66–70	62
B	71–75	68
C	76–80	73
D	81–85	75
E	86–90	80
F	91–95	82
G	96–100	90

Then our predictions would have been better differentiated and, in the case described, would have resulted in a smaller error variance. The scattergram in Fig. 16.10 shows the predictions that would be made following Table 16.2. Remember that only a few of the scores are shown.

249

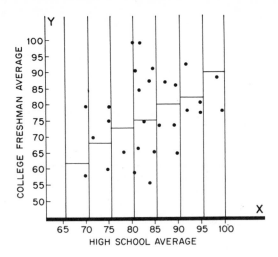

Fig. 16.10

Finally, suppose we were able to consider our predictor variable, high school averages, as actually continuous for purposes of prediction. That is, suppose we were able to go from each single value of our predictor variable to a specific prediction. Now the situation would be that of having a continuous curve to indicate which predictions we were to make. This is shown in Fig. 16.11.

This prediction curve is called a *regression curve*. Admittedly we never have enough information literally to draw a curve like it. No matter how many cases we collect, it would obviously always be possible to find values of the predictor variable which have not appeared even once. Actually we supply the continuity of the curve ourselves. In practice, we either draw the regression curve after first plotting the scattergram or we use mathematical methods to determine its shape and position. In chapter 18 we shall consider the construction of the regression curve in a particular case, but for now we shall assume that we have drawn this curve precisely so that we can examine some of its meanings.

To look back at Fig. 16.11, it should be clear that, for each value of the predictor variable, the regression curve tells what the prediction concerning the other variable must be. For instance, suppose a student arrives whose high school average was 82. The first step toward predicting what his college average as a freshman will be is to locate the value 82 on the horizontal axis in Fig. 16.11. Then we move straight up to the point where the regression curve is over the value 82. The height of the curve at this point, which is its value on the vertical axis, is 74. Thus, for the student whose high school average was 82, we predict that his freshman average at the state university will be 74. Our prediction is shown by the broken lines on

Fig. 16.11.

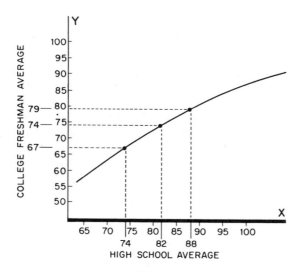

Fig. 16.11

The reader should verify using Fig. 16.11 that, for the student whose high school average was 74, the prediction is that his college freshman average will be 67. For the student whose high school average was 88, the prediction is that his college freshman average will be 79.

ind. var. *X*
It should be clear that a regression curve relates every value of the predictor variable to some value of the predicted variable. In other words, *Y -the dep. var.* for every single possible high school average, the curve identifies some prediction of a corresponding college freshman average. We might take any particular high school average and conceive of the multitude of students who attain that average over the long run. Then this particular collection of students would get some mean freshman grade at the state university. We may think of an infinitesimally narrow column directly over each point on the horizontal axis. The points in any given column represent performances of students with the particular high school average. As the curve passes through such a column, it takes on the value of the mean height of these points. In other words, the height of the curve as it passes through the particular column tells us what the students' mean college freshman grade would be.

In sum, the regression curve passes over the continuity of points on the horizontal axis. Its height at each point indicates what students with the particular high school average would get as their mean college freshman average. More generally, a regression curve indicates for each value on the X axis what the mean Y value is. Thus a prediction made using a regression curve is essentially the prediction of a mean. One begins with a particular X value and predicts for the individual who received it that his Y value will be the mean of all Y values obtained by those who got the same X value. When we used the regression curve in a previous illustration we were essentially saying: "John's high school average was 82. What is the mean college

251

freshman average of all those whose high school average is 82?" The regression curve by its height over the point 82 tells us that the answer is 74. Therefore we predict that John's freshman average will be 74.

We have considered the meaning of the regression curve and the procedure for making predictions using it. The subject of regression and its implications is quite broad and the purpose here is merely to introduce the reader to it. Having laid the groundwork, we shall next consider a very specific problem of regression—namely, that of correlation. Actually the correlation coefficient that we shall examine next is meaningful without making reference to the issue of regression. But as we use it we shall make reference to implications concerning regression, and they will be seen to throw much light on the whole issue of correlation.

PROBLEM SET A

The data in the table below are used in Problems 16.1 through 16.5.

Age of Child	Mean Mechanical-Aptitude Score
6	3
7	6
8	10
9	16
10	24

16.1 What is the predicted mechanical-aptitude score for a child who is (a) 7 years old? (b) 10 years old?

16.2 Suppose you tested four 8-year-old children and they received mechanical aptitude scores of 8, 10, 11, and 13. (a) What would you have predicted their scores to be before you tested them? (b) Compute the error variance if this value is used as a prediction of mechanical-aptitude scores for these four children.

16.3 Here are the actual test scores for some children in each age group.

6 years	7 years	8 years	9 years	10 years
1	3	7	13	19
2	5	10	15	21
4	7	11	17	25
5	8	12	18	26

Compute the error variance for each age group if you had used the mean mechanical-aptitude scores for prediction.

16.4 Draw the scattergram to represent the data in Problem 16.3. Using the mean mechanical-aptitude scores presented above Problem 16.1, draw the regression curve in the scattergram.

16.5 Using the regression curve in the scattergram, what is your estimated mechanical-aptitude score for a child who is (a) $6\frac{1}{2}$ years old? (b) $8\frac{1}{2}$ years old? (c) $9\frac{3}{4}$ years old?

The data in the table below are used in Problems 16.6 through 16.10.

Months of Typing Practice	Mean Number of Typing Errors
1	25
2	10
3	5
4	2

16.6 What is the predicted number of typing errors for an individual who has been practicing for (a) 1 month? (b) 3 months?

16.7 Suppose you tested six individuals who have had two months of typing practice and found that they made 6, 8, 10, 11, 13, and 15 errors. (a) What would you have predicted their error rates to be before you tested them? (b) Compute the error variance if this value is used as a prediction of the error rate for these six individuals.

16.8 Here are the actual error rates of some secretaries after various months of typing practice.

	Months of Practice		
1 month	*2 months*	*3 months*	*4 months*
30	5	2	0
27	8	5	1
23	12	8	3
15	16	11	5

Compute the error variance for each group if you had used the mean number of typing errors for prediction.

16.9 Draw the scattergram to represent the data in Problem 16.8. Using the mean number of typing errors presented above Problem 16.6, draw the regression curve in the scattergram.

16.10 Using the regression curve in the scattergram, what is your estimated typing-error rate for an individual who has been practicing for (a) $1\frac{1}{2}$ months? (b) $2\frac{1}{2}$ months? (c) $3\frac{1}{2}$ months?

PROBLEM SET B

The following mean grade-point averages according to selected entrance-examination scores are used in Problems 16.11 through 16.15.

Entrance-Examination Scores	Grade-Point Average
10	2.45
20	2.75
30	3.05
40	3.15

16.11 What is the predicted grade-point average for a student who has an entrance-examination score of (a) 10? (b) 30? (c) 40?

253

16.12 Suppose you selected four students who each have an entrance-examination score of 30 and found, from their transcripts, that their grade-point averages were 2.95, 2.00, 3.05, and 3.07. What would you have predicted their grade-point averages to be before you looked at their transcripts? (b) Compute the error variance if this value is used as a prediction of their grade-point averages.

16.13 Here are the actual grade-point averages for four groups of students according to their entrance-examination scores.

	Entrance-Examination Score			
	10	20	30	40
Grade-Point	2.21	2.70	2.97	3.11
Averages	2.37	2.74	3.00	3.15
	2.49	2.80	3.02	3.17
	2.51	2.82	3.04	3.18

Compute the error variance for each group, using the mean grade-point averages as presented above Problem 16.11.

16.14 Draw a scattergram to represent the data in Problem 16.13. Using the mean grade-point averages, draw the regression curve in the scattergram.

16.15 Using the regression curve in the scattergram, what is your estimate of grade-point average for a student with an entrance-examination score of (a) 15? (b) 28? (c) 42?

The data in the table below are used in Problems 16.16 through 16.20.

Age (in months)	Mean Reaction Time (in seconds)
1	1.5
2	0.8
3	0.5
4	0.4

16.16 What is the predicted reaction time of an infant who is (a) 2 months old? (b) 4 months old?

16.17 Suppose you measured the reaction time of five 3-month-old infants and determined their reaction times to be 0.2, 0.4, 0.5, 0.7, and 0.9 seconds. (a) What would you have predicted their reaction times to be before you measured them? (b) Compute the error variance if this value is used as a prediction of the reaction times of these five infants.

16.18 Here are the actual reaction times of four groups of infants.

	Age (in months)			
	1	2	3	4
	1.2	0.5	0.3	0.2
	1.5	0.9	0.5	0.3
	1.7	1.0	0.7	0.6
	2.0	1.3	0.9	0.8

Compute the error variance for each age group if you had used the mean reaction time for prediction.

16.19 Draw the scattergram to represent the data in Problem 16.18. Using the mean reaction times presented above Problem 16.16, draw the regression curve in the scattergram.

16.20 Using the regression curve in the scattergram, what is your estimated reaction time of an infant who is (a) $1\frac{1}{2}$ months old? (b) $2\frac{1}{2}$ months old? (c) $3\frac{1}{2}$ months old?

seventeen

correlation

17.1   The Concept of Correlation.   One particular type of relationship, called correlation, is of major interest to investigators.   Two variables may show a direct or positive correlation, or no correlation, or finally a negative correlation, as we shall see.   We shall illustrate correlation in its various forms before actually defining it; otherwise the formal definition will seem unnecessarily abstract and forbidding.   As we have been doing, we shall use simplified illustrations to make clear the basic notion.   In particular we shall consider populations made up of only five cases each.

Suppose the heights and weights of a population (A) of five men are as follows.   The shortest man is 5 feet 5 inches tall and weighs 140 pounds.   The next is 5 feet 6 inches tall and is five pounds heavier, and so on, as indicated in Table 17.1.

TABLE 17.1   POPULATION A

1	5′ 5″	140
2	5′ 6″	145
3	5′ 7″	150
4	5′ 8″	155
5	5′ 10″	165

The first thing we note is that for the five men there is a direct or positive relationship between their heights and weights.   As the heights increase, the weights increase correspondingly.   The shortest man is the lightest; the one following him is the next heavier, and so on.

But now let us look at the relationship more specifically.   Note that, as we begin to move downward in Table 17.1, each inch in height corresponds to exactly five pounds in weight. For instance, as we go from the first person to the second, the increase in height is one inch and the corresponding increase in weight is five pounds. The same is true as we go from the second person, who is 5 feet 6 inches, to the third, who is 5 feet 7 inches. Once again the increase of a single inch corresponds to a five-pound weight increase. Finally, even as we go from the fourth person to the fifth, the same correspondence exists. The increase of two inches corresponds to an increase of ten pounds, which is the same as saying that each inch in height corresponds to five pounds in weight.

In sum, not only is there a perfect positive relationship between height and weight in the population described, but the relationship is one of a very special kind. It is characterized by the fact that the correspondence between one inch and five pounds is uniform throughout. The notion of correlation refers not merely to a relationship, but more particularly to one in which each unit change on one variable corresponds to a designated change in the other, and the correspondence is uniform.

So far we have been talking about the case of perfect or exact correlation. Naturally one does not expect to find an exact correspondence, such as the one described, that will hold for all the cases in a population. Nearly always we find ourselves considering situations where some relationship exists and where there is some degree of correlation as we shall define it, but in these cases the correlation is far from perfect, as it was in the case described.

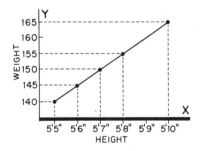

Fig. 17.1

The scattergram for the five cases given in Table 17.1 appears in Fig. 17.1. All five points lie on the same straight line, and this line has actually been drawn in the illustration. Correlation is often described as linear relationship because it has the geometric interpretation that points representing cases lie on the same straight line. The degree to which two variables correlate is the degree to which a single straight line comes close to (or contains) the points representing cases in the population. (Only when there is perfect correlation does a straight line actually contain all the points.) To see why, let us consider the case of positive correlation in particular. Here the 257 increase of a given amount on one variable always implies some specified increase on the other. Graphically, for the variable on the horizontal axis, X each unit increase means a shift of a particular distance to the right. Where there is perfect positive correlation, for each shift of this amount there must be a corresponding upward movement of some other particular amount.

We may start out with a line segment of any specified length (say length A) to represent a one-unit shift on the variable marked on the horizontal axis. We may pick any length (say length B) for our second line segment, which is to represent the corresponding change on the other variable. For instance, the two line segments might be those in Fig. 17.2.

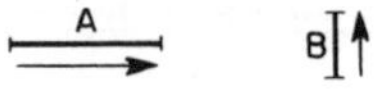

$$\xrightarrow{\quad A \quad} \qquad B\uparrow$$

Fig. 17.2

Now, beginning with our axes, suppose we make sure that for every horizontal shift of length A we make a vertical shift of length B. In other words, the only way we can move to the right a length A is to go up a length B. Following this rule, we would find that we always ended on the same straight line (see Fig. 17.3). Perfect positive correlation implies that for

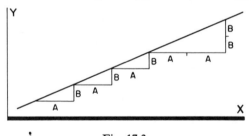

Fig. 17.3

every increase of length A on one variable, there must be a corresponding increase of length B on the other. Note that an increase of length 2A implies a corresponding increase of 2B. Similarly, an increase of A/8 would imply a corresponding increase of B/8.

To further highlight the notion of positive correlation, let us look at Population B, where the heights and weights of five men are as shown in Table 17.2. Here there is a positive relationship in the sense that the shortest man is the lightest, the next to shortest is the next to lightest, and so on. However, this time each increase in height does not have an identical meaning

TABLE 17.2 POPULATION B

1	5′ 5″	140
2	5′ 6″	145
3	5′ 7″	150
4	5′ 8″	170
5	5′ 10″	220

in terms of increase in weight. For the first three men, there exists a perfect positive correlation; or to put it another way, the relationship is positive and linear. As before, each increase of one inch corresponds to an increase of five pounds. But as we go from the third person to the fourth, the increase of one inch corresponds to an increase of 20 pounds. And as we go from the fourth person to the fifth, the increase of two inches corresponds to an increase of 50 pounds, so that now it is one inch per 25 pounds. The scatter-gram for the members of Population B appears in Fig. 17.4, and we can see that where the correspondence breaks down in the last two cases, the relationship between height and weight is no longer linear. Incidentally, the straight line coming closest to the five points has been drawn in Fig. 17.4. It no

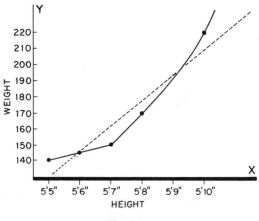

Fig. 17.4

longer contains them, but still we shall have use for it later when we consider it as the regression line.

Next we shall consider the case of a perfect negative correlation. This time each successive increase on one variable corresponds to some uniform decrease on the other. We shall say that each of five students was given first an adjustment scale and then a test to measure the degree of rage reaction that appeared when the student was thwarted in carrying out a defined task in front of the examiner. On the adjustment scale each student received a score indicating the adequacy of his personal adjustment. Each student's score on the frustration scale indicates his readiness to respond with an infantile rage reaction when thwarted. The data for Population C (Table 17.3) are

TABLE 17.3 POPULATION C

	ADJUSTMENT SCORE	RAGE REACTION SCORE
1	38	18
2	42	17
3	50	15
4	54	14
5	74	9

259

contrived, since in practice a perfect correlation of any kind would scarcely ever appear, but the data do illustrate what a perfect negative correlation would look like.

We may note first that there is a negative relationship between the two variables, which means that the higher the individual's adjustment score is, the lower his rage-reaction score becomes. The first person, whose adjustment score is 38, has the highest rage-reaction score. Each succeeding person has a somewhat higher adjustment score than his predecessor and also a somewhat lower rage-reaction score. Finally the person with the highest adjustment score, the fifth person, is the one with the lowest rage-reaction score.

So far we have described what might be called a negative or inverse relationship. However, it may also be seen that this negative relationship is uniform and linear in the sense already described. For each four-point increase in adjustment score, there is a one-point decrease in rage reaction score. As we go from the first person to the second, the increase in adjustment score is exactly four points and the decrease on the rage reaction variable is exactly one point. As we go from the second person to the third, the adjustment score increase is eight points and there is a corresponding drop of two points on the rage reaction score. Next, as we go from the third person to the fourth, the increase of four points on adjustment once again corresponds to a one point decrease on the rage reaction scale. Finally as we go from the fourth person to the fifth, the increase of twenty points on the adjustment score corresponds to a decrease of five points on the rage-reaction scale. Once again each four-point increase on the adjustment scale corresponds precisely to a drop of a single point on the rage-reaction scale.

In sum, the data are such that for every unit increase on the adjustment variable there is a specific decrease on the rage-reaction variable. Two variables that enter a relationship of this kind are described as being *perfectly negatively correlated*. Two variables are perfectly negatively correlated when for each unit increase on one of them there is a corresponding decrease of a specified amount on the other.

The data of Table 17.3 appear in the form of the scattergram in Fig. 17.5. Since there is a perfect negative correlation between the two variables, the points that represent the five cases lie on the same straight line just as they did when there was a perfect positive correlation between the two variables. This time, each increase (or shift to the right) on the adjustment variable is coupled with a decrease or drop on the rage-reaction variable. Thus, as the straight line moves to the right, it moves downward. The reader should compare the scattergram of Fig. 17.1, where there was perfect positive correlation, with that of Fig. 17.5, which illustrates perfect negative correlation. In each instance the points all lie on the same straight line, but in Fig. 17.1 as the line moved to the right it moved upward, whereas now as the line moves to the right it moves consistently downward.

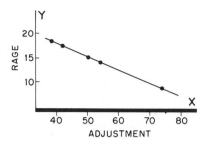

Fig. 17.5

We have already described the case of perfect positive correlation as an ideal and we have noted that we usually come upon instances where merely some degree of positive correlation exists. Our concern is to measure the degree of positive correlation that does exist, and we considered the most extreme case simply to fix the idea of positive correlation. The same has been true regarding this presentation of negative correlation and what we actually find in practice. In practice, we come upon instances where two variables are negatively correlated and our task is to measure the degree of correlation. The most extreme case—namely, that in which there is a perfect negative correlation—was presented merely to communicate the idea. However, we shall refer to these most extreme cases again to make clear some of the basic concepts.

Finally we shall consider the case of no correlation between two variables. Specifically, suppose the heights and IQ scores of ten people composing Population D are those listed in Table 17.4. In this table we see that, as the

TABLE 17.4 POPULATION D

HEIGHT	IQ
5' 5"	107
5' 5"	99
5' 6"	114
5' 7"	94
5' 8"	98
5' 8"	106
5' 9"	94
5' 10"	114
5' 11"	108
5' 11"	98

heights increase, the IQ scores do not show an apparent change in either direction. There is in fact no correlation between the variables, which means that as one variable increases, the other variable does not show even the slightest over-all tendency to increase or decrease uniformly. The scattergram for this case of no correlation between the variables, height and IQ, is shown in Fig. 17.6.

261

This time there is no correlation.   As we move to the right on the variable, height, there is no systematic upward or downward movement on the variable, IQ.   We have drawn in the straight line that comes closest to all of the points considered together, and it turns out that this line is horizontal.   We will see that this fact is important later on—that where two variables do not correlate, the straight line that is "best fitting" in the scattergram representing the case is perfectly horizontal.

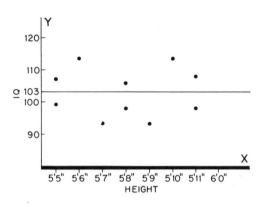

Fig. 17.6

The two variables, height and IQ, as found in Population D in Table 17.4, show no relationship of any kind.   We noted that in particular they do not correlate.   However, of all possible relationships, the linear relationship, called correlation, is only one type.   Thus it is possible for two variables to enter a complex relationship and for there still to be absolutely no correlation between them.

The data to be discussed are contrived once more for the sake of making the clearest possible illustration.   Our purpose is to illustrate how two variables may enter a complex relationship that is of such a nature that the correlation between them remains nonexistent.   Ten subjects who were members of a closely knit group were given a well-known anxiety scale and each received a score indicating what is called his chronic anxiety level.   A low score on this scale meant that the subject had a relatively low chronic anxiety level, and a high score meant that he had a high one.   The subjects were also each given a differentiation scale, which measured for each one of them the degree to which he was able to see personality differences among his fellow group members.   A relatively low score on this differentiation scale meant that a subject tended not to differentiate among his fellow group members; or, to put it another way, he tended to see them all as relatively similar.   A relatively high score meant that the subject tended to differentiate—that is, tended to perceive differences among his fellow group members.   The data for Population E appear in Table 17.5.

Note in Table 17.5 that the subjects are listed in order of their anxiety scores. The first subject is the one with the lowest anxiety score, the next subject is next lowest, and so on. Successive subjects after the first one show progressively increasing anxiety scores. As we go down the list, the differentiation scores start at their lowest and increase progressively up to the fifth subject. In other words, for the first five subjects there is a positive relationship between anxiety and differentiation. But now as we go from the fifth subject to the sixth, we find that our subjects no longer continue to increase in their ability to differentiate. Now as the anxiety scores continue to increase, the corresponding differentiation scores cease to increase and

TABLE 17.5 POPULATION E

	ANXIETY SCORES	DIFFERENTIATION SCORES
1	20	1
2	24	2
3	26	4
4	30	6
5	35	8
6	40	8
7	45	6
8	49	4
9	51	2
10	55	1

instead they actually begin progressively to decrease. The scattergram showing the data of Table 17.5 appears in Fig. 17.7.

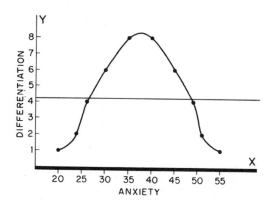

Fig. 17.7

In Fig. 17.7 the points indicating the ten cases have been joined by a smooth continuous curve. It may clearly be seen how, as we move to subjects with increasing anxiety, their differentiation scores first increase and then begin to decrease. Because the curve drawn in Fig. 17.7 depicts the

263

relationship so well, this type of relationship is often called a *curvilinear* one. We have also drawn the straight line that comes closest to all the points; *note that it is horizontal.* An analysis indicates that for the first five cases there is a positive correlation and for the last five there is a negative correlation. The two variables are specifically related, but their over-all relationship can by no means be defined as a correlation. We cannot say that as anxiety increases there is any kind of uniform change on the variable, differentiation. Loosely speaking, if we did we would necessarily be as wrong on some of the cases as we would be right on others.

We have presented two illustrations of how two continuous variables may be uncorrelated. In one case they may have no relationship whatsoever. In the other case, they may enter a very specific relationship, but we may not be able to conceptualize it when we restrict ourselves to thinking of a linear relationship. What has been said has been aimed at highlighting the fact that a correlation is a very specific kind of relationship between two variables —namely, a linear relationship.

We may now sum up what has been said with a formal definition.

Definition 17.1 *Two variables are said to correlate positively when, as one variable increases, the other shows some trend to increase correspondingly in a uniform way. Two variables are said not to correlate at all when, as one of them changes, the other shows absolutely no over-all trend to change in a uniform way with respect to it. Two variables are said to correlate negatively when, as one variable increases, the other shows some trend to decrease correspondingly in a uniform way.*

Whenever we have a population of individuals scored on each of two variables, the case of perfect positive correlation should be viewed as one extreme; the case of no correlation is a central point; and the case of perfect negative correlation should be viewed as the other extreme. If we think of a line representing possible correlations, as shown in Fig. 17.8, then we must conclude that the relationship between any two variables must be some point on that line.

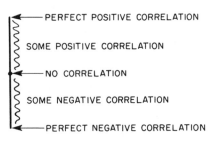

PERFECT POSITIVE CORRELATION

SOME POSITIVE CORRELATION

NO CORRELATION

SOME NEGATIVE CORRELATION

PERFECT NEGATIVE CORRELATION

Fig. 17.8

Remember that when we discuss correlation we are always talking about a trend that exists and not a necessary or invariable relationship. Also remember that the fact that two variables correlate does not at all imply that either variable causes the other to change. In any given case, the relationship may be coincidental or the consequence of some external cause that has not even been discovered.

Correlation refers to the degree to which two variables move uniformly with respect to each other, and it follows that two variables may show some correlation even when their precise relationship is not perfectly linear. That is, the conceptualization of them as moving together in a linear way may be informative even where it is not the best possible conceptualization of how they relate. To put it another way, the straight line describing their relationship may be informative but still only a meaningful approximation of some curve that would better describe it. Thus, where a correlation between two variables is found, we must remember that this correlation may not provide the ultimate statement of how the variables relate.

17.2 Correlation Construed as the Relationship between Two Sets of z Scores. The individuals composing a population are apt to be scored on each of two variables where the units of measurement are quite different. For instance, as we have seen, one variable may be height in inches and the other may be weight in pounds. The units of measurement of the two variables have absolutely nothing to do with whether or not the two variables correlate. In other words, the means and standard deviations of the variables may be entirely different, and still the variables may correlate moderately or perfectly in either direction. The point is that we may write down the scores of the individuals in ascending order on one variable and, once we have done so, the issue is whether the terms follow a linear order in one direction or the other on the other variable.

To illustrate, suppose each of nine students takes an IQ test and also another test, which we shall simply call Test A. On the IQ test the mean of their scores is 100 and the standard deviation is 5.16. On Test A the mean is 54 and the standard deviation is 12.9. For our first illustration, suppose that the IQ scores and those on Test A show a perfect positive correlation. The scores for the nine students are given in Table 17.6A.

TABLE 17.6A   IQ SCORES AND TEST A SCORES
SHOWING A PERFECT POSITIVE CORRELATION

STUDENT	IQ SCORE	TEST A SCORE
1	92	34
2	94	39
3	96	44
4	98	49
5	100	54
6	102	59
7	104	64
8	106	69
9	108	74

265

Examination of the data of Table 17.6A reveals that for each two-point increase in IQ there occurs among the students exactly a five-point increase in their scores on Test A. The means and standard deviations on the two variables are different, but their values have no bearing on the perfect positive correlation that exists between the two variables. There could have existed a perfect negative correlation between the variables, still assuming that both the IQ scores and the scores on the other variable had the same means and standard deviations as in Table 17.6A.

Suppose that the IQ scores of the nine students had been exactly the same but the order of their scores on the other variable had been exactly reversed. Then the correlation between the scores on the two tests would have been a perfect negative one. In Table 17.6B the IQ scores of the same nine students are shown along with their scores on a fictitious Test B, which has the same mean and standard deviation as Test A. This time there is a perfect negative correlation as the reader may verify.

TABLE 17.6B IQ SCORES AND TEST B SCORES
SHOWING A PERFECT NEGATIVE CORRELATION

STUDENT	IQ SCORE	TEST B SCORE
1	92	74
2	94	69
3	96	64
4	98	59
5	100	54
6	102	49
7	104	44
8	106	39
9	108	34

In Table 17.6C the same IQ scores are shown in correspondence with scores on a fictitious Test C. This time the IQ scores show no correlation with the test scores. Note that the mean and standard deviation of the scores on Test C are exactly the same as they were on Tests A and B.

TABLE 17.6C IQ SCORES AND TEST C SCORES
SHOWING NO CORRELATION

STUDENT	IQ SCORE	TEST C SCORE
1	92	69
2	94	34
3	96	49
4	98	64
5	100	54
6	102	44
7	104	59
8	106	74
9	108	39

In our illustrations, IQ shows a perfect positive correlation with Test A, a perfect negative correlation with Test B, and no correlation with Test C. These facts are true even though the means and standard deviations of the three tests were identical. The point is that the fact of correlation relates specifically to how two variables go together and is independent of the values of the mean and standard deviation of the variables in question. It follows that we preserve whatever correlation exists between two variables when we change the terms in each distribution to z scores. The z scores of the terms in a distribution are in a sense simplified versions of the original terms and do not reflect the size of the mean or standard deviation of the original terms.

Table 17.7A shows the z score correspondence indicating the perfect positive correlation between IQ scores and those on Test A. Both the IQ scores and the Test A scores have been changed to z scores, and these appear in Columns (2) and (3) of Table 17.7A. In essence the scores that appeared in Table 17.6A have been replaced by z scores that appear in Table 17.7A. The same has been done in Table 17.7B for the IQ and Test B scores. The z score relationship, indicating the perfect negative correlation between IQ and Test B scores, may be seen in this table. Finally the z score correspondence between IQ and Test C scores may be seen in Table 17.7C. In each case the z scores have been rounded off. Thus the value 8.98 is an approximation to 9.00, the value that would have been found were it not for rounding errors.

The substitution of z score values for the original terms in a distribution actually moves us forward so far as the issue of correlation is concerned. In fact we shall make reference to the z score data of Tables 17.7A, 17.7B, and 17.7C now as we proceed to discuss the classic indicator of correlation, which is called the *correlation coefficient*.

17.3 The Correlation Coefficient. The correlation coefficient (often called the *Pearson product-moment correlation coefficient*), signified by the

TABLE 17.7A z SCORES OF IQ AND TEST A
SHOWING PERFECT POSITIVE CORRELATION

(1) STUDENT	(2) IQ	(3) TEST A	(4) PRODUCT
1	-1.5	-1.5	2.25
2	-1.2	-1.2	1.44
3	$-.8$	$-.8$	.64
4	$-.4$	$-.4$	.16
5	.0	.0	.00
6	$+.4$	$+.4$	.16
7	$+.8$	$+.8$	.64
8	$+1.2$	$+1.2$	1.44
9	$+1.5$	$+1.5$	2.25

$$\Sigma = 8.98$$

267

TABLE 17.7B   z SCORES OF IQ AND TEST B
SHOWING PERFECT NEGATIVE CORRELATION

(1) STUDENT	(2) IQ	(3) TEST B	(4) PRODUCT
1	−1.5	+1.5	−2.25
2	−1.2	+1.2	−1.44
3	−.8	+.8	−.64
4	−.4	+.4	−.16
5	.0	.0	.0
6	+.4	−.4	−.16
7	+.8	−.8	−.64
8	+1.2	−1.2	−1.44
9	+1.5	−1.5	−2.25
			$\Sigma = -8.98$

TABLE 17.7C   z SCORES OF IQ AND TEST C
SHOWING NO CORRELATION

(1) STUDENT	(2) IQ	(3) TEST C	(4) PRODUCT
1	−1.5	+1.2	−1.80
2	−1.2	−1.5	1.80
3	−.8	−.4	.32
4	−.4	+.8	−.32
5	.0	.0	.00
6	+.4	−.8	−.32
7	+.8	+.4	+.32
8	+1.2	+1.5	+1.80
9	+1.5	−1.2	−1.80
			$\Sigma = 0.00$

letter r, is a precise measure of the way in which two variables correlate. Its value is such as to indicate both the direction (positive or negative) and the strength of the correlation between two variables. We are now going to discuss the correlation coefficient in terms of z score data, since it may be best understood when considering such data. However, as we shall see later, we do not actually need to substitute z score values for original terms in order to compute a correlation coefficient.

To begin with, an examination of Table 17.7A shows that corresponding to each negative z score on IQ is a negative z score on Test A. Corresponding to each positive z score on IQ is a positive z score on Test A. Quite naturally, this type of correspondence reflects a positive correlation between two variables. It means that those who are below the mean on one variable are also below the mean on the other, and that those who are above the mean on the first variable are also above it on the second one.

Suppose now that each individual's pair of z scores are multiplied together, giving what we shall call a *cross product* in each case. These are shown in Column (4) of Table 17.7A. Remember that the product of two negative numbers is positive, as is the product of two positive ones. Thus, in each case, the perfect positive correlation reflects itself in a positive or zero cross product. In Column (4) the nine cross products are indicated. They have been added and the mean found. The mean of these cross products is $+1.0$. (Were it not for rounding errors, the sum of the products would be 9.00 instead of 8.98. And $9.00/9 = 1.0$.)

It is the mean cross product of the z scores that is the correlation coefficient. Its sign, which was positive in the case just described, reflects the fact that the correlation was positive. Its size indicates the strength of the correlation in that the stronger the correlation is, the further away from zero is the value of the correlation coefficient. It may be shown mathematically that the maximum value possible for the correlation coefficient is $+1.0$, and that it takes this value only in the instance of a perfect positive correlation as in the case described. But let us go on to compute the value of the correlation coefficient in other cases.

Table 17.7B presents the z scores for IQ and Test B where the two variables had a perfect negative correlation. There, negative z scores on IQ went with positive ones on Test B, and positive z scores on IQ went with negative ones on Test B. Since the product of a positive and negative number is negative, the cross products in this case all turn out to be negative or zero. Therefore the mean of these cross products—which is the correlation coefficient—is also negative, reflecting the perfect negative correlation. These cross products are shown in Column (4) of Table 17.7B, and their mean is -1.0. The negative sign of the correlation coefficient indicates that the correlation is negative. The obtained correlation coefficient of -1.0 is the most extreme possible negative value that a correlation coefficient can take. It reflects the perfect negative correlation between the two variables.

Finally, where there is no correspondence, as between IQ and Test C scores, the cross products are in some cases positive and in others negative, and they tend to balance each other out. The illustration in Table 17.7C was contrived to show the situation of exactly no correlation between two variables, and thus it turns out that the correlation coefficient is exactly zero. The computation of this correlation coefficient is shown in Column (4) of Table 17.7C. A correlation coefficient of exactly zero reflects the fact that there is absolutely no correlation between two variables. Incidentally, as one might guess, even when two variables would be absolutely uncorrelated in the long run, there typically turns out to be some small correlation between them in a given sample. This correlation is as likely to be positive as negative. Just as even a perfectly fair coin would be unlikely to show exactly 500 heads in 1,000 flips, a correlation of exactly zero as in this illustration would be extremely unlikely.

It is meaningful that any departure from perfect correlation tends to reduce the size of the correlation coefficient, even where plus and minus signs are not involved. For instance, Table 17.7A shows the z scores and cross products where there is a perfect positive correlation. We may pretend that the z scores of the first two students on Test A are interchanged, in which case the table of z scores for IQ and Test A would appear as in Table 17.8. Note that the single interchange reduces the sum of the cross products. Hence this single lack of perfect correlation quite properly reflects itself by reducing the correlation coefficient. It is, in fact, true that any departure from perfect correlation, positive or negative, results in moving the correlation coefficient closer to zero, since only when the most extreme z scores on both variables are coupled does it turn out that the cross products add up to a maximum. Thus the correlation coefficient is precisely sensitive to the direction and degree of correlation that exists in each given case.

TABLE 17.8 ALTERED z-SCORE DISTRIBUTION FOR IQ AND TEST A

STUDENT	IQ	TEST A	PRODUCT
1	-1.5	-1.2	1.80
2	-1.2	-1.5	1.80
3	$-.8$	$-.8$	.64
4	$-.4$	$-.4$	.16
5	.0	.0	.00
6	.4	.4	.16
7	.8	.8	.64
8	1.2	1.2	1.44
9	1.5	1.5	2.25
			$\Sigma = 8.89$

Definition 17.2 *The correlation coefficient is the mean of the cross products of the z scores of two variables.*

$$r = \frac{\Sigma z_1 z_2}{N}$$

We shall see that one need not go through computing z scores to obtain the value of the correlation coefficient. The correlation coefficient is a measure of the degree to which two variables correlate. The sign of the correlation coefficient indicates whether the correlation is negative or positive. The correlation coefficient always takes some value between -1 (when there is a perfect negative correlation) and $+1$ (when there is a perfect positive correlation). It is zero when there is absolutely no correlation between two variables; and it departs from zero in one direction or the other depending upon the strength of the relationship between the two variables.

The vertical line in Fig. 17.9 indicates the range of possible values of the correlation coefficient. The different segments of the line are labeled as they were in Fig. 17.8.

17.4 Computing the Correlation Coefficient. Up to now we have substituted for the terms on each variable the appropriate z scores in order to get the "feel" of how each pair of variables went together. By translating the original terms in each case to *z* scores we were able to communicate the meaning of the correlation coefficient most easily. However, as mentioned, it is not necessary in practice to make the translation in each case to *z* scores. One can operate upon the original scores in such a way as to compute from them the correlation coefficient; and quite naturally, since the substitution of

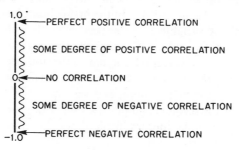

Fig. 17.9

z score values for the original terms is laborious, it is not usually done in practice.

We are going to illustrate now the usual method for computing the correlation coefficient. This method entails working with the original data and applying a formula that gives exactly the same result as if the data had first been put into *z* score form and then the average *z* score cross product had been found. In other words, just as we computed the value of *r* before, we are going to do so once more. Only this time we are going to work from original data, or what is sometimes called "raw data," and we shall not trouble ourselves to translate the data to *z* score form.

Let us say that we wish to compute the correlation coefficient and that our two variables are IQ scores and scores on a musical aptitude test. Specifically, 20 students have each been given both tests and their scores are shown in Table 17.9. From now on we shall not bother to number the students. Nor shall we arrange them in any particular order since it is not necessary to do so.

TABLE 17.9 IQ AND MUSICAL APTITUDE SCORES

IQ SCORE	MUSICAL APTITUDE TEST SCORE	IQ SCORE	MUSICAL APTITUDE TEST SCORE
95	30	102	30
106	37	105	30
110	35	111	32
104	28	106	33
98	17	110	38
114	45	143	42
123	33	106	40
100	18	98	20
98	24	104	23
88	18	97	25

In order to use the formula for the correlation coefficient, which will be introduced soon, we must make various separate computations. Our first step is to choose one of the sets of scores arbitrarily and call it the X variable. We shall call IQ scores values of the X variable and musical aptitude scores values of the Y variable.

To compute the correlation coefficient, we make a table of preparatory computations (see Table 17.10). In the first column under the heading X go the values of the X variable. More specifically, the values of the X variable in our present problem are the IQ scores and they are listed in the first column of Table 17.10. In the second column under the heading X^2 appear the squares of each of these scores. For instance, the first subject's X score is 95, which appears in the first column, and the square of this X score is 9025, which appears in the second column. ($95^2 = 9025$). Similarly, each subject's score on the X variable appears in the first column and the square of this score appears in the second one.

Each subject's score on the Y variable—the musical aptitude test—appears in Column (3). The square of each subject's score on the Y variable appears in the fourth column under the heading Y^2. For instance, the first subject's score on the musical aptitude test is 30, which appears in the third column, and the square of this score is 900, which appears in the fourth column.

TABLE 17.10   COMPUTATIONS FOR THE
CORRELATION COEFFICIENT WHERE IQ SCORES
ARE THE X VARIABLE AND MUSICAL APTITUDE
SCORES ARE THE Y VARIABLE

(1) X	(2) X^2	(3) Y	(4) Y^2	(5) XY
95	9025	30	900	2850
106	11236	37	1369	3922
110	12100	35	1225	3850
104	10816	28	784	2912
98	9604	17	289	1666
114	12996	45	2025	5130
123	15129	33	1089	4059
100	10000	18	324	1800
98	9604	24	576	2352
88	7744	18	324	1584
102	10404	30	900	3060
105	11025	30	900	3150
111	12321	32	1024	3552
106	11236	33	1089	3498
110	12100	38	1444	4180
143	20449	42	1764	6006
106	11236	40	1600	4240
98	9604	20	400	1960
104	10816	23	529	2392
97	9409	25	625	2425
$\sum X = 2118$	$\sum X^2 = 226854$	$\sum Y = 598$	$\sum Y^2 = 19180$	$\sum XY = 64588$

Each subject's score on the X variable has been multiplied by his score on the Y variable to yield the value for him in the fifth column. For instance, the first subject got an X value of 95 and a Y value of 30; the product of these two numbers, which is 2850, appears in the fifth column. The second subject's X score of 106 has been multiplied by his Y score of 37 to yield the value 3922, which appears in the fifth column. Note that the heading of the fifth column is XY, indicating that each value in this column is a cross product for a single subject.

Beneath each column at the bottom of Table 17.10 appears the total for that column. Remember that the symbol $\sum$ stands for "the sum of." The sum of the 20 X values, indicated as $\sum X$, is 2118. The sum of the 20 X^2 values, indicated as $\sum X^2$, is 226854. The sum of the 20 Y values, indicated as $\sum Y$, is 598. The sum of the 20 Y^2 values, indicated as $\sum Y^2$, is 19180. Finally, the sum of the 20 XY values, indicated as $\sum XY$, is 64588. The value of N, the number of individual scores, is of course 20.

Actually, it was to obtain the totals that we went through the entire procedure, for these totals are the important "pieces" that we use when we compute the correlation coefficient.

Now we are ready to put down the formula for working from original data and computing the correlation coefficient. In Formula (17.1) the correlation coefficient is indicated by the letter r. This is sometimes called the Pearson Product-Moment Correlation Coefficient.

$$(17.1) \qquad r = \frac{N \sum XY - (\sum X)(\sum Y)}{\sqrt{N \sum X^2 - (\sum X)^2} \; \sqrt{N \sum Y^2 - (\sum Y)^2}}$$

To solve Formula (17.1) we must insert six "pieces of information" in their appropriate places. The "pieces of information" are the following facts:

$$\sum X = 2118 \qquad \sum Y = 598 \qquad \sum XY = 64588$$

$$\sum X^2 = 226854 \qquad \sum Y^2 = 19180 \qquad N = 20$$

Now we insert each numerical value in place of its algebraic equivalent in Formula (17.1), and we get:

$$r = \frac{(20)(64588) - (2118)(598)}{\sqrt{(20)(226854) - (2118)^2} \; \sqrt{(20)(19180) - (598)^2}}$$

$$r = \frac{25196}{\sqrt{51156} \; \sqrt{25996}}$$

$$r = .69$$

Our ultimate finding is that the correlation coefficient is .69. Thus there is some degree of positive correlation between scores on the musical aptitude test and IQ scores for the students tested. We have gone through a rather laborious procedure, including the computations that we had to make in

273

Table 17.10. However, this procedure is a standard one and it must be thoroughly understood from beginning to end. The reader should be clear exactly how, from the data of Table 17.10, the various totals were found. These totals were the crucial "pieces" substituted in Formula (17.1), and it should be clear how this was done. Finally, it should be clear how the substitutions were made and the computations were done in solving Formula (17.1). The reader will very likely have occasion to repeat the entire procedure, and he will have to set up a table like Table 17.10 and to use Formula (17.1) to solve for the correlation coefficient.

Sometimes an experimenter already knows the population means and the standard deviations of both variables, in which case a different computing formula (also equivalent) is helpful.

$$(17.2) \qquad r = \frac{\dfrac{\sum XY}{N} - \mu_X \mu_Y}{\sigma_X \, \sigma_Y}$$

For the problem discussed, suppose we knew the following:

$$\sum XY = 64{,}588 \qquad \sigma_X = 11.31$$
$$\mu_X = 105.9 \qquad \sigma_Y = 8.06$$
$$\mu_Y = 29.9$$

Using Formula (17.2), the reader should verify that .69 is again obtained for the correlation coefficient.

Note that had we converted the original scores to z scores, we might simply have found the mean of the z score cross products. This mean, which is the correlation coefficient, would also have turned out to be .69. However, the task of making the conversions to z scores would have been much more laborious than our task was here. We might best describe the z score procedure as one that is important to consider in order to gain insight into the meaning of the correlation coefficient. On the other hand, the procedure just described is one that can be followed blindly for computing the value of r. The fact that one may operate blindly is a disadvantage in the sense that one does not gain insight, but it is an advantage for the worker who understands the meaning of the correlation coefficient and wishes to compute its value in a particular case as quickly as possible. The reader who is interested may demonstrate to himself that taking the mean z score cross product is algebraically equivalent to solving Formula (17.1) for the value of the correlation coefficient.

The correlation coefficient is an enormously important measure. Nearly every meaningful kind of relationship between the variables involves at least some degree of correlation between them. The first step for researchers when looking for a relationship is usually to compute the value of the correlation coefficient. The correlation coefficient is by far the simplest measure of relationship between two continuous variables, so far as computing is concerned. For data in grouped frequency distributions the same basic computational approach as in Table 17.10 may be used. However, as in Chapter 6, the midpoint of each interval on each variable is used to represent the value of each score that has been grouped in that interval. In practice one almost always uses commercially prepared charts for computing r from grouped frequency distributions. These charts come complete with a detailed list of steps and with a formula essentially the same as Formula (17.1), perhaps in some mathematically equivalent form.

The use of the correlation coefficient is evidenced in virtually every field where data are gathered. The reader should review the meaning of the correlation coefficient and understand it well, for it is hard to find a text relating to education, psychology, or the social sciences that is not replete with correlation coefficients.

PROBLEM SET A

17.1 Given the following five pairs of measures:

X	114	118	122	120	126
Y	140	95	155	125	185

(a) plot a scattergram; (b) compute r by using z score products; (c) compute r by using Formula (17.1).

17.2 Compute the correlation coefficient for the following pairs of scores.

Student	IQ Score	Test A Score
1	92	34
2	94	39
3	96	44
4	98	49
5	100	54
6	102	59
7	104	64
8	106	69
9	108	74

17.3 Compute the correlation coefficient for the following pairs of scores.

Student	IQ Score	Test C Score
1	92	69
2	94	34
3	96	49
4	98	64
5	100	54
6	102	44
7	104	59
8	106	74
9	108	39

17.4 The following table lists 15 pairs of college entrance-examination scores and college grade-point averages for four years:

Entrance Examination	Grade-Point Average	Entrance Examination	Grade-Point Average
419	2.18	579	2.79
465	2.34	591	2.90
495	2.23	616	2.65
522	2.47	643	3.38
530	2.51	659	2.97
532	2.14	687	3.32
554	2.42	704	3.46
569	3.04		

Compute the correlation coefficient.

17.5 The following are data on the index of retail food prices and the index of industrial production:

Year	1949	1950	1951	1952	1953	1954	1955	1956	1957	1958
Price Index	100	101	113	115	113	113	111	112	115	120
Production Index	64	75	81	84	91	85	96	99	100	93

Compute the correlation coefficient.

17.6 The following are data on the amount of consumer credit (in billions of dollars) and the number of commercial and industrial failures (in hundreds) in selected years:

Year	1950	1951	1952	1953	1954	1955	1956	1957	1958	1959
Credit	21	23	28	31	32	39	43	45	46	52
Failures	92	80	76	89	111	110	127	137	150	141

Compute the correlation coefficient.

17.7 The following are data on monthly normal temperature and precipitation in New York City:

Month	Jan.	Feb.	Mar.	Apr.	May	June
Temperature	33	33	41	50	61	70
Precipitation	3.5	3.1	3.6	3.2	3.5	3.7

Month	July	Aug.	Sept.	Oct.	Nov.	Dec.
Temperature	75	73	67	57	46	36
Precipitation	4.2	4.3	3.7	3.0	3.1	3.1

Compute the correlation coefficient.

17.8 The following are reading and arithmetic-reasoning scores:

Reading	43	58	45	53	37	58	55	61	46	64	46	62	60	56
Arithmetic Reasoning	32	25	28	30	22	25	22	20	20	30	21	28	34	28

Compute the correlation coefficient.

17.9 Compute the correlation coefficient for the following pairs of scores:

X	96	92	91	88	84	83	78	78	76	74
Y	96	96	90	88	85	80	78	76	76	75

17.10 Age, X, and systolic blood pressure, Y, of 10 women are shown in the table. Compute the correlation coefficient.

X	36	47	49	42	38	68	56	42	63
Y	118	128	145	125	115	152	147	140	149

17.11 The table below shows the weight, W, in grams of a certain chemical that will dissolve in 1,000 grams of water at temperature $T°$ Centigrade. Compute the correlation coefficient.

T	0	10	20	40	60	80	100	110
W	480	552	650	725	840	970	1,034	1,100

17.12 The table below gives the heights and weights of 15 college freshmen. Compute the correlation coefficient for these measures.

Height (in in.)	Weight (in lbs.)
65.8	166.0
68.3	115.2
72.7	157.8
66.1	152.5
73.1	149.3
71.8	181.0
73.1	173.2
66.5	120.4
69.3	124.5
73.4	163.2
67.3	125.2
73.6	173.3
67.9	146.7
69.7	158.9
68.7	134.5

PROBLEM SET B

17.13 Given the following five pairs of measures:

X	72	67	65	60	56
Y	86	84	85	80	75

(a) plot a scattergram; (b) compute r by using z score products; (c) compute r by using Formula (17.1).

17.14 Compute the correlation coefficient for the following pairs of scores.

Student	IQ Score	Test B Score
1	92	74
2	94	69
3	96	64
4	98	59
5	100	54
6	102	49
7	104	44
8	106	39
9	108	34

277

17.15 Following are the high school and actual college freshman averages for 10 students:

High School	66	70	75	80	82	85	90	92	95	98
College	60	78	65	87	74	70	78	95	88	90

Compute the correlation coefficient.

17.16 The following are 15 pairs of college grade-point averages and National Teacher Examination scores:

Grade-Point Average	Teacher Examination	Grade-Point Average	Teacher Examination
3.11	567	2.75	647
2.23	592	2.37	742
2.82	573	2.42	580
3.62	794	2.45	588
3.57	668	3.31	638
2.33	595	3.26	622
2.65	639	2.47	659
2.84	698		

Compute the correlation coefficient.

17.17 The following data show the number of commercial and industrial failures (in hundreds) and the amount of personal expenditures for recreation (in hundreds of millions of dollars) in selected years:

Year	1940	1945	1950	1955	1957	1959
Failures	136	8	92	110	137	141
Recreation Expenditure	38	61	113	142	161	183

Compute the correlation coefficient.

17.18 The following are data on the number of commissioned officers in the WAF and the gold reserve (hundred millions) of U.S. banks in selected years:

Year	1954	1955	1956	1957	1958	1959
Women Officers	789	704	634	630	672	733
Gold Reserves	218	218	221	229	206	195

Compute the correlation coefficient.

17.19 The following are data on monthly normal temperature and precipitation in San Francisco:

Month	Jan.	Feb.	Mar.	Apr.	May	June
Temperature	50	53	55	56	57	59
Precipitation	4.0	4.0	2.8	1.5	0.6	0.1

Month	July	Aug.	Sept.	Oct.	Nov.	Dec.
Temperature	59	59	62	61	57	52
Precipitation	0	0	0.1	1.1	2.3	4.1

Compute the correlation coefficient.

17.20 The following are vocabulary and spelling scores:

Vocabulary	73 86 80 73 67 65 74 65 76 73
Spelling	39 71 74 78 76 74 87 62 78 74

Vocabulary	70 88 60 79 63 56 80 92 65 74
Spelling	78 87 71 78 66 74 83 79 81 80

Compute the correlation coefficient.

17.21 The following table shows the heights, X, of a sample of fathers and the heights, Y, of their eldest sons.

X	67	63	70	66	68	71	65	62	69
Y	68	66	68	64	72	69	68	67	71

Compute the correlation coefficient.

17.22 A merchant keeps account of his weekly advertising expenditures, X, and weekly sales, Y, as shown in the following table. Compute the correlation coefficient.

X	$40	55	60	55	50	45	60	60	65	70
Y	$1,250	1,380	1,425	1,425	1,450	1,300	1,400	1,510	1,575	1,650

17.23 The mean height of a certain plant, in centimeters, X, is compared with the number of days of emergence above the ground, Y. Compute the correlation coefficient.

X	7	23	35	42	46	48	49
Y	10	20	30	40	50	60	70

17.24 The table below gives the chest girth in inches and lung capacity in cubic inches of college freshmen. Compute the correlation coefficient between these measures.

Chest Girth	Lung Capacity
38.9	311
35.0	305
31.3	330
30.3	210
38.0	269
31.5	238
30.8	305
33.7	219
37.6	226
34.5	278
32.6	310
37.5	275
34.3	220
34.4	219
37.2	265

eighteen

correlation and linear regression

18.1 Introduction. We are now going to consider what is by far the most important case in which we make predictions—namely, that which arises when we use a straight line as our regression curve. The correlation coefficient has crucial meaning in this case, as we shall see. We shall proceed as before, by a close analysis of the scattergram. The purpose is to use it here to demonstrate some basic relationships. Later we shall see how to make use of these relationships without even drawing the scattergram for the population of interest.

To use an illustrative situation, let us say that a vast number of students whose high school averages range from 65 to 95 have been going to the state university. Once again we shall say that there exists a relationship between their high school averages and their college freshman averages, and it is such that the higher a student's high school average, the better he tends to do as a college freshman. We shall again assume that a vast number of students' high school and college freshman averages have been recorded. The relationship has been consistent and there is every reason to believe that it will continue.

Fig. 18.1 is a scattergram showing the performance of a number of students. High school averages are recorded along the horizontal axis and freshman averages at the state university are recorded along the vertical axis. Note that each dot in Fig. 18.1 tells us by its location how a single student did both in his high school average and in his freshman average. For instance, the dot that has been encircled represents the performances of a student whose high school average was 70 and whose average at the state university was 74.

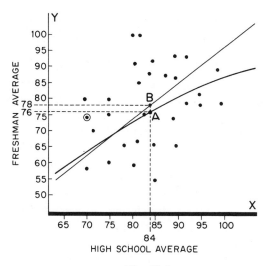

Fig. 18.1

The curve drawn in Fig. 18.1 is the regression curve as defined in Chapter 16.   In other words, one should refer to the height of the curve (based on past performances) to make the best possible prediction of what a student's freshman average at the state university will be.   One finds the point on the horizontal axis corresponding to the high school average attained by the particular student and locates the height of the curve as it passes over that point.   It is this height that becomes the predicted freshman average for the student.   For instance, suppose a student comes along whose high school average was 84.   The regression curve is at Point A in Fig. 18.1 as it passes over the value 84 on the horizontal axis.   The height at Point A is 76, as we may see by referring to the vertical axis.   Therefore, the best prediction is that the student's freshman average will be 76.

Next let us look at the straight line drawn in Fig. 18.1.   Though the curve might be somewhat more accurate, we might also make it a practice to use the straight line for predictive purposes.   For instance, in the case of the student whose high school average was 84, we would follow the straight line to Point B and predict that his freshman average at the state university would be 78.

We have defined the regression curve as the single continuous path that comes closest to all the dots.   Thus the errors made when predicting from the curve would, by definition, lead to an error variance smaller than that made when predicting from the line.   However, in practice there are enormous advantages to making predictions from a straight line.   For instance, there is a direct way to "fit" a straight line to data and to use it for predictive purposes.   One need not have a vast number of cases to "fit" a relatively accurate straight line, as we shall see.   On the other hand, a large number of cases are needed to fit an accurate curve that can be used for future cases.

281

It is easy to extend a straight line in either direction, but unless a curve is accurately drawn, it is not so easy to determine its path and so to extend it.

Thus, instead of using the regression curve, it is almost universal practice to set up as a stipulation that our prediction path be a straight line. In many instances, the best regression curve is a straight line anyhow and there is no problem; in fact, we are going to consider only these instances here. But the point is important that, even where there is some discrepancy between the best fitting curve and a straight line, it may be preferable to use the line. Thus, in practice, it is typical for the worker in the field of psychology or education to proceed immediately to find the best fitting straight line for his data and to use this line for making his predictions.

Having decided to use only a straight line for predictive purposes, what we must do quite naturally is to find the best-fitting straight line and use that one. More specifically, we are going to consider the problem of finding the line that comes closer to all the dots than any other line that might be drawn. It is this line that we call the *best-fitting straight line*.

To further distinguish this line we are discussing, we have reproduced the data of Fig. 18.1 in Fig. 18.2, where once again this best-fitting line has been drawn. Consider what happens when we use this line to make our prediction for each individual student. That is, given each student's high school average, we would refer to the appropriate point on the line to predict what his college freshman average would be. In each case, our error would be the distance between our chosen point on the line and the dot representing the actual performance of the student of interest.

For instance, the high school average of the student whose performance is encircled in Fig. 18.2 was 78. Suppose this student had come along and we knew only his high school average. Given only this fact, we would have gone ahead to find the height of the line over the point 78, and our prediction would have been that the student's freshman average was to be 72. Since the student's actual freshman average is 66, our error would have been six points, indicated by the length of the dotted line in Fig. 18.2. This dotted line connects the point on the regression line (which represented our prediction) to the dot indicating the student's actual performance. The length of the dotted line represents our error in the given case. The various vertical lines in Fig. 18.2 similarly stand for the errors that we would have made in individual cases.

Now suppose that we were to square each of these errors and then to find the mean of all of the squares thus obtained. The best-fitting line may be defined as the one that would make this mean of the squared errors less than if any other line had been used. We said that in Fig. 18.2 we have drawn the best-fitting straight line. This means that the mean of the squared errors in Fig. 18.2 is less than it would have been had we drawn any other line in the same scattergram and computed the error variance from it. Thus intuitively we call the best-fitting straight line in a scattergram the one that comes closest to all the dots. In more precise language, we may describe it as the one that makes the mean of the squared errors less than if any other line had been drawn. The two statements are equivalent. (This line is often called the "least-squares line" and in many cases the method we are discussing here is called the "least-squares method.") As we shall see, there is a computational way to determine exactly where this best-fitting line should go, and we will make use of it.

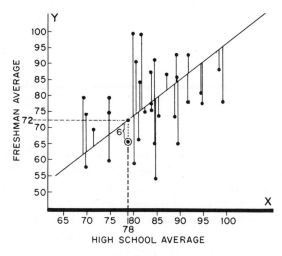

Fig. 18.2

18.2 The Geometric Meaning of the Correlation Coefficient. Suppose now that we were to take our data and transform the scores on each variable into z scores. For the variable, high school averages, we would have to find its mean and standard deviation. Then for each student we would have to determine how many standard deviations below or above the mean his high school average was, and the value obtained for each particular student would be his z score. For instance, if a student's high school average was one standard deviation above the mean, then his z score on this variable would be 1.0. Similarly, the student would have a z score on the variable, freshman average.

In essence, we would be able to represent the performance of each student by his pair of z scores. One would be the z score of his high school average and the other would be the z score of his freshman average at the state university. Suppose that, having computed for each student his pair of z scores, we were now to draw a scattergram. This time the units of measurement on each variable would be z scores. This scattergram is shown in Fig. 18.3 as it would appear. The z scores of the students' high school averages are indicated along the horizontal axis, and those of their freshman averages are indicated along the vertical axis. Once more the best-fitting straight line has been drawn in Fig. 18.3.

283

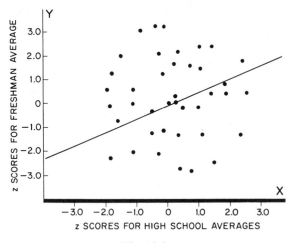

Fig. 18.3

As the reader may know, a line drawn on a graph generally has some slope. That is, a point moving on the line would always be rising (or dropping) at some speed in proportion to how fast it is moving horizontally. To determine the slope of a line, we may pick any point on the line at random. From that point we measure off some distance to the right. For instance, for the line in Fig. 18.4 we have randomly picked point *A*. We then draw a horizontal line any number of units that we decide upon. Let us say that in Fig. 18.4 we have chosen to measure off five units to the right, bringing us to point *B*. Then from this point we draw a line directly vertically to the original line in question. In Fig. 18.4 our vertical dotted line meets the original line at Point *C*, creating a triangle. We shall say that there the length of the line *BC* turns out to be two units.

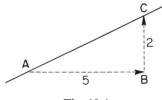

Fig. 18.4

We may conclude from our computations that for every five units that a point moves from left to right on our line, it ascends two units. The line *AB* is sometimes called the *run* of our original line, and the line *BC* is sometimes called its *rise*.

The slope of a line is its rise divided by its run, which in this case is 2/5. The slope of a line is the fraction that tells us how fast a point moving along the line is rising compared with the speed that it is moving horizontally. We may pick any point on the line as Point *A* and draw our horizontal line any length to Point *B*. Then we draw our vertical line *BC*. The ratio of the segment *BC* to the segment *AB* will always be the same. Thus if we had made *AB* ten units, line *BC* would have been four units, so that once again the ratio of *BC* to *AB* would be 2/5.

Now let us go back to the situation in which we drew our *z* score scattergram for the variables, high school average and freshman average. This scattergram was Fig. 18.3 and on it was drawn the best-fitting straight line. This best-fitting straight line has a slope; or, to put it another way, its rise divided by its run is some fraction. *The correlation coefficient for two variables is the same as the slope of the best-fitting line in the z score scattergram.* Thus suppose the correlation between high school averages and freshman averages is .40. Then the slope of the line in Fig. 18.3 would turn out to be .40 or 2/5.

Once again the translation into *z* scores reveals a meaning of the correlation coefficient. Naturally, one does not go to the trouble of translating the scores on both variables into *z* scores, then making up a *z* score scattergram, and finally drawing in the best-fitting straight line. The point is merely that if one went to this trouble, it would turn out that the exact correlation between the two variables would reveal itself as the slope of that line. As in Chapter 17, we have conceived of translating the scores into *z* score units simply for the purpose of making clear an insight.

There occurs a variation in terms of slope when we consider a negative correlation. Then, as scores increase on the *X* variable, they decrease on the *Y* variable. This means that even in *z* score terms, the best-fitting straight line moves downward instead of upward. As we move horizontally from our chosen Point *A*, we find that the best-fitting straight line for the data is dropping below us. As in Fig. 18.5 we must draw our vertical line *BC* downward in order to return to the best-fitting straight line. In this case we say that our best-fitting straight line has a negative slope.

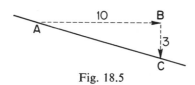

Fig. 18.5

Once more the ratio of *BC* to *AB* tells us the slope of the line, but this time we must put a minus sign in front of our fraction to indicate that the slope is negative. For instance, suppose that in Fig. 18.5 the ratio of *BC* to *AB* is 3/10. Then the slope of our line would be $-3/10$, meaning that the correlation between the variables of interest would be $-.30$.

Obviously we have not provided a mathematical proof that the correlation coefficient represents the slope of the best-fitting line in *z* score units. The proof would involve using theorems from an advanced topic in mathematics called vector theory. But, as we shall see, the fact not only provides insight into the meaning of the correlation coefficient but also becomes useful to us in deriving more practical facts.

18.3 The Best-Fitting Line in the Scattergram of Original Data. We shall begin this time by supposing that we have in our possession the z score scattergram for the data described. That is, we have taken the trouble to convert the students' high school grades and their university freshman grades into z scores and plotted them as in Fig. 18.3. We shall say further that the correlation between the two variables is .40, which means that the slope of the best-fitting line is .40, or 2/5. This best-fitting straight line is also indicated in Fig. 18.6.

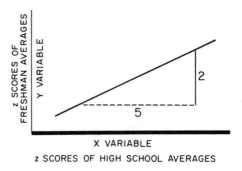

Fig. 18.6

As mentioned, a scattergram for the two variables in their original forms might also be constructed. We have presented this scattergram for our data in Fig. 18.1. In that illustration, high school averages in their original form are plotted along the X axis and are the X variable. The University freshman averages are plotted along the Y axis and are the Y variable. As opposed to the z score scattergram, we shall describe a scattergram of this kind as a *raw score scattergram*, since scores in their original form are called raw scores.

We know that in the z score scattergram the best-fitting straight line has a slope, which is the correlation coefficient. Our concern now shall be with using this information in particular to determine what the slope of the best-fitting straight line is in the raw score scattergram. We are about to compare the z score scattergram with the raw score scattergram to solve our problem.

In Fig. 18.3 the z scores of the X variable are plotted along the horizontal axis. Each z score unit represents a certain number of points. In particular, each z score unit represents exactly the number of points contained in one standard deviation of this X variable. We shall say that the standard deviation of high school averages, which is the X variable, is 9 points. Thus the high school average with a z score of 1.0, which is exactly one standard deviation above the mean, is 9 points above the mean. The high school average with a z score of 2.0 would be 18 points above the mean, and so on. Similarly, we shall say that the standard deviation of the Y variable is 10 points, which would mean that each z score unit along the vertical axis represents 10 points.

Now look again at the scattergram and the best-fitting line in Fig. 18.3. Remember that in z score units the slope of this best-fitting line for the two variables was 2/5. But we have seen that a shift of five z score units to the right means a shift of 45 points. Fig. 18.7 shows how five z score units on the X variable represent 45 points of original X scores.

FIVE z SCORE UNITS

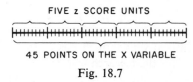

45 POINTS ON THE X VARIABLE

Fig. 18.7

Similarly each change of one z score unit on the Y variable represents a change of 10 points. Thus an increase of two z score units on the Y variable means an increase of 20 points in terms of original scores (see Fig. 18.8).

We have supposed that for high school averages and college freshman averages the correlation was .40. Thus the slope of the best-fitting line in terms of z scores is 2/5. Now we can interpret our "slope triangle" in terms of original units on each variable. We see that five z score units on the X

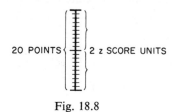

20 POINTS {   2 z SCORE UNITS

Fig. 18.8

variable stand for 45 original units, and that two z score units on the Y variable stand for 20. Thus, as we see in Fig. 18.9, the slope of the best-fitting line in the raw score scattergram for the two variables described is 20/45.

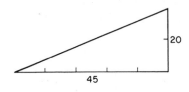

20

45

Fig. 18.9

We have seen how to go from the z score slope to the raw score slope. The z score slope was the fraction 2/5. The numerator represented z score change on the Y variable, so we multiplied it by σ_y. Since $\sigma_y = 10$, in effect we multiplied 2 by 10 to get our new numerator of 20. The denominator

287

represented z score change on the X variable, so we multiplied it by σ_z. Since $\sigma_z = 9$, in effect we multiplied 5 by 9 to get our new denominator. To put what we did in its simplest form, we need only to take the z score slope and multiply it by σ_y/σ_z in order to get the raw score slope.

The symbol b shall be used to stand for the slope of the best-fitting line in the raw score scattergram. This symbol will be satisfactory so long as we consistently use the letter X to refer to the variable measured along the horizontal axis and the letter Y to refer to the other variable. To summarize what we have said, the formula for b is as follows:

(18.1)
$$b = r \frac{\sigma_y}{\sigma_z}$$

It should be emphasized that though r is a straightforward measure of correlation, b is not. Thus the research worker may indicate the value of the correlation coefficient he has obtained to throw light on how his variables are related, but he may not use his obtained value of b to do the same thing. It should be noted that though r had a lower limit of -1 and an upper limit of $+1$, the same is not true of b. In cases where σ_y is much larger than σ_z, for instance, even where r is small the value of b may turn out to be considerably larger than 1.0. For instance, suppose $r = .10$, $\sigma_y = 40$ and $\sigma_z = 2$. Then when we solve for b using Formula (18.1), it turns out that $b = 2.0$.

Actually there is another formula for b that looks more forbidding than Formula (18.1) but does not necessitate the computing of r, σ_y, or σ_x. To use this formula, one must set up the same kind of table as Table 17.10. The totals that one derives from this table are the appropriate "pieces" not only for computing r but also for computing b. The formula (where X is the predictor variable and Y the predicted variable) is:

(18.2)
$$b = \frac{N \sum XY - (\sum X)(\sum Y)}{N \sum X^2 - (\sum X)^2}$$

To illustrate the use of the formulas just given, we shall look back at the data in Table 17.10 where the IQ scores and scores on a musical aptitude test for 20 subjects are listed. In order to solve for b using Formula (18.1) we would first have to compute the values of σ_z and σ_y. Using Formula (3.2), we find that $\sigma_z = 226.2$ and $\sigma_y = 161.2$. Thus using Formula (18.1) we get:

$$b = (.69) \frac{161.2}{226.2} = .49$$

Table 17.10 yields the totals that are the various "pieces" that go into Formula (18.2). Using this formula we get:

$$b = \frac{(20)(64588) - (2118)(598)}{(20)(226854) - (2118)^2}$$

$$b = .49$$

Incidentally, the reader who likes to work with symbols will find it rather easy to show that Formula (18.1) is exactly equivalent to Formula (18.2).

Finally, having computed the value of b for our situation, the scattergram of Fig. 18.10 is presented to illustrate its meaning.

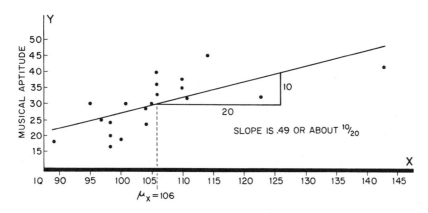

Fig. 18.10

18.4 The Best-Fitting Line and Prediction. Note that one needs only to possess raw data, and has no need of a scattergram, to compute the value of b. On the other hand, without having the scattergram in mind, it is unimaginable that one could have proper insight into the meaning of b. In the same way, one need not have drawn a scattergram in order to make predictions in the manner to be described. However, it is virtually necessary for us to go back to the scattergram once more in order to make clear the standard method for making predictions using the best-fitting straight line. Thus we shall go back to the scattergram to develop this method, though the reader need not actually draw a scattergram if and when he wishes to make use of the method.

Let us now consider a new situation and say that we have been giving an engineering aptitude test to a large population of students at the outset of their college careers as engineering majors. We have been recording these students' test scores and also their final college averages. In other words, each student's aptitude-test score and his final college average as an engineering major have been recorded. Our purpose is ultimately to use the aptitude-test scores to predict how students will fare so that we can make certain decisions regarding them as they enter.

In this situation the test scores obviously compose the predictor variable. In other words, we shall refer to the test scores as values of the X variable and record them along the horizontal axis. The students' engineering averages are values of the predicted variable. These values shall constitute our Y variable and shall be recorded along the vertical axis.

Suppose that for the population of students that we have tested the following facts are true. On the aptitude test the mean score is 80 and the standard deviation is 8 points. The students' engineering averages have a mean of 75 and a standard deviation of 6 points. The correlation between

289

the students' aptitude scores and their engineering averages is .80. From the data given we can compute by Formula (18.1) that the value of b is .60. We shall put these findings in a form that enables us to refer to them readily.

$$\mu_x = 80 \qquad \mu_y = 75 \qquad r = .80$$
$$\sigma_x = 8 \qquad \sigma_y = 6 \qquad b = .60$$

Now we are going to look at the scattergram for the data described in terms of the original scores and the best-fitting straight line. This scattergram appears in Fig. 18.11 and the slope of the best-fitting straight line is .60, as indicated. Note that the values $X = \mu_x$ and $\mu_y = 75$ have been labeled. A horizontal line passing through the point on the vertical axis where $Y = \mu_y$ has been drawn covering all points in the scattergram at which $Y = 75$. Similarly, a vertical line passes through all points at which $X = \mu_x$. Thus the horizontal line passes through all points at which the Y variable is at its mean, and the vertical line passes through all points at which the X variable is at its mean.

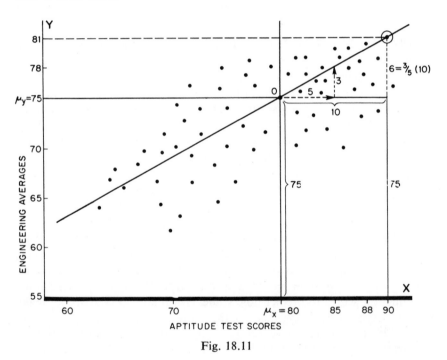

Fig. 18.11

Most important is the fact that these two lines representing the means of the two variables and the best-fitting straight line intersect at a single point. We shall call this point the origin. It is marked O in Fig. 18.11. It is always the case that all three lines—the regression line and the lines indicating the means of the two variables—coincide at a single point.

We shall assume that the data recorded in Fig. 18.11 are typical of what we may expect for students in the future. As each new entering freshman receives a score on our aptitude test, our task will be to predict what this student's ultimate engineering average will be.

So far we are merely referring to the scattergram. As explained, we make our prediction in each case by locating the student's test score on the X axis and then finding the point on the best-fitting line directly above it. For instance, suppose a student gets a 90 on our test. To make a prediction of what his engineering average will be, we look directly upward to the point on the best-fitting line over the value of 90 on the horizontal axis. This point is encircled. Having found this point, we now look over at the Y axis where we see that it has a Y value of 81. Thus our prediction for this student is that his college engineering average will be 81.

Our procedure is identical in each case. Given a student's score on the X variable, we simply refer to the point on the best-fitting line directly over it. The height of this point becomes our prediction of his score on the Y variable. Now we are going to substitute for this procedure a numerical way of doing the same thing. In other words, given any X value (for example, 90), there is a numerical way of determining the corresponding Y value on the best-fitting line. We shall make use of the scattergram to explain this procedure.

Referring to Fig. 18.11, the reader should keep in mind that the slope, b, of the best-fitting line has the value 3/5. This fact means that for every five-point shift on the X variable, there is a three-point shift on the Y variable. The other crucial fact, already mentioned, is that there necessarily exists some point on the line where X is at its mean and Y is at its mean. We have called this point the origin.

With reference to Fig. 18.11, consider once more the student whose X score is 90. Beginning at the origin, where X is at its mean, we must shift 10 points to the right to get to 90. But the slope of 3/5 tells us that, whatever amount X increases, Y increases 3/5 of that amount. Thus, having calculated the increase in X, we must multiply it by 3/5 to get the increase in Y. In essence, as far as X is above its mean, the increase in Y is 3/5 of that amount. For the X value of 90, the shift is $(90-80)=10$ points. The value of $b=3/5$. Thus, to determine the increase in Y over its mean, we multiply 3/5 by $(90-80)$. $(3/5)(90-80)=6$. For the individual in question, our prediction is that his Y score will be μ_y+6. In other words, it will be $75+6=81$. Note that all the statements made in this paragraph are illustrated in Fig. 18.11.

The procedure just described may be used to compute for any X value its corresponding Y value on the best-fitting straight line. To consider another example, suppose that the value of X is 88, meaning that a student's score on the aptitude test was 88. Our task is to find the appropriate Y value on the best-fitting straight line, and this value is to become our prediction of what his engineering average will be.

291

This time it is from the X score of 88 that we subtract 80 (see Fig. 18.12). The difference, 8, tells us that the X score represents a shift of eight points. Next we multiply the slope, b, by this shift to determine how much Y has ascended above its mean. Since $b = 3/5$ we multiply 3/5 by (88−80). Our product, 4.8, represents the shift of Y above its mean. To the mean of Y, which is 75, we add this increase of 4.8 points and get 79.8, or 80.

Thus, where X is 88, the corresponding Y value on the best-fitting straight line is 79.8. In other words, for the student whose test score is 88, we predict that his engineering average will be 80, as we would have had we been able to make a perfect visual reading of the appropriate point on the best-fitting line. Note that in Fig. 18.12, where $X = 88$, the height of the line at Point A is 79.8.

In each instance thus far we have started out with a supposedly obtained value of X and have computed the Y value, or height, of the best-fitting line at the appropriate point. This Y value, which is always a point on the best-fitting line, is in each case our prediction. The usual symbol for a prediction of this kind is $\tilde{Y}$, which we read "predicted Y."*

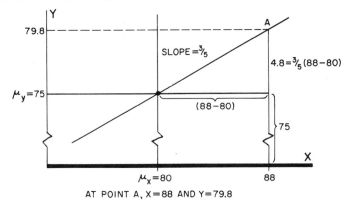

AT POINT A, X=88 AND Y=79.8

Fig. 18.12

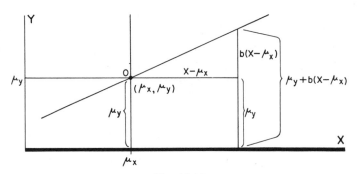

Fig. 18.13

---

* The mark above the Y is called a "tilde." $\tilde{Y}$ is sometimes called " Y tilde" as well as "predicted Y."

We have pointed out that, given any value of X, there is the implication that the appropriate $\tilde{Y}$ value is a distance of $b\,(X-\mu_x)$ from the mean of the Y values. Thus at any value of X the height of the line is $\mu_y+b(X-\mu_x)$ (see Fig. 18.13). We have seen that for any value of X we make the prediction $\tilde{Y}$. Now we see that given any value of X, the appropriate value of $\tilde{Y}$ is given by Formula (18.3).

$$(18.3) \qquad\qquad \tilde{Y} = \mu_y+b(X-\mu_x)$$

Formula (18.3) holds even if the given value of X is less than μ_x. In this case, the "shift" on the X variable is negative and reflects in the prediction. The value of $\tilde{Y}$ yielded by Formula (18.3) is still the appropriate value on the line. Look back at Fig. 18.11. Suppose that a student comes along whose test score was 60. Where X is 60, our prediction from the scattergram is that his $\tilde{Y}$ score will be about 63. Applying Formula (18.3) we get

$$\tilde{Y} = 75+\frac{3}{5}(60-80)$$
$$= 75-12$$
$$= 63$$

This time the value $(X-\mu_x)$ turned out to be negative, indicating that the shift in X from its mean was negative or to the left. As a consequence the formula led us to take the value μ_y and diminish it rather than add to it.

In practice, where we wish to make a vast number of predictions we can first do some simplifying of Formula (18.3). First we note that $\mu_y=75$, $b=3/5$, and $\mu_x=80$. It is only the value of X that will be new each time when we come to make predictions. Therefore we may write Formula (18.3) once more and then put in the values that we already have:

$$\tilde{Y} = \mu_y+b(X-\mu_x)$$
$$= 75+\frac{3}{5}(X-80) = 75+\frac{3}{5}X-48$$
$$= 27+\frac{3}{5}X$$

The formula above for $\tilde{Y}$ in its final form would undoubtedly be simpler to work with than the original formula, especially if we had many successive predictions to make. Therefore, a simplification of the kind just done is often carried out, especially where many predictions are to be made. However, Formula (18.3) remains the crucial one so far as understanding is concerned, and this last simplification is not absolutely necessary.

Formula (18.3) is sometimes called a *prediction equation* and sometimes a *regression equation*. It is extremely important. It enables one to do arithmetically with perfect precision what one might try to do visually with a scattergram. That is, it is aimed at enabling a worker to take scores on one variable and to make rapid-fire predictions of scores on another. One assumption of course is that a definite linear relationship has been discovered. 293

Another is that the facts, such as the correlation coefficient, derived from the previous collection of cases shall be largely the same for the new cases about which predictions are to be made.

One more illustration is in order to make clear how Formula (18.3) is used. A large number of U.S. Army Lieutenants were given a leadership ability scale on which they received scores. The top score possible was 100. They were also given scores on their leadership ability after having served one year. Here the top score possible was 25. The test scores of course shall be said to constitute the X variable and the leadership scores the Y variable. The findings for this group were as follows:

$$\mu_x = 74 \qquad \mu_y = 14 \qquad r = .48$$
$$\sigma_x = 8 \qquad \sigma_y = 3 \qquad b = .18$$

Now assuming that what was true for this collection of subjects will be largely true for others like them, we may wish to make some predictions for future cases. Suppose in particular that a subject gets 70 on the leadership ability scale. We want to predict what his leadership ability score will be. To proceed, we solve Formula (18.3).

$$\tilde{Y} = \mu_y + b(X - \mu_x)$$
$$= 14 + .18(70 - 74)$$
$$= 13.28$$

The reader should verify that, for the individual whose test score is 90, our prediction is that his leadership ability score will be 16.88. For the individual whose test score is exactly 74, our prediction is that his leadership ability score will be exactly 14. It is always true, when we are making linear predictions, that for the subject whose score was exactly the mean on one variable, we end by predicting that his score will be exactly the mean on the other.

18.5 Correlation and the Accuracy of Linear Prediction. An obvious question is how much advantage do we gain in the long run by making predictions from the best-fitting straight line? The answer is that the advantage that we gain in accuracy depends completely on the strength of the correlation. Where the correlation is trivial, then it hardly pays to use the best-fitting straight line to make predictions. But the stronger the correlation, the more precise is the best-fitting line in coming close to all the dots. Thus the stronger the correlation, the more it pays to use the best-fitting line for predictive purposes.

There is a specific relationship between the magnitude of the correlation coefficient (and its sign is not relevant here) and the degree to which our enlightened predictions are improvements in accuracy. Remember that with no information the best predictive device is to ascertain the mean of the Y variable and to use this value as the prediction for each individual case. Let us consider, for instance, the situation where the X variable and the Y

variable have a correlation of .60.   The only facts we need are that the mean of the Y variable is 44 and the standard deviation of the Y variable is 10. In sum, $r = .60$, $\mu_y = 44$, and $\sigma_y = 10$.

The scattergram for this situation has been drawn in Fig. 18.14A, where μ_y is represented by the horizontal line at the height of 44.   With no information concerning a relationship, it is this value 44 that would be the best one to use as the prediction for each new case.   As mentioned, it is most efficient to use μ_y as one's guess concerning each new case when one has no outside information.   This guess would of course lead to errors, indicated in Fig. 18.14A.   As mentioned, we are saying that the variance of the population of Y terms is 100 ($\sigma_y^2 = 100$).   When we guess the mean each time, this value becomes the variance of our errors.

In Fig. 18.14B the scattergram has been reproduced with the best-fitting line drawn in.   It is based on the correlation of .60, which we supposed. Once again using this line for predictions, there would be errors, which would have an error variance.   The question is: based on the correlation of .60, how much did we reduce the error variance by shifting from the use of μ_y as a predictor to the use of values on the best-fitting line?   Specifically, we said that the error variance in Fig. 18.14A is 100.   By how much do we reduce this error variance when we make predictions using the best-fitting straight line, as in Fig. 18.14B?

Remember that $r = .60$.   It may be shown that what one must do is to square this value. $r^2 = (.60)^2 = .36$.   The value of $r^2 = .36$ tells us that when we use the best-fitting straight line, the variance of our errors will be less than the original variance by .36 of that original variance.   Here the error variance in Fig. 18.14A was 100.   The fact that $r^2 = .36$ tells us that 36 per cent of this variance no longer exists when we predict as in Fig. 18.14B.

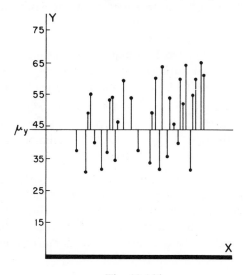

Fig. 18.14A

Of the original 100 units, there remain only 64. Using the best-fitting line as in Fig. 18.14B, the predictions have an error variance of 64.

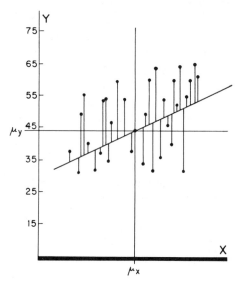

Fig. 18.14B

The value of r^2 is sometimes described as indicating the *proportion of variance accounted for* by a correlation. As mentioned, the error variance when we use the mean as a repeated prediction (as in Fig. 18.14A) is σ_y^2. What remains of the error variance when we use knowledge of an X variable to predict is designated by $\sigma_{y \cdot x}^2$. The subscript is to indicate that the variance is of the Y variable, but that predictions are based on knowledge of the X variable. The value of $\sigma_{y \cdot x}^2$ in Fig. 18.14B is 64.

Thus the original error variance (as in Fig. 18.14A) is σ_y^2. The proportion of this variance eliminated by using the best-fitting line is r^2 and the actual amount of original variance eliminated is $r^2\sigma_y^2$. Thus, using the best-fitting straight line, we find that the remaining error variance is given by

$$(18.4) \qquad \sigma_{y \cdot x}^2 \qquad = \qquad \sigma_y^2 \qquad - \qquad r^2\sigma_y^2$$

error variance	original	the amount
using the best	error	eliminated
fitting line	variance	

Suppose that in some community the average high school graduating mark is 82 and the variance of marks is 80. If we make a prediction for each entering student, knowing nothing about the person, our best approach is to predict that the student will get an 82. By this procedure, our errors will have a variance equal to the variance of the marks themselves. The error variance is 80.

$\sigma_y^2 = 80$

Now suppose we know each student's IQ score upon entering. IQ scores on our test correlate .6 with graduating marks. This time, for each student we make a special prediction based on the student's IQ score. Our predictions are better in the long run. How much better? $r^2\sigma_y^2$ is the amount of variance eliminated. Here,. $r^2\sigma_y^2 = (.6)^2(80) = 28.80$. Our new error variance $\sigma_{y \cdot x}^2$ is less than the old one by 28.80.

$$\sigma_{y \cdot x}^2 = \sigma_y^2 - r^2\sigma_y^2$$

$$\sigma_{y \cdot x}^2 = 80.00 - 28.80$$

$$\sigma_{y \cdot x}^2 = 51.20$$

Formula (18.4) reduces to

$$\sigma_{y \cdot x}^2 = \sigma_y^2(1 - r^2)$$

The name for the square root of the error variance is the *standard error of estimate* $\sigma_{y \cdot x}$.

The major fact that should be understood is that the higher the correlation, the greater the proportion of reduction in error variance when we use the best-fitting straight line. The reader may verify the following two statements, using Formula (18.4). When the correlation is perfect (meaning that r is either $+1.0$ or -1.0), then using the best-fitting line leads to an error variance of zero. Also, when the correlation is zero, then the original error variance is not at all reduced by using knowledge of the X variable.

The first statement means that knowledge of scores of a variable perfectly related to the Y variable leads to predictions that are without error. The second statement means that knowledge of scores of an uncorrelated variable does not improve predictions made by use of the method described. In fact, when the X variable is uncorrelated with the Y variable, then, even when X scores are known, the best-fitting straight line turns out to be the horizontal line as drawn in Fig. 18.14A.

18.6 The Test of Significance for a Correlation Coefficient. A frequent question is whether two variables have a correlation other than zero in some vast population from which a sample is drawn. For instance, one might wish to determine whether scores on an anxiety scale are correlated with IQ scores among ten-year-olds. In this case, one has in mind the vast population of ten-year-olds to whom the anxiety scale and the IQ test might be given. There exist methods for using a correlation coefficient obtained from a sample in order to test the hypothesis that the correlation is zero in the population from which the sample is drawn. We shall describe one such procedure by illustrating it.

Suppose that a researcher has given both the anxiety scale and the IQ test to 30 ten-year-olds, and his obtained value of r is .40. This time it is not the value $r = .40$ that is of ultimate concern to him: rather, he wishes to use it to make a decision about whether the correlation is zero between anxiety and IQ in the population that yielded the 30 cases. He knows that, even were this correlation exactly zero, there would likely be a correlation of some value other than zero in a sample such as his. The question is whether his obtained value is sufficiently far from zero to indicate that he should reject the null hypothesis that the correlation is zero in the population.

As usual, our researcher must pick a significance level, and we.shall say it is the .05 level. Next he asks: Under the hypothesis that the correlation is zero in the population, what would be the probability of obtaining a sample value of r as far from zero as mine? In other words, our experimenter conceives of his sample as one of a vast number of random samples from the same population. He conceives of his obtained r value as one of a collection of r values, each of which would have been yielded by a different sample.

Under the hypothesis that the correlation is zero in the population, the obtained values of r in different samples would vary around zero. There would be r values below zero and others above it. The question is whether the value of .40 is far enough away from zero to indicate rejecting the hypothesis.

We shall consider here a procedure applicable when the assumption can be made that the variables of interest are roughly normally distributed. In Chapter 19 we shall consider a different approach that is applicable when the assumption cannot be made that the variables are normally distributed.

When the variables are roughly normally distributed the following procedure may be followed. The researcher who has computed his r value should compute t from Formula (18.5).

$$(18.5) \qquad\qquad t = r \frac{\sqrt{N-2}}{\sqrt{1-r^2}}$$

This t value has $N-2$ d.f.

To illustrate, in the case described, $r=.40$ and $N=30$. Therefore,

$$t = \frac{.40\sqrt{30-2}}{\sqrt{1-(.40)^2}} = \frac{.40\sqrt{28}}{\sqrt{.84}} = 2.31$$

The obtained t value has exactly the same meaning as the t values that we computed in the past. Formula (18.5) is sometimes called a *conversion*. Earlier our researcher had to think of each of his fellow researchers as obtaining a value of r from his sample. Now he must think of each of them as having gone through the next step, which means having computed a value of t. Under the hypothesis, the distribution of the obtained t values would have 28 degrees of freedom (see Fig. 18.15).

As for our researcher, using his knowledge of the t distribution for 28 d.f., he must locate his obtained t value in this t distribution. (This t value is a representative of the r value that he computed.) At the .05 significance level, his obtained t value of 2.31 is significant. Therefore he is able to reject the null hypothesis, which was that the correlation in the population is zero. He may conclude that anxiety and IQ are at least somewhat related in the population from which the sample was drawn.

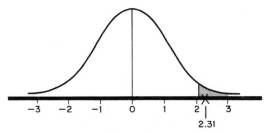

Fig. 18.15

Had the obtained value of r been .18, for instance, then the value of t would have been .97, as the reader may verify. The null hypothesis could not be rejected for 28 d.f. Note that a negative correlation would lead to a negative value of t but the procedure for testing the hypothesis would be the same. Also, the desire to prove that the correlation in the population differs from zero in a particular direction would lead to a one-tail test, exactly as in the situations described earlier. For instance, the desire to prove that the population correlation is less than zero would lead to a one-tail test where only an extreme negative value would lead a researcher to reject the null hypothesis.

The technique described involves what statisticians call *the use of a transformation.* Formula (18.5) was used to transform the statistic r into the statistic t since the latter has a distribution that is familiar. Thus the test of the hypothesis was directly involving t values, but the implication of the finding was immediate with regard to the original issue, which was whether the population correlation was zero.

PROBLEM SET A

18.1 Presented below are pairs of measures on two variables.

X	114	118	122	120	126
Y	140	95	155	125	185

(a) Obtain the prediction formula for $\tilde{Y}$; (b) determine the values of $\tilde{Y}$ for $X = $ 105, 117, 130; (c) state the proportion of variance in Y that is accounted for by the correlation between X and Y.

18.2 Presented below are pairs of measures on two variables.

X	66	70	75	80	82	85	90	92	95	98
Y	60	78	65	87	74	70	78	95	88	90

(a) Obtain the prediction formula for $\tilde{Y}$; (b) determine the values of $\tilde{Y}$ for $X = $ 75, 82, 95; (c) state the proportion of variance in Y that is accounted for by the correlation between X and Y.

18.3 Presented below are pairs of measures on two variables.

X	Y	X	Y
3.11	567	2.75	647
2.23	592	2.37	742
2.82	573	2.42	580
3.62	794	2.45	588
3.57	668	3.31	638
2.33	595	3.26	622
2.84	698	2.47	659
2.65	639		

(a) Obtain the prediction formula for $\tilde{Y}$; (b) determine the values of $\tilde{Y}$ for $X = 2.80, 3.40, 4.10$; (c) state the proportion of variance in Y that is accounted for by the correlation between X and Y.

18.4 The following data show the number of commercial and industrial failures (in hundreds) and the amount of personal expenditures for recreation (in hundreds of millions of dollars) in selected years:

Year	1940	1945	1950	1955	1957	1959
Failures	136	8	92	110	137	141
Recreation Expenditures	38	61	113	142	161	183

Let X represent failures and Y represent recreation expenditures. (a) Obtain the prediction formula for Y; (b) determine the values of Y for $X = 60, 90, 130$; (c) state the proportion of variance in Y that is accounted for by the correlation between X and Y.

18.5 In a sample of 83 pairs of X and Y values, the correlation coefficient is 0.75. Use the .05 level of significance to test the hypothesis that the population correlation coefficient is zero.

18.6 Consider the pairs of data given in Problem 18.1 as random samples from approximately normal populations, and test the null hypothesis that the correlation in the population is zero. Use a two-tail test at the .05 significance level.

18.7 Refer to the pairs of measures given in Problem 18.2. Consider the data as random samples from approximately normal populations, and test the null hypothesis that the correlation in the population is zero. Use a one-tail test at the .05 significance level. *|

18.8 Refer to the pairs of measures given in Problem 18.3. Consider the data as random samples from approximately normal populations, and test the null hypothesis that the correlation in the population is zero. Use a two-tail test at the .01 significance level.

18.9 Refer to the data shown in Problem 18.4. Consider the pairs of data as random samples from approximately normal populations, and test the null hypothesis that the correlation in the population is zero. Use a one-tail test at the .01 significance level.

*In all one-tail tests in this and subsequent problem sets, assume that the obtained value (of O or $\bar{X}$) differs from the expectancy in the direction necessary to permit rejection of the hypothesis.

18.10 Consider the following pairs of reading and arithmetic test scores as random samples from approximately normal populations, and test the null hypothesis that the correlation in the population is zero. Use a two-tail test at the .05 significance level.

Reading	43	58	45	53	37	58	55	61	46	64	46	62	60	56
Arithmetic	32	25	28	30	22	25	22	20	20	30	21	28	34	28

PROBLEM SET B

18.11 Presented below are pairs of measures on two variables.

X	72	67	65	60	56
Y	86	84	85	80	75

(a) Obtain the prediction formula for $\tilde{Y}$; (b) determine the values of $\tilde{Y}$ for $X = 55, 68, 75$; (c) state the proportion of variance in Y that is accounted for by the correlation between X and Y.

18.12 Presented below are pairs of measures on two variables.

X	Y	X	Y
419	2.18	579	2.79
465	2.34	591	2.90
495	2.23	616	2.65
522	2.47	643	3.38
530	2.51	659	2.97
532	2.14	687	3.32
554	2.42	704	3.46
569	3.04		

(a) Obtain the prediction formula for $\tilde{Y}$; (b) determine the values of $\tilde{Y}$ for $X = 450, 500, 625$; (c) state the proportion of variance in Y that is accounted for by the correlation between X and Y.

18.13 The following are data on the index of retail food prices and the index of industrial production in a 10-year period.

Year	1949	1950	1951	1952	1953	1954	1955	1956	1957	1958
Price Index	100	101	113	115	113	113	111	112	115	120
Production Index	64	75	81	84	91	85	96	99	100	93

Let X represent the price index and Y the production index. (a) Obtain the prediction formula for $\tilde{Y}$; (b) determine the values of $\tilde{Y}$ for $X = 100, 113, 119$; (c) state the proportion of variance in Y that is accounted for by the correlation between X and Y.

18.14 In a sample of 30 pairs of X and Y values, the correlation coefficient is calculated to be 0.350. Does this indicate a significant correlation at the .05 significance level?

18.15 A correlation coefficient of 0.62 is calculated for 20 pairs of observations. Should this value be considered significantly different from $r = 0$ on the basis of the .05 significance level? .01?

18.16 Consider the following pairs of data as random samples from approximately normal populations, and test the null hypothesis that the correlation in the population is zero. Use a two-tail test at the .05 significance level.

X	72	67	65	60	56
Y	86	84	85	80	75

18.17 Refer to the pairs of measures given in Problem 18.12. Consider the pairs of data as random samples from approximately normal populations, and test the null hypothesis that the correlation in the population is zero. Use a one-tail test at the .05 significance level.

18.18 Refer to the data given in Problem 18.13. Consider the pairs of data as random samples from approximately normal populations, and test the null hypothesis that the correlation in the population is zero. Use a two-tail test at the .01 significance level.

18.19 The following are data on the number of commissioned officers in the WAF and the gold reserve (in hundreds of millions of dollars) of U.S. banks in selected years:

Year	1954	1955	1956	1957	1958	1959
WAF Officers	789	704	634	630	672	733
Gold Reserves	218	218	221	229	206	195

Consider the pairs of data above as random samples from approximately normal populations, and test the null hypothesis that the correlation in the population is zero. Use a one-tail test at the .01 significance level.

18.20 Consider the following pairs of vocabulary and spelling scores as random samples from approximately normal populations, and test the null hypothesis that the correlation in the population is zero. Use a two-tail test at the .05 significance level.

Vocabulary	73	86	80	73	67	65	74	65	76	73
Spelling	39	71	74	78	76	74	87	62	78	74

Vocabulary	70	88	60	79	63	56	80	92	65	74
Spelling	78	87	71	78	66	74	83	79	81	80

nineteen

nonparametric statistical tests

19.1 Introduction. We have discussed procedures for testing various hypotheses. Some of these hypotheses concerned population means, and the one in the last chapter was the hypothesis that the correlation coefficient in a population is zero. Remember that when testing hypotheses regarding means, we have in each case assumed an approximately normal distribution in the population from which our samples were drawn. As mentioned, we have also assumed normality of the variables when testing the hypothesis that a population correlation coefficient is zero.

In contrast, when we worked with contingency tables and did the chi square test in Chapter 15, we made no assumption of this kind. That is, we did not stipulate that the distributions of the variables involved had to be normal, or had to have any other specified shape, for that matter.

There is an important distinction between the procedures that necessitate making some assumption regarding the shape of a distribution and those for which we need not make any such assumption to proceed. In contrast with the first type of procedure, the latter type is called a *nonparametric* or *distribution-free* method of testing a hypothesis. Although the only instance of a nonparametric test discussed so far has been the chi square test, there exist many other such tests used for different purposes. The inability to assume normality in various situations is becoming increasingly of concern to researchers. As substitutes for the older parametric tests, which had been used where they were not indicated, the distribution-free hypothesis tests are becoming more and more popular.

To use a nonparametric test is a necessity when the distribution of interest is known not to be normal and when one can infer absolutely nothing regarding its shape. Where an assumption is not warranted, it is obviously inappropriate to make it implicitly by using a test that follows from it. On the other hand, where the assumption of normality is warranted, to renounce making this assumption and to use a nonparametric test would be a serious mistake. As in other instances, to renounce making a warranted assumption is to hold oneself back from drawing a proper conclusion. Where normality exists, the refusal to assume it and the decision to use a nonparametric test is the surrendering of a powerful tool for a weaker and more general one. It is true that nonparametric tests require fewer assumptions than the methods described earlier. But for this very reason they give less precise results than parametric tests when they are used in situations where parametric tests are applicable.

Nonparametric tests tend as a group to be easy to apply. Where such tests are applicable they are a great convenience. But especially where the actual data gathering has been painstaking, the fact that a nonparametric test is easy to carry out should not commend it to use. The careful research worker always asks himself first whether he can use a parametric test, since it would give him the most accuracy, and if it appears reasonable to do so he uses one. It is only after he has carefully ruled out the use of a parametric test that he uses a nonparametric one. Having issued these words of caution, we shall move on to discuss several nonparametric methods. Those presented here are among the more important and widely used of the nonparametric tests.

19.2 The Rank Correlation Test. The oldest and for many years the most widely used of all nonparametric methods is that known as *rank correlation*. It was originally devised as a short-cut method for estimating the value of a correlation coefficient. As the reader will soon see, it has the advantage of circumventing much of the computation involved in the calculation of the Pearson product-moment correlation coefficient. But in view of what has been said, when it is appropriate to compute a product-moment correlation coefficient and to test it for significance, it is often not satisfactory merely to compute a rank correlation coefficient instead.

In the application of this method we make no assumptions whatsoever about the distributions of the underlying populations. We do not assume normality either to compute the rank correlation coefficient or to test the hypothesis that there is zero correlation in the population. It is only necessary for us to be able to arrange the sample observations in rank order. Sometimes the ranking is actually done by judges, and at other times we must do the actual ranking of numerical scores we have obtained—when we are unable to assume that the scores are normally distributed. Each situation will be illustrated to show how we proceed to compute the rank correlation.

First we shall consider the situation in which the actual ranks are given to us, and later we shall consider the one where we must rank the scores ourselves. Table 19.1 shows the ratings given to ten contestants by two judges

TABLE 19.1   RANK CORRELATION FOR
BEAUTY CONTEST RATINGS

(1) CONTESTANT	(2) JUDGE A x	(3) JUDGE B y	(4) $x-y$	(5) $(x-y)^2$
M	1	3	−2	4
N	2	2	0	0
O	3.5	1	2.5	6.25
P	3.5	8	−4.5	20.25
Q	5	7	−2	4
R	6	9	−3	9
S	7	4	3	9
T	8	6	2	4
U	9	5	4	16
V	10	10	0	0

$$\sum (x-y)^2 = \overline{72.50}$$

in a beauty contest.   The ten contestants are numbered in Column (1).   The judges are designated A and B.   Each contestant's rating received from Judge A appears in Column (2) and from Judge B in Column (3).   For instance, the girl who is numbered as the first contestant received a rating of 1 from Judge A and a rating of 3 from Judge B.   Similarly, each contestant's pair of ratings appears in Columns (2) and (3).

Our purpose now is to measure the correlation between Columns (2) and (3).   In other words, our purpose is to measure the degree to which Judges A and B correlate in their ratings of the beauty of the ten contestants.

Let us look briefly at the ratings by Judge A, which appear in Column (2). The two ratings of 3.5 appearing in Column (2) indicate that Judge A was unable to decide between two of the contestants and was willing to divide third and fourth ranks between them $\left(\frac{3+4}{2}=3.5\right)$.   The ranks given by Judges A and B have been designated as variables x and y, respectively. These are our variables of interest in this example.

The formula for the rank correlation coefficient requires the value of the square of the difference of the ranks given to each contestant.   Accordingly, the differences, $x-y$, have been found and are entered in Column (4) of Table 19.1.   These differences have been squared and the squared values are shown in Column (5) of the table.

The formula that has been devised mathematically for the rank correlation coefficient is:

(19.1) $$R = 1 - \frac{6 \sum (x-y)^2}{N(N^2 - 1)}$$

In Formula (19.1), R represents the coefficient of rank correlation that we are to find. $\sum (x-y)^2$ represents the sum of the squares of the differences in ranks.   That is, $\sum (x-y)^2$ represents the sum of the terms in Column (5) of Table 19.1.   Finally, N stands for the number of individuals who have

305

been ranked. We see that in our problem $\sum (x-y)^2$ turns out to be 72.50, and $N=10$. Thus, we have

$$R = 1 - \frac{6(72.50)}{10(10^2 - 1)}$$
$$= 1 - .43$$
$$= .57$$

We have found the value of R, the rank correlation coefficient for the ratings given by Judge A and Judge B, to be .57. Note that in this particular problem our original data were ranks. When this is the case—that is, when we do not have scores but only ranks to begin with—then the formula for r given in Chapter 17 would yield the same value as Formula (19.1),* except that Formula (19.1) is easier to apply. However, we still cannot consider our finding of .57 as a value of r, but we must think of it as a value of R. The distinction becomes meaningful when we come to test whether our finding is significantly different from zero.

Where our original data are ranks, we must of course compute a rank coefficient. Where our data are scores, and we suspect that the scores on even one variable are far from normal, we must hesitate before interpreting the correlation coefficient. A chance pairing of extreme scores might have accounted for much of the correlation (positive or negative). In such a case we may proceed by converting our scores to ranks and proceeding. (See Section 19.3.)

Without assuming normality, we could not ultimately take our obtained correlation coefficient and use it to test the hypothesis that the correlation in the population from which the sample came is zero. On the other hand, the rank coefficient has the advantage of leading to a test of this hypothesis that does not necessitate the assumption of normality. Therefore, when normality seems unlikely and we wish to correlate two variables, what we do is rank the scores on each variable ourselves in order to compute the rank coefficient.

As an illustration, consider the scores made by a class of fifteen students on each of two tests, A and B. Let us say that the scores on Test A tended to be either very low or very high, since students were either well or poorly prepared for it. Clearly the scores on this test were far from normally distributed. Therefore, in order to measure the correlation between scores on the two tests, we are led to compute a rank correlation coefficient.

The scores of the 15 students on Tests A and B are given in Columns (1) and (3) of Table 19.2. (This time we have omitted a left-hand column listing the students themselves.) Our first step is to assign a rank to each score. The ranks for the scores on Test A are given in Column (2). These have been obtained by giving the highest score the rank 1, the next highest score the rank 2, and so on. Similarly, the ranks for the scores on Test B have been found

---

* The values are identical only provided that there are no ties in rank.

and are given in Column (4). The reason for the appearance of rank 7 three times in Column (4) is that three students had scores of 81 on Test B. These three students occupy ranks 6, 7, and 8, which have been divided among them $\left(\frac{6+7+8}{3}=7\right)$. The values in Columns (2) and (4) have been labeled x and y. They are the two variables of interest in our problem.

·Next the values of $x-y$, the differences in rank, have been found and entered in Column (5) and the squares of these values are in Column (6) of Table 19.2. The sum of the numbers in Column (6) is $\sum (x-y)^2 = 164$. Applying Formula (19.1) we have

$$R = 1 - \frac{6\sum(x-y)^2}{N(N^2-1)} = 1 - \frac{6(164)}{15(225-1)} = 1 - \frac{984}{15(224)}$$

$$= 1 - \frac{984}{3360} = 1 - .29 = .71$$

This is the value of the rank correlation coefficient between scores on Tests A and B as found for the fifteen students.

TABLE 19.2 RANK CORRELATION FOR TEST SCORES

(1) TEST A	(2) RANK x	(3) TEST B	(4) RANK y	(5) $x-y$	(6) $(x-y)^2$
99	1	96	2	−1	1
98	2	95	3	−1	1
96	3	100	1	2	4
94	4	81	7	−3	9
88	5	88	5	0	0
70	6	65	12	−6	36
62	7	81	7	0	0
54	8	58	14	−6	36
47	9	90	4	5	25
41	10	81	7	3	9
35	11	72	10	1	1
33	12	50	15	−3	9
28	13	80	9	4	16
25	14	60	13	1	1
22	15	70	11	4	16
					$\sum (x-y)^2 = 164$

19.3 Significance of Rank Correlation Coefficient.   In Section 18.6 we discussed the test for significance of a value of the Pearson product-moment correlation coefficient, r.   The test presented there was subject to the assumption that the variables of interest could be taken as samples derived from normal distributions.   However, the rank correlation coefficient, R, is calculated directly from ranks, and hence there is no assumption about the distributions of the variables of interest.   Remember that this is true of all the nonparametric methods that we shall discuss in this chapter.

It may be shown that, even when the variables are not normal, we may test R for significance as we have tested r. That is, we may use the transformation:

$$(19.2) \qquad t = \frac{R\sqrt{N-2}}{\sqrt{1-R^2}}$$

where $\text{d.f.} = N-2$.

Note that had we computed r instead of R in our second problem, we would not have been correct in using this transformation. In other words, when one or both variables are not normal, it is not correct to compute r and then to use Formula (18.5). One must first rank the variables and then compute R, which is apt to have quite a different value. Then one can use Formula (19.2), which transforms the obtained value of R into a t value, to test the hypothesis that the correlation in the population is zero.

For example, suppose we consider the test-score data of the previous section as constituting a sample. Our theory is that there is a positive correlation between the two tests. Therefore we test the hypothesis of no correlation between scores on Tests A and B. We shall use the .05 level of significance. We have $R=.71$ and $N=15$. Using Formula (19.2), we have $t=.71\sqrt{15-2}/\sqrt{1-(.71)^2}=3.64$. Remember, we are predicting that the correlation in the population departs from zero in the positive direction. Therefore we are doing a one-tail test for 13 d.f. Since the necessary value of t for significance at the .05 level for a one-tail test is 1.77, we are able to reject our hypothesis and consider our theory as proved. The reader should verify that the value of R that we obtained is large enough so that it would have led to a t value that is significantly different from zero at the .01 level for 13 d.f. also.

The reader should also test the null hypothesis for the value $R=.57$ obtained for the beauty contest ranks in the first example of the previous section. He should verify that this hypothesis cannot be rejected at the .01 level for 8 d.f. The obtained value of $R=.57$ is one that would occur too often by chance to be considered significant at that level.

19.4 The Sign Test—"Before" and "After" Data. In Sections 14.3 and 14.4 we discussed the t-test for the difference between means of two distributions. We saw that this test has many important applications in practice. The use of this test was predicated on the assumptions of normality and the independence of the samples. In other words, we assumed that each sample was chosen in such a way as to have no effect on the choice of the other sample. This assumption is of course implicit in the definition of "random sample" and it must be kept in mind whenever data obtained from random samples are used.

The assumption of independence is not satisfied in one very important type of experiment—namely, the "before" and "after" type of study.

Suppose, for example, that a researcher has only one class and wishes to determine whether a particular lecture that he gives this class actually increases their incentive to perform well in arithmetic. He gives the class a pre-test of their performances on handling arithmetic concepts; then he gives his lecture, after which he gives the class a post-test. He would like to use the results of the two tests to determine whether his lecture actually improved their scores.

In essence, each student gets a pre-test score and a post-test score, as indicated in the first two columns of Table 19.3 for the ten students whom we shall say compose his class. Note that here the scores in Column (1) are not independent of those in Column (2). Rather, each score in one group is coupled with a score in the other—namely, that produced by the same person. Because of this natural pairing, we cannot do the *t*-test described in Section 14.3; but we can use the one in Section 14.5.

TABLE 19.3   SIGN TEST FOR PRE-TEST AND POST-TEST SCORES

(1) PRE-TEST	(2) POST-TEST	(3) SIGN OF DIFFERENCE
65	75	+
80	79	−
50	90	+
92	90	−
85	80	−
75	85	+
45	68	+
89	89	0
60	82	+
65	75	+

We might take each student's post-test score and subtract from it his pre-test score to obtain what is sometimes called his "change" score. For the case in which we have reason to believe that these "change" scores are normally distributed, there does exist a variation of the *t*-test. This test has frequent application and it is described in Section 14.5. More often it occurs that because of individual characteristics, some persons change drastically whereas others change hardly at all. In such a case, one drastic change in a negative direction, for instance, might be enough to balance out many changes in the positive direction. Where such a result is anticipated, there may be reason to believe that the obtained differences will not be normally distributed. In fact, an experimenter might properly wish to avoid using a measure that equates the units of change for a relatively few people with those of others whose scores do not seem to be so amenable to change. Specifically, he is interested only in the proportion of individuals who change in one direction as opposed to the other. That is, he would like to give each individual equal weight regardless of how much he changed.

In such a case, the appropriate procedure is for the researcher to record for each individual only the sign, plus or minus, that indicates the direction

309

in which he changed. Thus, where a subject increases his score in going from pre-test to post-test, this fact is recorded by a plus sign. The fact of a decrease is recorded by a minus sign. The next step for the experimenter is to test whether the discrepancy between plus and minus signs is extreme enough to reach the decision that the discrepancy was not the result of chance.

The appropriate test to be used is known as the sign test, since it deals with only the signs of the individual changes and not with their magnitudes. Column (3) of Table 19.3 shows only the sign of the difference in performance of each student on the two tests. This has been recorded as + if the student did better on the post-test than on the pre-test and as − if his score went down from the first test to the second. In one case where there was no change, a 0 has been recorded in Column (3). This case will not be included in our treatment of the data. Thus we find that there are 6 plus signs and 3 minus signs.

The theory to be proved is that the lecture given is effective in increasing students' performances on the arithmetic test. The hypothesis to be tested therefore is that the lecture produces no change in students' achievement. It follows that if the method produces no meaningful change in students, then in any given sample we should expect no more students to receive better scores than receive poorer scores on the post-test as compared to the pre-test. In other words, our hypothesis is the same as the hypothesis that we are as likely to obtain a + sign as a − sign. The reader will recognize this as equivalent to the statement that we are as likely to obtain a head as a tail when tossing a fair coin. The situation in our example is somewhat equivalent to testing the hypothesis that a coin is fair if we have obtained six heads and three tails in a sample consisting of nine tosses.

However, our situation differs in one important respect from that involved in testing the fairness of a coin. We shall use a *one-tail test*, because we are interested in the specific alternative that the lecture is effective—that is, that it aids student achievement. Let us also decide in advance to use the .05 significance level.

Since we are dealing with a dichotomous variable, we now use the procedure described in Section 11.3. We suppose that our sample is one of a vast number of equal-sized samples obtained in the same way. Let us designate the plus sign as the one of interest, so that the outcome of our sample is 6. Under the hypothesis the expectancy is for an equal number of pluses and minuses. Thus, in this vast number of samples of 9 cases each, some would contain no pluses, others would contain one plus, others two pluses, and so on. In other words, the number of pluses in each sample is the outcome of interest, and in different samples this number would range from zero to nine.

Our concern is with locating our outcome of 6 in the distribution of outcomes that would be found, assuming that the hypothesis is true. Applying Theorem 11.1, we calculate the mean and standard deviation of this distribution of outcomes. Under the hypothesis, the probability, P, of a plus is $P=1/2$. $N=9$. The mean,

$$\mu = NP = 9(1/2) = 4.5; \qquad \sigma = \sqrt{NP(1-P)} = \sqrt{9(1/2)(1/2)} = 1.5$$

This distribution is shown in Fig. 19.1.

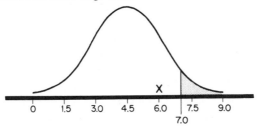

Fig. 19.1

Recalling our discussion of one-tail tests in Section 12.6, we now set up our rejection region in the right-hand tail of the distribution. Referring to Appendix III, we find that the z value, 1.64, places 5 per cent of the area in the extreme right-hand tail. For our example, 1.64 standard deviations amount to 2.46 points. Thus, where we are at the mean of 4.50 and add 1.64 standard deviations, what we do in effect is add 4.50 and 2.46, giving us approximately the point 7.0 as the point at which to draw the vertical line that marks the boundary of the rejection region that is shaded in Fig. 19.1. But our observed number of plus signs is six and this does not fall in the rejection region. Hence, we cannot reject the hypothesis that the teaching method produces no difference in student achievement.

We could have proceeded without reference to Fig. 19.1 by simply computing the z score for our observed value. Using Formula (12.1) we have,

$$z = \frac{O-NP}{\sqrt{NP(1-P)}} = \frac{6-4.5}{1.5} = \frac{1.5}{1.5} = 1.0$$

Since this z value is smaller than the 1.64 required for one-tailed rejection of our hypothesis at the .05 level, we fail to reject the hypothesis. We cannot conclude that the lecture effectively improved students' performances.

As another example, suppose that the experiment had been carried out with 30 students and had resulted in 22 students receiving better scores, 3 showing no change, and 5 receiving poorer scores on the post-test as compared to the pre-test. We would then have 22 plus signs and 5 minus signs for the 27 students for which there was an observable change. If we again test the hypothesis of no change at the .05 level, following the same procedure as before, we have,

$$z = \frac{O-NP}{\sqrt{NP(1-P)}} = \frac{22-27(1/2)}{\sqrt{27(1/2)(1/2)}} = \frac{22-13.5}{(1/2)\sqrt{27}} = \frac{8.5}{(1/2)(5.2)}$$

$$= \frac{8.5}{2.6} = 3.26$$

311

Since this exceeds the value of 1.64 needed for rejection of the hypothesis, we accordingly do reject the hypothesis. This, then, proves the theory that the teaching method is effective, using the .05 level.

19.5 Runs. In all of our previous discussions of sampling and tests based on samples, it has been assumed that the samples were random. This assumption may not always be justified. For instance, samples of the prices of stocks on the New York Stock Exchange taken at different times over a period of several months would be influenced by any general trends of the market over that period of time. As another example, samples of the speeds of automobiles passing a given point that are taken by recording the speeds of several cars in succession may very well be suspected of lacking randomness. Drivers who are caught behind a "slow poke" will also be driving slowly, while drivers following a speeder may also be exceeding the speed limit, perhaps unconsciously. In each of these examples, we see that there would be a tendency for successive items included in the sample to have similar values: high stock prices or slow speeds. Such a succession of similar values is known as a "run." The examples just cited would be suspected of producing long runs and, hence, relatively few runs in a given sample.

Statistical methods of the nonparametric type have been devised for testing data for the presence of an abnormal number of runs. A sample with too many runs is just as suspect as one with too few runs. Too many runs tend to indicate that the data occur in definite cycles. The tests based on runs, then, provide us with information for drawing conclusions about the randomness of a sample. Such information is particularly important in instances where it is necessary to utilize data recorded over a period of time in order to make predictions about future events. Such would be the case, for example, in certain aspects of meteorology, in studying the incidence of infectious diseases, and in studying fatalities resulting from drowning or other types of accidents.

To start with a simple illustration of a test based on runs, let us suppose that the speeds of cars on a certain residential street have been checked and compared with the speed limit, which is 25 miles per hour. For each car an "a" has been recorded if it was traveling at a speed above 25 m.p.h., and a "b" if it was traveling at a speed below 25 m.p.h. Cars traveling at exactly 25 m.p.h. have been omitted from the list. Let us suppose that 18 successive entries have been obtained as follows: a, b, b, b, a, a, a, a, a, b, b, a, a, b, b, b, b, b. We now state:

Definition 19.1   *A run is a sequence of identical symbols that is preceded and followed by a different symbol (or no symbol at all).*

Thus the first run in the sequence above consists of the single letter "a"; the next run consists of three "b's"; the next of five "a's"; and so on. In all there are six runs in the sequence of 18 letters.

Now, we may suspect that our sample has too few runs for it to be a random sample. The influence of "slow pokes" and speeders may have caused long runs so that the total number of runs in the sample is relatively small. We need some way to test the relative size of the number of runs in comparison to the sample. The mathematicians have come to our aid again, and have devised such a test. The application of this test does not depend on the lengths of the individual runs; but it does depend on the total number of times that each symbol appears in the sequence. We observe, in our example, that there are eight "a's" and ten "b's." Our theory is that the sample is not random, so the hypothesis to be tested is that the sample *is* random.

The method of approach to our test based on runs should be familiar by now. We assume that each of a vast number of investigators obtains a random sample consisting of 8 "a's" and 10 "b's". Under the hypothesis, there is no tendency either toward or away from runs, but of course the obtained number of runs would vary from one sample to the next. This number of runs per sample is commonly designated by the letter u. Remember that we are assuming the hypothesis is true and conceiving of a collection of random samples, each consisting of 8 symbols of one kind and 10 of another, and it is under this condition that we are considering the distribution of u values that would be obtained.

Had there been different numbers of symbols involved in our sample, we would be concerned with a different distribution of u values, or what we call a different sampling distribution of u. That is, we do not think merely of a sample size, N, but rather of a sample of n_1 symbols of one kind and n_2 symbols of another kind. Here $n_1 = 8$ and $n_2 = 10$ (though we might have made it the reverse). In other words, in our case we conceive of our sample as one of a vast collection of similar random samples, in each of which there are 8 symbols of one kind and 10 symbols of another. This fact is important, since for each pair of values n_1 and n_2 we are led to consider a unique distribution of u values that would be found in different samples.

Our next step is to determine whether our obtained u value is sufficiently below or above the expectancy to indicate rejecting the hypothesis. In our particular problem where $n_1 = 8$ and $n_2 = 10$, our obtained u value turned out to be 6. Specifically, we ask whether this obtained value is sufficiently far from the expectancy to indicate rejecting the hypothesis. Appendix VI is set up for testing our hypothesis at the .05 level of significance. For each particular pair of values of n_1 and n_2 it gives the critical values of u. Note that each cell entry in Appendix VI contains two numbers. The smaller number is called the left-tail critical value of u, and the larger number is the right-tail critical value of u, for a two-tail test at the .05 level. This means that the rejection region consists of values at or below the smaller number and the values at or above the larger number.

Specifically, for any pair of values of n_1 and n_2, we find the appropriate cell in Appendix VI, which means the cell whose row is given by the smaller of the values of n_1 and n_2 and whose column is given by the larger of these

313

values. Once we have found that cell, we reject the hypothesis only if our obtained value of u is either equal to or less than the smaller number in the cell or equal to or greater than the larger number in the cell. In our example, $n_1 = 8$ and $n_2 = 10$, and the observed value of u is 6. In using the table, since 8 is the smaller of n_1 and n_2, we find the row with 8 in the left margin; and since 10 is the larger we find the column with 10 at the top. We thus find the cell that is in this row and column. The two numbers in this cell are 5 and 15. This means that we would reject as not random a sample having 5 or fewer runs or a sample having 15 or more runs. Since in our sample $u = 6$, we fail to reject at the .05 level the hypothesis that our sample is random, and thus we have not proved our theory.

As a variation on our example, suppose that the first "a" had not been present in the sequence based on speed data. The sequence would then contain 7 "a's" and 10 "b's" and would have 5 runs. We now have $n_1 = 7$, $n_2 = 10$, and $u = 5$. In the table in Appendix VI, we look in the cell that appears in the column with 10 at the top and in the row with 7 in the left-hand margin. The critical values given for u in this cell are 5 and 14. Since our observed value is 5, we would now reject the hypothesis that the sample is random and thus consider proved the theory that it is not a random sample. In this new example there are too few runs and the reason may involve the "slow pokes" and speeders previously cited.

The test using runs as we have presented it may be used in any situation where each item of the data is given as one or the other of two symbols. Another example would be the situation in which there are data on defective and satisfactory articles produced by a manufacturing process. The runs test may be done as a one-tail test and also at other levels of significance, though of course in each case appropriate tables would be needed. Note that we might use Appendix VI to do the runs test as a one-tail test at the .025 level of significance. In such a case we would pick one of the critical values in the appropriate cell as our theory dictated, and we would then reject the hypothesis only upon obtaining a u value at or beyond that critical value. To prove that there are too few runs for randomness, we would reject the hypothesis only upon obtaining a u value at or smaller than the lesser critical value given. To prove that there are too many runs for randomness, it would be correct to reject the hypothesis only upon obtaining a u value at or larger than the greater critical value given.

19.6 Runs Relative to the Median. In the example of the previous section, there was a "built-in" reference point. The speed limit was 25 miles per hour and so it was natural to classify the speeds as above and below this value. In some situations where it is desirable to apply the runs test, there is no such "built-in" reference point. In such cases we will have to supply a reference point before we can classify the data as "above" and "below" and check on runs.

Some examples of the type of data just mentioned are weather observations and stock market prices. These do not appear in a form containing only two symbols. Before we can apply the runs test to them, we will have to convert the data into such a form. Then we will be able to proceed in the same manner as in the previous section and we can obtain a check on the randomness of the data.

We shall use an example from weather observations. Let us suppose that the number of inches of snowfall in 25 successive winters in our city has been recorded. The data appear as follows: 32, 45, 49, 42, 30, 67, 72, 73, 79, 25, 28, 36, 49, 52, 21, 42, 49, 67, 43, 75, 21, 17, 44, 42, 96. We are interested in whether the data appear in essentially random fashion or in patterns of some sort. Too few runs would indicate that the data tend to fall in rather long cycles of heavy and light snowfall, thus resulting in long runs. Too many runs would indicate that winters of heavy and light snowfall tended to alternate, thus resulting in short runs.

Before proceeding we must decide on a reference point. The median suggests itself as a good choice, owing to its definition as a value that is larger than or equal to half of the other values in the distribution and equal to or smaller than half of them. In the set of data above there are 12 terms that are greater than 44 and 12 terms that are less than 44, so the median is 44.

We now replace each actual term by one of the symbols "a" and "b," depending on whether the term is above or below the median. However, we omit values at the median—in this case the one term that is 44. After each value above the median has been replaced by an "a" and each value below the median by "b," the resulting series is: b, a, a, b, b, a, a, a, a, b, b, b, a, a, b, b, a, a, b, a, b, b, b, a. We have now reduced the data to the form used in the previous section and thus are ready to apply the runs test precisely as we did there. Our theory is that the data do not fall in a random pattern. Our hypothesis to be tested is that the data do constitute a random sample.

We have 12 "a's" and 12 "b's" and there are 12 runs in the sequence. Hence, we are ready to use the table in Appendix VI with $n_1 = 12$, $n_2 = 12$, and our observed value of u is 12. We look in the cell that appears in the column with 12 at the top and in the row with 12 in the left margin. The critical values given for u in this cell are 7 and 19. Hence we cannot reject our hypothesis of randomness at the .05 level for a two-tail test. Note that were we using a one-tail test, whether right-hand or left-hand, we still could not reject the hypothesis of randomness, this time at the .025 level.

PROBLEM SET A

19.1 Consider the data below as coming from populations not necessarily normal. Compute the rank correlation coefficient and test it for significance using a two-tail test at the .05 significance level.

X	114	118	122	120	126
Y	140	95	155	125	185

19.2 Consider the following high school and college freshman averages for 10 students as coming from populations not necessarily normal. Compute the rank correlation coefficient and test it for significance using a two-tail test at the .01 significance level.

High School	66	70	75	80	82	85	90	92	95	98
College	60	78	65	87	74	70	78	95	88	90

19.3 Consider the following 15 pairs of college grade-point averages and National Teacher Examination scores as coming from populations not necessarily normal. Compute the rank correlation coefficient and test it for significance using a one-tail test at the .05 significance level.*

Grade-Point Average	Teacher Examination	Grade-Point Average	Teacher Examination
3.11	567	2.75	647
2.23	592	2.37	742
2.82	573	2.42	580
3.62	794	2.45	588
3.57	668	3.31	638
2.33	595	3.26	622
2.65	639	2.47	659
2.84	698		

19.4 Consider the following vocabulary and spelling scores as coming from populations not necessarily normal. Compute the rank correlation coefficient and test it for significance using a one-tail test at the .01 significance level.

Vocabulary	73	86	80	73	67	65	74	65	76	73
Spelling	39	71	74	78	76	74	87	62	78	74

Vocabulary	70	88	60	79	63	56	80	92	65	74
Spelling	78	87	71	78	66	74	83	79	81	80

19.5 A reducing diet has produced the following weight changes with 12 people:

Before	176	157	166	202	178	179	170	181	190	192	140	138
After	160	157	167	193	172	171	174	187	185	187	144	130

Test the null hypothesis that the diet is not effective. Apply the sign test at the .05 significance level for a two-tail test.

19.6 The weights of 12 people before they stopped smoking and six weeks after they stopped smoking are as follows:

Before	127	157	145	162	181	172	169	156	187	115	118	173
After	136	161	137	164	190	180	162	164	165	118	117	180

Test the null hypothesis that to quit smoking has no effect on a person's weight. Apply the sign test at the .01 significance level for a one-tail test.

19.7 Student spectators waiting in line for admission to a high school athletic contest are arranged as follows (where "b" denotes boy and "g" denotes girl): g, g, b, b, b, b, b, b, g, g, b, b, b, b, b, b, b, g, g, g, g, b, b, b, b, b. Test these data for randomness using a two-tail test at the .05 significance level.

* See the footnote on page 150.

19.8 The arrangement of adults and children entering a certain movie theater is as follows: a, c, a, c, c, a, c, c, c, a, c, a, c, c, a, a, c, c, a, c, c, a, c, c, a, c, a, c, a, a, c, a, c, c, a, c, a, c, a. Test these data for randomness using a one-tail test at the .025 significance level.

19.9 The speeds of cars passing a given point are as follows: 27, 38, 45, 21, 29, 52, 38, 46, 18, 17, 55, 50, 32, 31, 31, 31, 58, 57, 38, 49, 27, 39, 31. Test these data for randomness using a two-tail test at the .05 significance level.

19.10 Successive weights (in milligrams) of a certain chemical precipitate were recorded as follows: 17, 15, 18, 15, 18, 18, 17, 17, 15, 19, 14, 15, 18, 17, 15, 14, 14, 17, 14, 15. Test these data for randomness using a one-tail test at the .025 significance level.

PROBLEM SET B

19.11 Consider the data below as coming from populations not necessarily normal. Compute the rank correlation coefficient and test it for significance using a two-tail test at the .05 significance level.

X	72	67	65	60	56
Y	86	84	85	80	75

19.12 Consider the following 15 pairs of college entrance-examination scores and college grade-point averages for four years as coming from populations not necessarily normal. Compute the rank correlation coefficient and test it for significance using a two-tail test at the .01 significance level.

Entrance Examination	Grade-Point Average	Entrance Examination	Grade-Point Average
419	2.18	579	2.79
465	2.34	591	2.90
495	2.23	616	2.65
522	2.47	643	3.38
530	2.51	659	2.97
532	2.14	687	3.32
554	2.42	704	3.46
569	3.04		

19.13 Consider the following pairs of reading and arithmetic scores as coming from populations not necessarily normal. Compute the rank correlation coefficient and test it for significance using a one-tail test at the .05 significance level.

Reading	43	58	45	53	37	58	55	61	46	64	46	62	60	56
Arithmetic	32	25	28	30	22	25	22	20	20	30	21	28	34	28

19.14 A certain reducing diet has produced the following weight changes in 10 people:

Before	129	129	122	127	145	129	147	132	139	136
After	121	126	115	121	141	125	153	130	130	141

Test the null hypothesis that the diet is not effective. Apply the sign test at the .05 significance level for a one-tail test.

317

19.15 A certain method of teaching for an understanding of the concept of correlation produces the following results:

Pre-test   67 72 79 87 92 81 96 79 85 88
Post-test 81 83 91 85 90 80 98 83 87 91

Test the null hypothesis that the method is not effective. Apply the sign test at the .05 significance level for a two-tail test.

19.16 A series of manufactured parts is tested for correct dimension. The resulting sequence of above and below readings is as follows: a, b, b, a, a, a, b, a, b, b, b, a, a, b, b, a, a, a, b, a. Test these data for randomness using a two-tail test at the .05 significance level.

19.17 In a row of mixed marigolds the yellow and orange flowers are arranged as follows: y, o, y, o, y, y, y, o, o, y, y, y, y, o, y, o, y, o, o, o, o, y, y, y, o, y, o. Test these data for randomness using a one-tail test at the .025 significance level.

19.18 The cases of scarlet fever reported in successive years in a certain residential district are as follows: 5, 8, 6, 3, 1, 1, 5, 7, 9, 10, 8, 7, 4, 2, 7, 1, 5, 6, 8, 1, 7. Test these data for randomness using a two-tail test at the .05 significance level.

19.19 In a quality-control process, manufactured items are removed intermittently from a production line and scored by inspection. The scores for a particular time interval are 7.2, 7.3, 7.0, 6.9, 7.1, 7.2, 7.2, 7.3, 7.2, 7:0, 6.9, 6.8, 6.9, 7.0, 7.1, 7.2, 7.2, 7.3, 7.0, 6.9, in consecutive order. Test for randomness using a two-tail test at the .05 significance level.

19.20 A coin is tossed 12 times with the following sequence of heads and tails: H T T H T T H H T T H T. Test for randomness using (a) a two-tail test at the .05 significance level, (b) a one-tail test at the .025 significance level.

twenty

the F test and analysis of variance

20.1 Introduction. We sometimes want to test the hypothesis that two populations have the same, or equal, variances. For instance, two methods of teaching arithmetic are tried with sixth grade students, and samples of scores of students taught by each method are gathered. We shall assume that the distribution of arithmetic scores is known to be normal. We compute the variances of the two samples and designate them $S_1{}^2$ and $S_2{}^2$. Our null hypothesis is that the two teaching methods do not produce different effects on the variability of students' scores; under this hypothesis, our two sample variances would be independent estimates of the same unknown σ^2, the variance of scores in the larger population.

Having obtained our values of $S_1{}^2$ and $S_2{}^2$, we wish to compare them in some way to test our null hypothesis. But how? If the two obtained values were identically equal, we could not possibly reject the hypothesis. Almost certainly though, the obtained values would be different. The question is, "How are we to tell whether the obtained discrepancy is sizeable enough to allow us to reject our null hypothesis at whatever significance level we choose?" The F test answers this question for us.

We make reference to the F distribution whenever we test the hypothesis that two sample variances, $S_1{}^2$ and $S_2{}^2$, are independent estimates of a single population variance, σ^2. We assume normality in the parent population, though departures from normality tend not to invalidate the procedure unless they are drastic. After considering the F distribution, we shall do several problems testing the null hypothesis that a pair of sample variances are both estimates of a common σ^2.

319

20.2   The Theoretical Model of the F Distribution.   Picture a giant wooden crate with an inexhaustible supply of slips of paper, each with a number written on it.   Let us say that the distribution of numbers on the slips is normal, with unknown mean and variance, which we call μ and σ^2, respectively.   We choose two numbers arbitrarily, which we shall use to determine the sizes of samples to be drawn.   The numbers, let us say, are $N_1 = 21$ and $N_2 = 25$.

We now draw a pair of random samples from the crate, the first of 21 slips and the second of 25.   The variance of the sample of 21 slips is, let us say, $S_1^2 = 35$, and that of the second sample is $S_2^2 = 20$.   We divide the variance of our first sample by that of the second.

$$\frac{S_1^2}{S_2^2} = \frac{35}{20} = 1.75$$

We repeat the process, again drawing random samples of sizes 21 and 25, and again dividing the variance of the sample of 21 slips by that of the sample of 25.   This time the variances are $S_3^2 = 10$ and $S_4^2 = 30$, where S_3^2 was computed from the sample of 21 slips and S_4^2 from that of 25 slips.

$$\frac{S_3^2}{S_4^2} = \frac{10}{30} = .333$$

Our next pair of samples give us a ratio with a value of 1.25.

$$\frac{S_5^2}{S_6^2} = \frac{25}{20} = 1.25$$

Let us say we repeat the process of drawing a pair of random samples, the first of size 21 and the second of size 25, and each time divide the variance of our 21 slip sample by that of our 25 slip sample.   Imagine, for instance, that the two hundredth time we do it, our ratio turns out to be .20:

$$\frac{S_{399}^2}{S_{400}^2} = \frac{4}{20} = .20$$

We record the computed ratios by piling up squares on a line.   After we've obtained 500 ratios, our graph looks like that shown in Fig. 20.1.

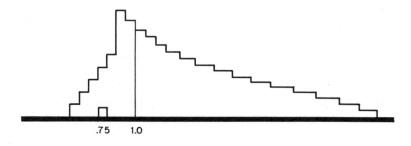

.75   1.0

Fig. 20.1

Remember that the variance of our sample of 21 cases has $N_1 - 1$ degrees of freedom, which is 20; and that the variance computed from our sample of 25 cases has $N_2 - 1$ degrees of freedom, which is 24. We have obtained and graphed a collection of ratios, each of which has a numerator based on 20 degrees of freedom and a denominator based on 24 degrees of freedom. In each fraction, both numerator and denominator are estimates of the unknown σ^2, and the numerator and denominator of each fraction are independent.

It might be argued that the numerator of any given fraction is as likely to be bigger than the denominator of that fraction as it is to be smaller than the denominator; and thus as we increase the number of ratios in our collection, it should become increasingly likely that about half of the ratios would have values less than 1.000 and half of them more than 1.000. They would tend to cluster around 1.000. Actually, it can be shown mathematically that only when N_1 and N_2 are the same will the median be exactly 1.000. If N_1 and N_2 are nearly equal, as in our example, the median will be close to 1.000. The true median is .994 in our case.

Unlike the normal curve, the long run distribution we are approaching this time is not symmetrical. Since variances are always positive, no ratio can be less than zero, whereas some ratios are very large. As we continue adding new ratios, the graph of their frequencies approaches the smooth curve in Fig. 20.2.

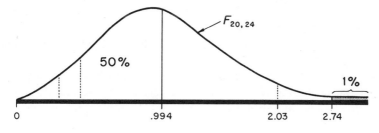

Fig. 20.2

We have generated a new distribution, the F distribution, with degrees of freedom 20 and 24. The first number, 20, indicates the degrees of freedom that went into each numerator estimate of σ^2; and the second number, 24, reports the d.f. in each denominator estimate. Figure 20.2 shows that as we continue our procedure, half of our obtained ratios are expected to be less than .994 and half more than .994. Half the area is to the left of the vertical line at .994. Five per cent of the ratios are expected to be bigger than 2.03 and one per cent are expected to be bigger than 2.74 in the long run.

Suppose, instead, that we had arbitrarily chosen pairs of samples of 16 and 10 cases each, and had each time divided the variance of our 16 slip sample by that of our 10 slip sample. This time the degrees of freedom in each ratio would be 15 and 9. The F distribution generated by continued collection of such samples would appear as in Fig. 20.3.

321

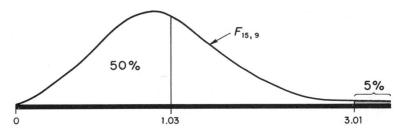

Fig. 20.3

This time it can be shown mathematically that the median is 1.03, so the axis at 1.03 divides the area in half, reflecting the fact that ratios are as likely to be less than 1.03 as more than 1.03. The curve is again nonsymmetrical, but slightly different in shape from the last one. Five per cent of the area is to the right of 3.01.

When we talk about F curves, we are talking about a family of distributions. Two members of this family were graphed in Figs. 20.2 and 20.3. Each is a long run distribution of ratios, in which the numerator and denominator are independent estimates of a common σ^2. The d.f. in the numerator and that in the denominator must both be given to designate a member of the F distribution. We denote a member of the F family by writing $F_{r,s}$ where r is the d.f. in the numerator and s in the denominator. Thus we have discussed $F_{20,24}$ and $F_{15,9}$. We use our knowledge of the F distribution whenever we test the hypothesis that two independent samples we have drawn from a population assumed to be normal are each providing unbiased estimates of the same σ^2.

When testing hypotheses and when we need to divide one variance by another, we shall always be dividing the bigger of our two sample variances by the smaller one, and then using a form of the F table which is constructed for use with one-tail tests. This practice, which assures that ratios which we look up in the F table when doing problems will always be bigger than one, simplifies computations and allows us to work with less elaborate tables than would be needed if we did two-tail tests. Later, when we do F tests in connection with analysis of variance problems, use of the one-tail test will be a logical consequence of our procedure.

Are students' arithmetic scores more variable when arithmetic is taught by Method A than by Method B? Let us choose, for example, the .05 level of significance to test the null hypothesis that there is *no* difference in variability of performance using the two methods. This null hypothesis assumes that the obtained sample variances will be independent unbiased estimates of the unknown σ^2.

Assuming a normal distribution of scores among all the students, suppose we sample 21 taught by Method A and 25 taught by Method B. If it turns out that the variance for the group taught by Method A is *not* bigger than the variance for the group taught by Method B, we need proceed no further, since we cannot possibly reject our null hypothesis in this situation. However, if it turns out that the group taught by Method A does have a bigger variance than the group taught by Method B, we must see whether that variance is enough bigger to permit us to reject our null hypothesis. Let us suppose that S_1^2, the variance of the group taught by Method A, is 46 and that S_2^2, the variance of the group taught by Method B, is 20. Since there were 21 people taught by Method A, S_1^2 is based on $N_1 - 1 = 21 - 1 = 20$ d.f. S_2^2 is based on $N_2 - 1 = 25 - 1 = 24$ d.f.

We divide the bigger sample variance by the smaller:

$$\frac{S_1^2}{S_2^2} = \frac{46}{20} = 2.30$$

Appendix Table VII shows the 1 and 5 per cent critical points for the F distribution where it is assumed that the bigger of two obtained estimates of a common σ^2 has been made the numerator. We must find the critical point relevant to our problem, where the numerator and denominator have 20 and 24 d.f. respectively. In Table VII, the column values pertain to the d.f. in the numerator. Here $n_1 = 20$. The row value in the table, indicating the d.f. in the denominator, is $n_2 = 24$. In Table VII, the number in Roman type in the box in row 24, and the column headed 20, is 2.03. This number gives us the beginning of the rejection region when the .05 level of significance is used. Since our obtained value of 2.30 is bigger than 2.03, we reject the hypothesis that S_1^2 and S_2^2 are each providing independent estimates of the same unknown σ^2.

If the experiment was conducted properly (and this implies, among other things, that the students were assigned the different methods randomly), we may conclude that the teaching methods exert different effects on the students' variability of performance, with Method A producing the more variable results. Intuitively, it should be seen that since we are always putting the bigger variance on top when we need to compute a ratio, then the bigger the ratio we find, the less likely it is that the two sample variances were estimates of the same σ^2. Since, under the hypothesis, a ratio as big as ours would be found less than five per cent of the time by chance, we were able to reject the hypothesis in our one-tail test. Note that if we had used the .01 level of significance, the ratio would not have been big enough to allow us to reject the hypothesis. Our obtained value was less than the .01 value (of 2.74) given in boldface type. For our one-tail tests, a bigger ratio is always needed to reject at the .01 level than at the .05 level.

We should like to consider one more problem, beginning with raw scores this time, in order to emphasize notation that will be needed shortly. An experimenter wants to prove that at age 21, men are more variable in mechanical aptitude than women are. He gives his aptitude test to 41 men and 31 women in this age group, planning to reject the null hypothesis only if S_1^2/S_2^2 is significantly bigger than one, where S_1^2 is the estimate of σ^2

323

based on the men's scores and $S_2{}^2$ is the estimate of σ^2 based on the women's scores. This means that if $S_2{}^2$ turns out to be bigger than $S_1{}^2$, the null hypothesis cannot be rejected. As before, the logic of our experiment dictates that the test be one-tailed. Our theory says to reject only if the

TABLE 20.1

	GROUP ONE MEN			GROUP TWO WOMEN	
	X	X^2		Y	Y^2
1.	14	196	1.	28	784
2.	29	841	2.	31	961
3.	37	1,369	3.	20	400
4.	20	400	4.	17	289
5.	38	1,444	5.	25	625
6.	29	841	6.	3	9
7.	52	2,704	7.	27	729
8.	32	1,024	8.	31	961
9.	34	1,156	9.	8	64
10.	31	961	10.	19	361
11.	23	529	11.	14	196
12.	31	961	12.	27	729
13.	34	1,156	13.	23	529
14.	29	841	14.	25	625
15.	8	64	15.	24	576
16.	16	256	16.	25	625
17.	37	1,369	17.	38	1,444
18.	32	1,024	18.	34	1,156
19.	36	1,296	19.	16	256
20.	25	625	20.	33	1,089
21.	29	841	21.	19	361
22.	38	1,444	22.	21	441
23.	30	900	23.	27	729
24.	51	2,601	24.	17	289
25.	58	3,364	25.	43	1,849
26.	29	841	26.	38	1,444
27.	25	625	27.	40	1,600
28.	35	1,225	28.	25	625
29.	29	841	29.	27	729
30.	30	900	30.	30	900
31.	32	1,024	31.	20	400
32.	28	784	$N_2 = 31$	$\Sigma\,Y = 775$	$\Sigma\,Y^2 = 21,775$
33.	27	729			
34.	28	784			
35.	9	81			
36.	27	729			
37.	21	441			
38.	28	784			
39.	33	1,089			
40.	30	900			
41.	26	676			
$N_1 = 41$	$\Sigma\,X = 1,230$	$\Sigma\,X^2 = 40,660$			

numerator is sufficiently larger than the denominator. The two sets of aptitude scores are shown in Table 20.1. The squares of the individual scores have been computed and will be used in the final computations. Our values of S_1^2 and S_2^2 will be computed by Formulas (13.2).

$$\text{d.f.} = N_1 - 1 = 40 \qquad\qquad \text{d.f.} = N_2 - 1 = 30$$

$$S_1^2 = \frac{\sum X^2 - \frac{(\sum X)^2}{N_1}}{N_1 - 1} \qquad\qquad S_2^2 = \frac{\sum Y^2 - \frac{(\sum Y)^2}{N_2}}{N_2 - 1}$$

$$= \frac{40,660 - \frac{(1,230)^2}{41}}{40} \qquad\qquad = \frac{21,775 - \frac{(775)^2}{31}}{30}$$

$$= \frac{40,660 - 36,900}{40} \qquad\qquad = \frac{21,775 - 19,375}{30}$$

$$= \frac{3,760}{40} \qquad\qquad = \frac{2,400}{30}$$

$$= 94 \qquad\qquad = 80$$

If the null hypothesis is true, the ratio of these two estimates of σ^2 is a member of $F_{40,30}$.

$$\frac{S_1^2}{S_2} = \frac{94}{80} = 1.175$$

We cannot reject our hypothesis at the .05 level of significance, since Appendix VII shows that a value as big as 1.79 or bigger is needed; and thus we cannot conclude that men are more variable than women in mechanical aptitude, at least not in the population sampled by our experimenter.

The *F* distribution was named after the late Sir Ronald Fisher, already mentioned, a genius at mathematical statistics and in particular at solving problems by intuitive geometrical devices when algebraic methods were opaque. We refer to the *F* distribution whenever we test the null hypothesis that the variances of two samples are estimates of the same σ^2 (or of σ^2's in different populations that, under our hypothesis, are identical in value).

Note that as a formula we have

(20.1)
$$F = \frac{S_1^2}{S_2^2}$$

20.3 Comparing More Than Two Sample Means. Suppose that children in the fourth grade of a school are randomly assigned to six groups, each to be taught spelling by a different method. At the end of the year, these children take a city-wide spelling test. The question we want to answer is, "May we conclude that the methods differ in effectiveness, as measured by scores on the city-wide spelling test?" We shall assume that, ordinarily, fourth grade spelling scores are distributed normally, and that 30 children have been taught by each of the six methods. At the end of the school year, 325 when the test is given to the six groups, the mean scores of these groups are:

GROUP

	Method A	Method B	Method C	Method D	Method E	Method F
MEAN SCORE	$\bar{X}_1 = 75$	$\bar{X}_2 = 72$	$\bar{X}_3 = 76$	$\bar{X}_4 = 79$	$\bar{X}_5 = 82$	$\bar{X}_6 = 72$

Are the differences wide enough for us to conclude that the methods really differ in effectiveness? Consider first how many t tests might be done involving pairs of sample means. Table 20.2 shows that there are 15.

TABLE 20.2

$\bar{X}_1$ with $\bar{X}_2$	$\bar{X}_2$ with $\bar{X}_3$	$\bar{X}_3$ with $\bar{X}_4$	$\bar{X}_4$ with $\bar{X}_5$	$\bar{X}_5$ with $\bar{X}_6$
$\bar{X}_1$ with $\bar{X}_3$	$\bar{X}_2$ with $\bar{X}_4$	$\bar{X}_3$ with $\bar{X}_5$	$\bar{X}_4$ with $\bar{X}_6$	
$\bar{X}_1$ with $\bar{X}_4$	$\bar{X}_2$ with $\bar{X}_5$	$\bar{X}_3$ with $\bar{X}_6$		
$\bar{X}_1$ with $\bar{X}_5$	$\bar{X}_2$ with $\bar{X}_6$			
$\bar{X}_1$ with $\bar{X}_6$				

With all those t tests to do, the experimenter is giving himself many more opportunities of finding a significant t value than would be the case if he had collected only two samples and therefore could make only one comparison. If instead of six, we had collected 50 samples and computed their means, and then performed t tests between all pairs of means, then even if the samples really came from exactly the same population, it is almost certain that *some* differences between pairs of means would be significant. For the very reason that we have given ourselves too much opportunity for a significant finding, the t test becomes inappropriate as soon as it is clear that more than two sample means are going to be compared with each other. When a comparison is to be made involving more than two sample means, another technique is needed to test the null hypothesis that these sample means, the $\bar{X}$'s, are *all* estimates of a common population mean, μ.

We are working toward a test that takes into account three or more group means *simultaneously* and tests the hypothesis that all of them are estimates of the same population mean, μ. In the case we are considering, our null hypothesis is that the different teaching methods were equally effective. It would follow from this hypothesis that the spread among our sample means was a chance phenomenon. Our test must answer the question, "How unlikely would it be to encounter as wide a spread among sample means as this one, assuming that the hypothesis is true?"

20.4   Introduction to Analysis of Variance. Table 20.3 shows the means, variances, and sample sizes of the six groups. Our hypothesis is that all six means have come from the same population, with mean μ and variance σ^2 both unknown.

TABLE 20.3

$\bar{X}$'s	$\bar{X}_1 = 75$	$\bar{X}_2 = 72$	$\bar{X}_3 = 76$	$\bar{X}_4 = 79$	$\bar{X}_5 = 82$	$\bar{X}_6 = 72$
S^2's	$S_1{}^2 = 173.2$	$S_2{}^2 = 168.7$	$S_3{}^2 = 170.1$	$S_4{}^2 = 169.8$	$S_5{}^2 = 172.0$	$S_6{}^2 = 167.6$
N	$N_1 = 30$	$N_2 = 30$	$N_3 = 30$	$N_4 = 30$	$N_5 = 30$	$N_6 = 30$
d.f.	29	29	29	29	29	29

Is the scatter among our six means big enough for us to reject the hypothesis, say at the .05 level of significance? If, under the hypothesis, our means would show at least as much scatter as or more frequently than five times in 100, we cannot reject the hypothesis; if, under the hypothesis, scatter this big would be found less frequently than five times in 100, we shall reject the hypothesis, concluding that the sample means were *pushed* apart by differences in the teaching methods.

To begin with, we need a way of representing the amount of scatter we actually found among our sample means. We compute our measure of scatter, the variance of the means, which we denote $S_M{}^2$. That is, using Formula (13.2) we calculate this variance of our six means, considering them as six members of a theoretical distribution of means.* Since means instead of individual cases are elements, $\bar{X}$ replaces X, and K, the number of means being considered, replaces N. Formula (20.2) is then obtained from Formula (13.2).

(20.2)
$$S_M{}^2 = \frac{\sum \bar{X}^2 - \dfrac{(\sum \bar{X})^2}{K}}{K-1}$$

Computations done with the data in Table 20.3 are as follows:

$$\sum \bar{X} = 75 + 72 + 76 + 79 + 82 + 72 = 456$$
$$(\sum \bar{X})^2 = (456)^2$$
$$\sum \bar{X}^2 = 75^2 + 72^2 + 76^2 + 79^2 + 82^2 + 72^2 = 34{,}734$$
$$K = 6$$

To find the variance of the sample means, $S_M{}^2$, we write

$$S_M{}^2 = \frac{34{,}734 - \dfrac{(456)^2}{6}}{5} = 15.6$$

Our question is becoming more specific. Now we can ask how often would the means of six samples of 30 cases each, drawn randomly from our population, have a variance $S_M{}^2$ as great as 15.6?

---

* $S_M{}^2$ can be obtained in this problem, as always, by subtracting each of the terms from their mean, squaring the differences, and then dividing by one less than the number of terms.

In the form presented, the question baffles us. But we can translate it into a form in which we can answer it. Notice that if the null hypothesis is true and the scatter among our means is random, then in Formula (20.3), which is obtained by squaring both sides of Formula (14.2), and replacing s by S, we have,

(20.3) $$S_M{}^2 = \frac{S^2}{N}$$

Here, N is the number of cases in each of our samples, and S^2 is an estimate of σ^2, the variance of scores in the population of fourth graders. Formula (20.3) implies that

$$NS_M{}^2 = S^2$$

In our problem, where $N = 30$,

$$NS_M{}^2 = (30)(15.6) = 468 = S^2$$

On the assumption that our six means came from random samples drawn from the same population, we have converted our variance of 15.6 into an estimate of σ^2, the variance of the spelling scores of all the fourth graders considered as individuals in one population. This estimate of σ^2, derived entirely from the K sample means, was based on $K-1$ d.f. We designate this estimate by $S_B{}^2$; the subscript stands for *between* and is intended to remind us that *this estimate of σ^2 is based entirely on the scatter between the group means*. We can now write as a formula

(20.4) $$S_B{}^2 = NS_M{}^2$$

In our example we have

$$S_B{}^2 = 468$$

remembering that $S_B{}^2$ is based on $K-1=5$ d.f.

Suppose that our null hypothesis were false—that is, the methods really exerted different effects. As experimenters, we would not know this fact, but it would already have manifested itself by increasing the value of $S_M{}^2$ and now the value of $S_B{}^2$, which is an estimate of σ^2. If the null hypothesis were false, $S_B{}^2$ would tend to be too big. In effect (when we multiplied by N), we have shifted from asking whether $S_M{}^2$ is "too big" to asking whether $S_B{}^2$ is "too big."

We have done so because we have a way of testing a hypothesis about $S_B{}^2$. We shall use an F test to answer the question, "Is our obtained $S_B{}^2$ significantly bigger than another unbiased estimate of σ^2?" If the answer is yes, we shall conclude that the six sample means did *not* come from the same population with mean, μ, or, in other words, that the six methods used were not all identical in their effects.

For our F test to be meaningful, our second estimate of σ^2 must be independent of $S_B{}^2$. We are going to divide $S_B{}^2$ by this second estimate, which we shall call S_W^2, and if the resulting F ratio is big enough, we are going to reject the null hypothesis and conclude that the sample means were not all estimates of a common mean, μ. Since the statistic we shall use to test the null hypothesis is to be a ratio, this second estimate of σ^2, which will be the denominator, must not suddenly increase if the hypothesis about a common mean, μ, is false, for then it would spuriously reduce the quotient; and it must not shrink, for then it would spuriously increase the quotient. In effect, we want our second estimate of σ^2 to be unaffected by the truth or falsity of the hypothesis about μ. Our inference is to be made essentially about the "bigness" of S_B^2, and a second estimate of σ^2 is, in a sense, a yardstick helping us make this inference.

To discover this second estimate, reconsider Table 20.3. We want an unbiased estimate of σ^2, which would not be affected if there were real differences between groups. $S_1^2 = 173.2$ is such an estimate. It provides us with an estimate of σ^2, which since it was based wholly on scores from Group One, cannot possibly have been influenced by differences between groups; and thus $S_1^2 = 173.2$ is the sort of value we want to put in our denominator, to help us reach a decision as to whether S_B^2 is unusually big under the null hypothesis. $S_2^2 = 168.7$ would also provide an unbiased estimate, and so would the sample variance of each of the groups. To find our best estimate, the one based on the most degrees of freedom, we shall use the data in all six sample variances. When samples are all the same size, we can add up the S^2's and divide by the number of groups, in this case six; when the samples differ in size, we must give proportional representation when computing our average variance. We weight each sample variance by its degrees of freedom, add, and then divide their sum by the total degrees of freedom. If there are N_1 scores in the first group, N_2 in the second, and so on, then S_1^2 is based on $N_1 - 1$ d.f., S_2^2 on $N_2 - 1$ d.f., etc. The weighted mean of these sample variances is the estimate of σ^2 we are looking for. We designate it S_W^2.

$$S_W^2 = \frac{(N_1 - 1)S_1^2 + (N_2 - 1)S_2^2 + \cdots + (N_K - 1)S_K^2}{(N_1 - 1) + (N_2 - 1) + (N_3 - 1) + \cdots + (N_K - 1)}$$

where K is the number of groups. (It may be recognized that the expression under the first square root sign in Formula (14.5) is the special case of this based on two samples.)

It is convenient to let $N_T = N_1 + N_2 + N_3 + \cdots + N_K$, so that we may write equivalently,*

(20.5)
$$S_W^2 = \frac{(N_1 - 1)S_1^2 + \cdots + (N_K - 1)S_K^2}{N_T - K}$$

This estimate is based on $N_T - K$ d.f.

---

* $(N_1 - 1) + (N_2 - 1) + \cdots + (N_K - 1) = N_T - K$.

From the values in Table 20.3, we can compute

$$S_W{}^2 = \frac{(30-1)(173.2)+(30-1)(168.7)+(30-1)(170.1)+(30-1)(169.8)+(30-1)(172.0)+(30-1)(167.6)}{180-6}$$

$$= \frac{(29)(1,021.4)}{174}$$

$$= \frac{1,021.4}{6}$$

$$= 170.2$$

This estimate is based on $(180-6)=174$ d.f.

$S_W{}^2$ is the estimate of σ^2 we have been looking for. It is *based entirely on the scatter among scores within the groups.* At no time did a distance between scores in different groups enter the calculations, and thus even if the sample means had been pushed very far apart by the teaching methods exerting different effects on the students' performances, this estimate, $S_W{}^2$, would not have been in the slightest altered on that account. $S_W{}^2$ is independent of $S_B{}^2$.

If the null hypothesis is true, both $S_B{}^2$ and $S_W{}^2$ are unbiased estimates of σ^2. If it is false, we may expect that $S_B{}^2$ will be inflated, whereas even under that condition $S_W{}^2$ will remain an unbiased estimate of σ^2.

We may now test the null hypothesis by computing our ratio $F=S_B{}^2/S_W{}^2$ using Formula (20.1). Under the null hypothesis, this ratio would have an F distribution with $K-1$ d.f. in the numerator and N_T-K d.f. in the denominator. We shall test the hypothesis at the .05 level of significance. We have

$$F = \frac{S_B{}^2}{S_W{}^2} = \frac{468}{170.2} = 2.75$$

Here the d.f. are $K-1=5$ and $N_T-K=174$.

Although the exact entry does not appear in Appendix Table VII, we see that if the d.f. had been 5 and 150, a ratio as big as 2.27 would have been significant. With the same d.f. in the numerator and even more d.f. in the denominator, a ratio of 2.27 must surely be significant. Any discrepancy becomes more meaningful as the number of cases from which it is computed increases. Thus, we can safely conclude that our obtained value of 2.75 represents significance at the .05 level. We reject the hypothesis that the six sample means were unbiased estimates of μ, and we conclude that something other than sampling error must have inflated the numerator estimate of σ^2. In general, if a finding would be signficant with fewer d.f. in the denominator than we have collected, then it must be significant in the problem at hand, where there are more degrees of freedom.

Note that we cannot conclude that all differences between pairs of means are significant. So far, we can draw no conclusions regarding any particular pair of means. Strictly speaking, a significant finding does not even allow us to say that the biggest of our K means is significantly bigger than the smallest! Other tests of significance, discussed in more advanced texts, show us how to make individual comparisons. The F test we have done allows us to conclude only that *some* factor is scattering the means; and if our experiment was properly controlled, we may conclude that this factor was inherent in the teaching methods used.

20.5   The Technique of Analysis of Variance.   The technique of analysis of variance is designed for testing the hypothesis that the means of a set of K samples are all estimates of the same population mean, μ.* We test the hypothesis by comparing an estimate of σ^2 derived from the scatter among means with an estimate of σ^2 stable enough not to undergo change if the hypothesis is false.   The F ratio obtained is the statistic used to test the hypothesis that the variable being studied had no effect on the group means— that is, did not push them apart.

This technique may be extended to problems where there are many variables and hence different sets of means to be studied.   It allows us to examine separately the influence of different variables operating in a single experiment, and even to study the interactions among the different variables. No matter how intricate the design, the form of each and every F test of a hypothesis is such that the numerator is subject to being increased, if the hypothesis is false, and the denominator is left uninfluenced.   This means that every F test we do, in simple designs and in complex ones, is one-tailed and that we reject the hypothesis if the numerator of our obtained F ratio is too big.

The aim so far has been to convey the logic behind the technique of analysis of variance, using information already given to the reader in earlier discussions.   In practice, when we must work from raw data, not having already computed the parts that go into our F ratio, it serves us well to put them into a table designed to give us particular parts when we need them and to help us interpret our findings.   The procedure to be presented is widely used; it has the advantage of being easily extended for use with more complex analysis of variance problems than are to be presented in this book.

To make sure this procedure is conveyed clearly, we shall use numbers that make the operations easier to perform than they would be in practice, and we shall keep the sample sizes smaller than they would be ordinarily.

Suppose we wish to show that there are differences among three groups in "personal defensiveness."   Our null hypothesis is that the three groups are the same in personal defensiveness and that whatever scatter we encounter among the sample means will be due to chance.   This is the hypothesis that the means of the samples of the three groups are all unbiased estimates of a common unknown μ.   We shall use the .05 significance level.   The personal defensiveness scores and their squares appear in Table 20.4.   Study the calculations closely.   This is the time to make sure you are familiar with the notation.   Though notation itself can be messy and should be kept to a

---

* Experimenters sometimes envision themselves drawing samples from different populations where $\mu_1 = \mu_2 = \mu_3 = \cdots = \mu$.   The hypothesis that this is so is statistically identical to the one stated.

minimum, a good notational system may be all we have to save us when computations become numerous and difficult to handle.

TABLE 20.4

TABLE OF COMPUTATIONS FOR PERSONAL
DEFENSIVENESS SCORES

	GROUP A		GROUP B		GROUP C		
	X_1	$X_1{}^2$	X_2	$X_2{}^2$	X_3	$X_3{}^2$	
	2	4	5	25	5	25	
	4	16	4	16	2	4	
	6	36	0	0	1	1	
	8	64	5	25	0	0	
			2	4			ROW TOTALS
1. $\sum X$	$\sum X_1 = 20$		$\sum X_2 = 16$		$\sum X_3 = 8$		$\sum X_T = 44$
2. $\sum X^2$	$\sum X_1{}^2 = 120$		$\sum X_2{}^2 = 70$		$\sum X_3{}^2 = 30$		$\sum X_T{}^2 = 220$
3. N	$N_1 = 4$		$N_2 = 5$		$N_3 = 4$		$N_T = 13$
4. $(\sum X)^2$	$(\sum X_1)^2 = 20^2$		$(\sum X_2)^2 = 16^2$		$(\sum X_3)^2 = 8^2$		
5. $\dfrac{(\sum X)^2}{N}$	$\dfrac{(\sum X_1)^2}{N_1} = \dfrac{20^2}{4}$		$\dfrac{\sum X_2{}^2}{N_2} = \dfrac{16^2}{5}$		$\dfrac{(\sum X_3)^2}{N_3} = \dfrac{8^2}{4}$		$G = \dfrac{(\sum X_1)^2}{N_1} + \dfrac{(\sum X_2)^2}{N_2} + \dfrac{(\sum X_3)^2}{N_3}$
6. $\dfrac{\sum X}{N} = \bar{X}$	$\bar{X}_1 = \dfrac{20}{4} = 5$		$\bar{X}_2 = 3.2$		$\bar{X}_3 = 2.0$		$G = \dfrac{20^2}{4} + \dfrac{16^2}{5} + \dfrac{8^2}{4}$ $G = 167.2$

The computations made at the bottom of Table 20.4 have each been labeled. In the first row, the sum of each group is given; the subscript indicates the group. In the second row appears the sum of squares of each group. For instance, in Group B the sum of the squared terms, $\sum X_2{}^2$, is 70. The third row shows the number of terms in each group. The fifth row gives the value in the fourth row divided by the value in the third row. The sixth row gives the mean of each group. Each time, the group number is indicated by the subscript. The subscript T, indicating *total*, refers to the entire collection of observations in the experiment. For instance, $\sum X_T{}^2$ stands for the sum of the squares of all the observations made. $\sum X_T{}^2$ will soon be important. N_T shall hereafter be used to indicate the total number of observations made; here 13. The terms in row four do not have immediate value for us but were needed to calculate the terms in row five. The total of the terms in row five is a very important quantity, which we shall call G.

$$G = \frac{(\sum X_1)^2}{N_1} + \frac{(\sum X_2)^2}{N_2} + \cdots + \frac{(\sum X_K)^2}{N_K}$$

In more complex problems, we need to compute several such quantities, each used in a different part of the analysis. It is always useful to examine the means of our different groups to get some sense of what we are asking. Here the means are 5, 3.2, and 2.0; we are testing the hypothesis which says that the scatter among these three estimates of μ is due to chance and not to influences specific to particular groups.

We can now compute our two estimates of σ^2 from the data in Table 20.4. We use the following computational formulas. Remember that K is the number of groups; here $K=3$.

$$S_B{}^2 = \frac{G - \dfrac{(\sum X_T)^2}{N_T}}{K-1}$$

so that

$$S_B{}^2 = \frac{167.2 - \dfrac{(44)^2}{13}}{2} = 9.15$$

And

$$S_W{}^2 = \frac{\sum X_T{}^2 - G}{N_T - K}$$

so that

$$S_W{}^2 = \frac{220 - 167.2}{13 - 3} = 5.28$$

In these fractions, the denominators represent the numbers of d.f. that went into the estimates. $S_B{}^2$ is based on $K-1$ d.f. and $S_W{}^2$ on $N_T - K$ d.f. Here $S_B{}^2$ is based on 2 d.f. and $S_W{}^2$ on 10 d.f.

Next we divide $S_B{}^2$ by $S_W{}^2$

$$\frac{S_B{}^2}{S_W{}^2} = \frac{9.15}{5.28} = 1.73$$

We must now make reference to Table VII and in particular to what it tells us about the distribution of $F_{2,10}$. Our value of 1.73 is not significant, being less than 4.10, which marks the beginning of the .05 rejection region. We have failed to reject our hypothesis. The evidence does not allow us to conclude that these groups differ in personal defensiveness.

The method described would enable us to do any "one-way analysis of variance" problem—that is, any problem in which scores on only one variable were studied and the question asked, "Did the means show significant scatter on that variable?" In the future, we shall compute the numerator and denominator of each of our variance estimates separately, and enter each as we find it.

The reader should study the form of Table 20.5 and learn the new names for the column headings, which will be discussed shortly.

P stands for the probability of our finding an *F* ratio as big as, or bigger than, the one we actually found, assuming that the null hypothesis is true. Here, our probability of having found one as big as 1.73 is larger than .05, and therefore we cannot reject the hypothesis.

We have based our computations on the following formulas, which you should study carefully. The essence of the analysis of variance work discussed in this chapter is given in Table 20.5. The entries have been made using the following formulas.

$$(20.6) \qquad G = \frac{(\sum X_1)^2}{N_1} + \frac{(\sum X_2)^2}{N_2} + \cdots + \frac{(\sum X_K)^2}{N_K}$$

$$(20.7) \qquad SS_B = G - \frac{(\sum X_T)^2}{N_T}$$

$$(20.8) \qquad SS_W = \sum X_T^2 - G$$

$$(20.9) \qquad S_B^2 = \frac{SS_B}{K-1}$$

$$(20.10) \qquad S_W^2 = \frac{SS_W}{N_T - K}$$

In our problem we find $G = 167.2$, and the results of the other computations are given in Table 20.5, culminating in

$$F = \frac{S_B^2}{S_W^2} = 1.73$$

TABLE 20.5

ANALYSIS OF VARIANCE TABLE

SOURCE OF VARIATION	NUMERATOR— SUM OF SQUARES	DENOMINATOR— d.f.	ESTIMATE OF σ^2	OBTAINED F RATIO	P
Between Group Means	SS_B 18.3	$K-1$ 2	S_B^2 9.15	1.73	larger than .05
Within Groups	SS_W 52.8	$N_T - K$ '10	S_W^2 5.28		
Total	71.1	12			

The reader should verify the computations done in connection with the following problem. The raw data are given in Table 20.6, and the information needed for the analysis of variance is entered in Table 20.7. This time, five procedures are to be compared for their effects on students' motivation to learn history. Fifty students, all of whom were considered to have low motivation to begin with, have been randomly assigned to five groups. Two of the students did not finish the semester, so that N_T, the total number of subjects, is 48. The scores are listed in Table 20.6.

This time we are testing the null hypothesis that the procedures do not differ in their effects on students' motivation. First we calculate the five sample means. They are $\bar{X}_1 = 33.0$, $\bar{X}_2 = 28.7$, $\bar{X}_3 = 29.7$, $\bar{X}_4 = 22.0$, and $\bar{X}_5 = 31.4$. It seems impossible to tell, by merely inspecting the terms in Table 20.6, whether the scatter among these five means is wide enough to suggest that these means were pushed apart by the difference in procedure. To test our null hypothesis, we shall see whether our obtained ratio,

TABLE 20.6

COMPUTATION OF ANALYSIS OF VARIANCE FOR HISTORY
MOTIVATION SCORES

PROCEDURES

1	2	3	4	5
32	23	27	26	27
33	32	34	17	29
36	20	31	30	34
34	19	22	18	37
32	32	28	26	19
30	39	30	14	25
32	32	37	30	47
35	30	27	15	34
35	31	26		32
31	29	35		30
$\sum X_1 = 330$	$\sum X_2 = 287$	$\sum X_3 = 297$	$\sum X_4 = 176$	$\sum X_5 = 314$

Computations reveal that $\sum X_1{}^2 = 10{,}924$; $\sum X_2{}^2 = 8{,}585$; $\sum X_3{}^2 = 9{,}013$; $\sum X_4{}^2 = 4{,}186$; $\sum X_5{}^2 = 10{,}370$; $G = 41{,}679.4$; $(\sum X_T)^2 = (1{,}404)^2$; and $\sum X_T{}^2 = 43{,}078$.

TABLE 20.7

ANALYSIS OF VARIANCE SUMMARY TABLE FOR
MOTIVATION SCORES

SOURCE OF VARIATION	SUM OF SQUARES SS	d.f.	ESTIMATE OF σ^2	OBTAINED RATIO	P
Between Groups	612.4	4	153.1	4.71	less than .01
Within Groups	1,398.6	43	32.5		
Total	2,011.0	47			

$F = S_B{}^2 / S_W{}^2$, is significant at the .01 level.   You should show that application of Formulas (20.6) through (20.10) yields the entries in our analysis of variance table.

Our obtained F ratio of 4.71 is bigger than 3.80, the .01 significance F value for 4 and 42 degrees of freedom as shown in Appendix VII.   With more observations involved in our denominator (one more in this case), our obtained F ratio is certainly significant.   Consequently, we reject the null hypothesis and conclude that the procedures did exert some effect that pushed the sample means apart.

Notice that the total d.f., which was 47 in this problem, is always one less than the total number of observations made.   Here, there were $N_T = 48$ observations, and if we tried to make a single variance estimate, considering them all as comprising one group, we would also have had 47 d.f. available. However, such an estimate would have been biased if the null hypothesis was false, which we have concluded was the case here.   We can compute the total sum of squares directly by Formula (20.11), which can be obtained by adding Formulas (20.7) and (20.8).

335

(20.11) $$SS_T = \sum X_T{}^2 - \frac{(\sum X_T)^2}{N_T}$$

$$SS_T = 43{,}078 - \frac{(1{,}404)^2}{48} = 43{,}078 - 41{,}067 = 2\,011$$

In complex problems of analysis of variance, the value of SS_T serves as a check on other computations. Here it is also a check but a weak one, since if any of the pieces of SS_B or SS_W were computed incorrectly but properly handled after that, the flaw is likely to appear in SS_T, and the totals will be internally consistent even though a mistake was made. In more complicated problems, SS_T is more useful, though it is never a perfect check, since many of the same computations that go into the component sums of squares are carried over to SS_T.

20.6 Partitioning the Sums of Squares.

To designate any single score in an analysis of variance problem, we need two subscripts, the first showing its group and the second indicating which score in that group we are talking about. For instance, the fifth score in the second group would be designated $X_{2,5}$. The difference between that score and the mean of its group is $(X_{2,5} - \bar{X}_2)$. If we designate the mean of the scores of all groups considered together by $\bar{X}_T$, then the difference between our particular score and the mean of all the scores is $(X_{2,5} - \bar{X}_T)$. The difference between the mean of the sample from which our particular score was taken and the grand mean is $(\bar{X}_2 - \bar{X}_T)$. We may subdivide the distance between our particular score and the grand mean into two parts and write:

$$(X_{2,5} - \bar{X}_T) = (X_{2,5} - \bar{X}_2) + (\bar{X}_2 - \bar{X}_T)$$

Imagine that we repeated the procedure for each of the 48 scores in our previous example. For the sixth score in group two, we would have the following equivalence:

$$(X_{2,6} - \bar{X}_T) = (X_{2,6} - \bar{X}_2) + (\bar{X}_2 - \bar{X}_T)$$

And for example, for the seventh score in group four, we have,

$$(X_{4,7} - \bar{X}_T) = (X_{4,7} - \bar{X}_4) + (\bar{X}_4 - \bar{X}_T)$$

The reader should verify these equivalences.

If we use the subscript i to stand for the group number, and j to stand for the number of the individual in that group, then

$$(X_{i,j} - \bar{X}_T) = (X_{i,j} - \bar{X}_i) + (\bar{X}_i - \bar{X}_T)$$

where $\bar{X}_i$ stands for the mean of the ith group.

The square of the distance between the score and the grand mean does not necessarily equal the square of its distance from its group mean plus the square of the distance from the group mean to the grand mean any more than

$7^2 = 3^2 + 4^2$ because $7 = 3 + 4$. However, suppose we carry out the process for each of the 48 scores, and then compute the totals.

$(X_{1,1} - \bar{X}_T)^2$ $\vdots$	$(X_{1,1} - \bar{X}_1)^2$ $\vdots$	$(\bar{X}_1 - \bar{X}_T)^2$ $\vdots$
$(X_{1,10} - \bar{X}_T)^2$	$(X_{1,10} - \bar{X}_1)^2$	$(\bar{X}_1 - \bar{X}_T)^2$
$(X_{2,1} - \bar{X}_T)^2$ $\vdots$	$(X_{2,1} - \bar{X}_2)^2$ $\vdots$	$(\bar{X}_2 - \bar{X}_T)^2$ $\vdots$
$(X_{5,10} - \bar{X}_T)^2$	$(X_{5,10} - \bar{X}_5)^2$	$(\bar{X}_5 - \bar{X}_T)^2$

The reader may verify that this leads to:

(20.12)
$$\sum_{i,j} (X_{i,j} - \bar{X}_T)^2 = \sum_{i,j} (X_{i,j} - \bar{X}_i)^2 + \sum_{i,j} (\bar{X}_i - \bar{X}_T)^2$$

It may be shown mathematically that Formula (20.12) is correct and works in *all* cases, not just in our example.

Formula (20.12) tells us that the sum of squared distances of all the scores from the grand mean equals the sum of their squared distances from their own group means plus the sum of squared distances between their group means and the grand mean. The left-hand side of Formula (20.12) is a statement of SS_T, which we computed without going through so many steps. The first term on the right-hand side is SS_W and the second term is SS_B. Thus, the partitioning leads to the statement of an equivalence that we have already used.

$$SS_T = SS_W + SS_B$$

In complex problems of analysis of variance, it is helpful to construe the different sums of squares as results of partitioning, and though the approach is not necessary here, it ought to be studied, for it adds insight. Remember that in the problems we have done:

SUM OF SQUARES	THEORETICAL MEANING	COMPUTATIONAL FORMULA
SS_W	$\sum_{i,j} (X_{i,j} - \bar{X}_i)^2$	$\sum X_T^2 - G$
SS_B	$\sum_{i,j} (\bar{X}_i - \bar{X}_T)^2$	$G - \dfrac{(\sum X_T)^2}{N_T}$
SS_T	$\sum_{i,j} (X_{i,j} - \bar{X}_T)^2$	$\sum X_T^2 - \dfrac{(\sum X_T)^2}{N_T}$

where $G = \dfrac{(\sum X_1)^2}{N_1} + \dfrac{(\sum X_2)^2}{N_2} + \cdots + \dfrac{(\sum X_K)^2}{N_K}$

337

PROBLEM SET A

20.1 Two samples of rats were randomly selected and given different diets for two weeks. The sample sizes and variance of their speeds on a maze completion test are given below. Using the .05 significance level, test the null hypothesis that there is no difference in variability of their speeds between the two groups.

$$N_1 = 15 \qquad N_2 = 22$$
$$S_1^2 = 35 \qquad S_2^2 = 14$$

20.2 Samples of two metal alloys were measured for yield strength. The sample sizes and variances are given below. Using the .01 significance level, test the null hypothesis that there is no difference in variability of yield strengths between the two alloys.

$$N_1 = 10 \qquad N_2 = 10$$
$$S_1^2 = 20 \qquad S_2^2 = 25$$

20.3 The variances in sociometric ratings of two randomly selected groups are presented below. Using the .01 significance level, test the null hypothesis that there is no difference in variability of ratings between the two groups.

$$N_1 = 76 \qquad N_2 = 27$$
$$S_1^2 = 87.56 \qquad S_2^2 = 41.35$$

20.4 Below are the variances of manual-skills scores obtained from two random samples of fourth-grade children. Using the .05 significance level, test the null hypothesis that there is no difference in variability of scores between the two groups.

$$N_1 = 25 \qquad S_1^2 = 215$$
$$N_2 = 41 \qquad S_2^2 = 72$$

20.5 Three pharmacologists share equally in a random selection of 15 drug samples. The coded results of their tests are shown here.

Pharmacologist

A'	B	C
5	4	5
6	4	6
7	5	6
8	7	6
9	5	7

Test for significant differences between the means, using the .05 significance level.

20.6 The bacterial count in a certain food product is observed after three samples are subjected to a process designed to destroy bacteria. Is there a significant difference in the treatments? Use the .05 significance level.

Treatment

I	II	III
135	105	120
170	92	118
154	89	109

20.7 In a cost-of-living survey, four cities were selected in which several shops sold a particular commodity. The prices noted for this commodity are shown below. Are the prices for the four cities significantly different? Use the .01 significance level.

City

A	B	C	D
$1.30	1.40	1.44	1.27
1.27	1.30	1.40	1.05
1.21	1.28	1.28	1.24
1.09	1.27	1.28	1.22
1.03	1.24	1.06	
1.01	1.08		
1.09			

20.8 At the .05 significance level, what do you conclude about the differences of the means of the five samples shown here:

Sample

I	II	III	IV	V
5	9	8	4	7
7	13	11	8	10
6	10	12	10	9
8	11	7	6	8
4	7	7	7	6

20.9 Thirty freshman students were divided into three socioeconomic groups and were administered a reading-comprehension test. The results of this test are presented below. Are the scores for the three groups significantly different? Use the .01 significance level.

Group A	Group B	Group C
21	20	24
18	20	23
17	19	23
17	18	22
17	17	20
16	17	20
15	14	18
13	14	17
12	9	16
8	7	
7		

20.10 In a certain experiment, four samples of unequal sizes produce these results:

I	II	III	IV
8	9	16	9
12	11	9	12
8	7	12	8
10	11	14	10
	8		7
	10		14
			12

Test for significant differences between the means, using the .05 significance level.

PROBLEM SET B

20.11 Two samples of students were given different types of spelling instruction. The sample sizes and variances of their final scores are given below. Using the .01 significance level, test the null hypothesis that there is no difference in variability of scores between the two types of instruction.

$$N_1 = 15 \qquad N_2 = 22$$
$$S_1^2 = 35 \qquad S_2^2 = 29$$

20.12 Samples of two types of diet foods were tested for caloric content. The sample sizes and variances are given below. Using the .05 significance level, test the null hypothesis that there is no difference in variability in caloric content between the types of diet food.

$$N_1 = 35 \qquad N_2 = 19$$
$$S_1^2 = 63 \qquad S_2^2 = 72$$

20.13 Samples of two types of cable insulation were tested for resistance. The sample sizes and variances are given below. Using the .05 significance level, test the null hypothesis that there is no difference in resistance between the types of insulation.

$$N_1 = 20 \qquad N_2 = 15$$
$$S_1^2 = 100 \qquad S_2^2 = 125$$

20.14 Two samples of guinea pigs were randomly selected and fed different types of grain for a month. The sample sizes and variances of their weight gains are given below. Test the null hypothesis that there is no difference in variability of weight gains between the two groups, using the .01 significance level.

$$N_1 = 15 \qquad N_2 = 13$$
$$S_1^2 = 115 \qquad S_2^2 = 28$$

20.15 Three samples of students were given a typing test after receiving different styles of typing instruction. The numbers of typing errors they made are given below. Using the .05 significance level, test for significant differences between the means.

Type I	Type II	Type III
6	4	6
7	5	6
7	5	6
7	5	6
8	6	6

20.16 In an agricultural experiment, 35 plots of nearly equal fertility are planted with seven different varieties of a grain, five plots for each variety. The distribution of the varieties among the various plots is assumed to be random. The yields in bushels per acre are shown below, each column corresponding to one of the seven varieties. Is there a significant difference in the varietal yields? Use the .01 significance level.

Grain Variety

I	II	III	IV	V	VI	VII
10	15	9	17	12	16	13
13	14	19	15	16	12	10
11	17	17	19	9	15	14
15	9	16	11	13	11	14
12	18	14	13	15	17	19

20.17 Listed below are the final grades obtained by five students in the subjects shown. At the .05 level of significance, test the hypothesis that the students are equally proficient in all three subjects—that is, that the means are equal.

Statistics	Psychology	Music
81	73	65
73	91	87
67	94	84
90	87	93
60	83	89

20.18 The Centerville school wished to determine whether certain methods of teaching arithmetic were better than others. It devised three different methods of teaching arithmetic and divided 21 children into three groups, giving each group a different method. The three groups were designated Group A, Group B, and Group C. These three groups of children did not differ initially on arithmetic achievement. At the end of the school year an arithmetic achievement test was given to each of the children in the three groups. The results of this test are presented below. Are the scores for the three groups significantly different? Use the .01 significance level.

Group A	Group B	Group C
94	90	86

92	86	84
90	84	80
90	82	78
86	82	78
86	80	72
84	78	
82		

20.19 Observers at three different locations on the earth's surface make coded measurements of radio signals received from outer-space sources. The results are compared with similar signals received on a space vehicle orbiting the earth. Test the hypothesis of equal means of A, B, C, and D. Use the .05 significance level.

Earth Observers			Space Observers
A	B	C	D
3	3	4	7
2	4	4	6
3	4	5	4
2	3	4	6
3	4	4	4
1	3	3	7

20.20 In Problem 20.19, test the hypothesis of equal means for A, B, and C, using the .05 significance level. What do you infer about space observations?

There are a few points to watch in using the table on the following pages. We shall discuss these briefly. First of all, observe that the first column lists all numbers, n, from 1.00 through 10.00. Each number in the second column is the square, n^2, of the corresponding number, n, in the first column. For example, $(1.78)^2 = 3.1684$ and $(7.17)^2 = 51.4089$.

The second column can also be used to obtain the squares of other numbers having the same succession of digits as the numbers given in the first column. For example, the square of 17.8 will also have the same succession of digits, 31684, as the square of 1.78. However, the position of the decimal point is not the same and $(17.8)^2 = 316.84$. This can be explained by the fact that we must multiply 1.78 by 10 to get 17.8. When the number is squared, the 10 is also squared $[(17.8)^2 = (10)^2 \cdot (1.78)^2]$. Thus the answer is 100 times 3.1684 or 316.84. Similarly, $(717)^2 = 514,089$ because 7.17 must be multiplied by 100 to give 717 and the square of 7.17 is then multiplied by the square of 100, which is 10,000. Note that in each case *the decimal point is moved twice as many places in the square as in the number that is squared.* Consider now the effect of moving the decimal point in the opposite direction. $(.178)^2$ will again contain the digits 31684 but this time the correct answer is .031684. The explanation is that 1.78 must be multiplied by .1 to give .178. The answer is then multiplied by $(.1)^2$ or .01 thus giving .031684. As another example, $(.0717)^2 = .00514089$. Note that the italicized statement holds regardless of the direction that the decimal point is moved.

The operation of taking the square root is the inverse of the operation of squaring, just as division is the inverse operation of multiplication. The discussion in the preceding paragraph is, therefore, also helpful in understanding the use of the third column of the table. This column gives the square root, $\sqrt{n}$, of the corresponding number, n, in the first column of the table. To further simplify the use of the table for finding square roots, the fourth column, $\sqrt{10n}$, has been added.

Since the first column contains all numbers from 1.00 to 10.00, we know that the third column contains the square roots of all these numbers. For example, $\sqrt{1.78} = 1.33417$ and $\sqrt{7.17} = 2.67769$. The fourth column enables us also to find directly the square roots, $\sqrt{10n}$, of all numbers from 10(1.00) to 10(10.00), that is, from 10.0 to 100.0, where now each number is given only to the nearest tenth. For example, we find from the fourth column opposite 1.78 that $\sqrt{17.8} = 4.21900$ and opposite 7.17 that $\sqrt{71.7} = 8.46759$. Hence from the third and fourth columns we can read directly the square roots of all numbers from 1.00 through 100.0. However, just as we extended the use of the table for squares, so we can extend its use for square roots.

Suppose we want $\sqrt{717}$. From the table we can read both $\sqrt{7.17}$ and $\sqrt{71.7}$. Which should we use? This question is answered when we consider the placement of the decimal point. Remembering that taking the square root is the inverse of squaring, and looking back at the italicized statement earlier in the discussion, we see that *the decimal point is moved half as many places in the square root as in the number.* Now, half of an odd number isn't a whole number and a decimal point can't be moved a fraction of a place. So we must move the decimal point an *even number* of places to begin with when converting a number in order to apply the table and find its square root. Therefore, in our example we want the digits in $\sqrt{7.17}$ that are 267769 and by application of the last italicized statement we have $\sqrt{717} = 26.7769$, since the decimal point is moved two places from 7.17 to 717 and half of two is one. Had the problem been to find $\sqrt{7170}$, we would again have moved the decimal point an *even* number of places in order to obtain a number whose square root we could read directly from the table. In this case we would have the digits in $\sqrt{71.7}$ or 846759. Observing the rule for placement of the decimal point would give us the answer, 84.6759. The problem of finding $\sqrt{.00717}$ leads to the same sequence of digits (moving the decimal point four places), but the answer this time is .0846759. Again, the last italicized statement holds regardless of the direction that the decimal point is moved.

n	n^2	$\sqrt{n}$	$\sqrt{10n}$	n	n^2	$\sqrt{n}$	$\sqrt{10n}$
1.00	1.0000	1.00000	3.16228	1.50	2.2500	1.22474	3.87298
1.01	1.0201	1.00499	3.17805	1.51	2.2801	1.22882	3.88587
1.02	1.0404	1.00995	3.19374	1.52	2.3104	1.23288	3.89872
1.03	1.0609	1.01489	3.20936	1.53	2.3409	1.23693	3.91152
1.04	1.0816	1.01980	3.22490	1.54	2.3716	1.24097	3.92428
1.05	1.1025	1.02470	3.24037	1.55	2.4025	1.24499	3.93700
1.06	1.1236	1.02956	3.25576	1.56	2.4336	1.24900	3.94968
1.07	1.1449	1.03441	3.27109	1.57	2.4649	1.25300	3.96232
1.08	1.1664	1.03923	3.28634	1.58	2.4964	1.25698	3.97492
1.09	1.1881	1.04403	3.30151	1.59	2.5281	1.26095	3.98748
1.10	1.2100	1.04881	3.31662	1.60	2.5600	1.26491	4.00000
1.11	1.2321	1.05357	3.33167	1.61	2.5921	1.26886	4.01248
1.12	1.2544	1.05830	3.34664	1.62	2.6244	1.27279	4.02492
1.13	1.2769	1.06301	3.36155	1.63	2.6569	1.27671	4.03733
1.14	1.2996	1.06771	3.37639	1.64	2.6896	1.28062	4.04969
1.15	1.3225	1.07238	3.39116	1.65	2.7225	1.28452	4.06202
1.16	1.3456	1.07703	3.40588	1.66	2.7556	1.28841	4.07431
1.17	1.3689	1.08167	3.42053	1.67	2.7889	1.29228	4.08656
1.18	1.3924	1.08628	3.43511	1.68	2.8224	1.29615	4.09878
1.19	1.4161	1.09087	3.44964	1.69	2.8561	1.30000	4.11096
1.20	1.4400	1.09545	3.46410	1.70	2.8900	1.30384	4.12311
1.21	1.4641	1.10000	3.47851	1.71	2.9241	1.30767	4.13521
1.22	1.4884	1.10454	3.49285	1.72	2.9584	1.31149	4.14729
1.23	1.5129	1.10905	3.50714	1.73	2.9929	1.31529	4.15933
1.24	1.5376	1.11355	3.52136	1.74	3.0276	1.31909	4.17133
1.25	1.5625	1.11803	3.53553	1.75	3.0625	1.32288	4.18330
1.26	1.5876	1.12250	3.54965	1.76	3.0976	1.32665	4.19524
1.27	1.6129	1.12694	3.56371	1.77	3.1329	1.33041	4.20714
1.28	1.6384	1.13137	3.57771	1.78	3.1684	1.33417	4.21900
1.29	1.6641	1.13578	3.59166	1.79	3.2041	1.33791	4.23084
1.30	1.6900	1.14018	3.60555	1.80	3.2400	1.34164	4.24264
1.31	1.7161	1.14455	3.61939	1.81	3.2761	1.34536	4.25441
1.32	1.7424	1.14891	3.63318	1.82	3.3124	1.34907	4.26615
1.33	1.7689	1.15326	3.64692	1.83	3.3489	1.35277	4.27785
1.34	1.7956	1.15758	3.66060	1.84	3.3856	1.35647	4.28952
1.35	1.8225	1.16190	3.67423	1.85	3.4225	1.36015	4.30116
1.36	1.8496	1.16619	3.68782	1.86	3.4596	1.36382	4.31277
1.37	1.8769	1.17047	3.70135	1.87	3.4969	1.36748	4.32435
1.38	1.9044	1.17473	3.71484	1.88	3.5344	1.37113	4.33590
1.39	1.9321	1.17898	3.72827	1.89	3.5721	1.37477	4.34741
1.40	1.9600	1.18322	3.74166	1.90	3.6100	1.37840	4.35890
1.41	1.9881	1.18743	3.75500	1.91	3.6481	1.38203	4.37035
1.42	2.0164	1.19164	3.76829	1.92	3.6864	1.38564	4.38178
1.43	2.0449	1.19583	3.78153	1.93	3.7249	1.38924	4.39318
1.44	2.0736	1.20000	3.79473	1.94	3.7636	1.39284	4.40454
1.45	2.1025	1.20416	3.80789	1.95	3.8025	1.39642	4.41588
1.46	2.1316	1.20830	3.82099	1.96	3.8416	1.40000	4.42719
1.47	2.1609	1.21244	3.83406	1.97	3.8809	1.40357	4.43847
1.48	2.1904	1.21655	3.84708	1.98	3.9204	1.40712	4.44972
1.49	2.2201	1.22066	3.86005	1.99	3.9601	1.41067	4.46094

n	n^2	$\sqrt{n}$	$\sqrt{10n}$	n	n^2	$\sqrt{n}$	$\sqrt{10n}$
2.00	4.0000	1.41421	4.47214	2.50	6.2500	1.58114	5.00000
2.01	4.0401	1.41774	4.48330	2.51	6.3001	1.58430	5.00999
2.02	4.0804	1.42127	4.49444	2.52	6.3504	1.58745	5.01996
2.03	4.1209	1.42478	4.50555	2.53	6.4009	1.59060	5.02991
2.04	4.1616	1.42829	4.51664	2.54	6.4516	1.59374	5.03984
2.05	4.2025	1.43178	4.52769	2.55	6.5025	1.59687	5.04975
2.06	4.2436	1.43527	4.53872	2.56	6.5536	1.60000	5.05964
2.07	4.2849	1.43875	4.54973	2.57	6.6049	1.60312	5.06952
2.08	4.3264	1.44222	4.56070	2.58	6.6564	1.60624	5.07937
2.09	4.3681	1.44568	4.57165	2.59	6.7081	1.60935	5.08920
2.10	4.4100	1.44914	4.58258	2.60	6.7600	1.61245	5.09902
2.11	4.4521	1.45258	4.59347	2.61	6.8121	1.61555	5.10882
2.12	4.4944	1.45602	4.60435	2.62	6.8644	1.61864	5.11859
2.13	4.5369	1.45945	4.61519	2.63	6.9169	1.62173	5.12835
2.14	4.5796	1.46287	4.62601	2.64	6.9696	1.62481	5.13809
2.15	4.6225	1.46629	4.63681	2.65	7.0225	1.62788	5.14782
2.16	4.6656	1.46969	4.64758	2.66	7.0756	1.63095	5.15752
2.17	4.7089	1.47309	4.65833	2.67	7.1289	1.63401	5.16720
2.18	4.7524	1.47648	4.66905	2.68	7.1824	1.63707	5.17687
2.19	4.7961	1.47986	4.67974	2.69	7.2361	1.64012	5.18652
2.20	4.8400	1.48324	4.69042	2.70	7.2900	1.64317	5.19615
2.21	4.8841	1.48661	4.70106	2.71	7.3441	1.64621	5.20577
2.22	4.9284	1.48997	4.71169	2.72	7.3984	1.64924	5.21536
2.23	4.9729	1.49332	4.72229	2.73	7.4529	1.65227	5.22494
2.24	5.0176	1.49666	4.73286	2.74	7.5076	1.65529	5.23450
2.25	5.0625	1.50000	4.74342	2.75	7.5625	1.65831	5.24404
2.26	5.1076	1.50333	4.75395	2.76	7.6176	1.66132	5.25357
2.27	5.1529	1.50665	4.76445	2.77	7.6729	1.66433	5.26308
2.28	5.1984	1.50997	4.77493	2.78	7.7284	1.66733	5.27257
2.29	5.2441	1.51327	4.78539	2.79	7.7841	1.67033	5.28205
2.30	5.2900	1.51658	4.79583	2.80	7.8400	1.67332	5.29150
2.31	5.3361	1.51987	4.80625	2.81	7.8961	1.67631	5.30094
2.32	5.3824	1.52315	4.81664	2.82	7.9524	1.67929	5.31037
2.33	5.4289	1.52643	4.82701	2.83	8.0089	1.68226	5.31977
2.34	5.4756	1.52971	4.83735	2.84	8.0656	1.68523	5.32917
2.35	5.5225	1.53297	4.84768	2.85	8.1225	1.68819	5.33854
2.36	5.5696	1.53623	4.85798	2.86	8.1796	1.69115	5.34790
2.37	5.6169	1.53948	4.86826	2.87	8.2369	1.69411	5.35724
2.38	5.6644	1.54272	4.87852	2.88	8.2944	1.69706	5.36656
2.39	5.7121	1.54596	4.88876	2.89	8.3521	1.70000	5.37587
2.40	5.7600	1.54919	4.89898	2.90	8.4100	1.70294	5.38516
2.41	5.8081	1.55242	4.90918	2.91	8.4681	1.70587	5.39444
2.42	5.8564	1.55563	4.91935	2.92	8.5264	1.70880	5.40370
2.43	5.9049	1.55885	4.92950	2.93	8.5849	1.71172	5.41295
2.44	5.9536	1.56205	4.93964	2.94	8.6436	1.71464	5.42218
2.45	6.0025	1.56525	4.94975	2.95	8.7025	1.71756	5.43139
2.46	6.0516	1.56844	4.95984	2.96	8.7616	1.72047	5.44059
2.47	6.1009	1.57162	4.96991	2.97	8.8209	1.72337	5.44977
2.48	6.1504	1.57480	4.97996	2.98	8.8804	1.72627	5.45894
2.49	6.2001	1.57797	4.98999	2.99	8.9401	1.72916	5.46809

n	n^2	$\sqrt{n}$	$\sqrt{10n}$	n	n^2	$\sqrt{n}$	$\sqrt{10n}$
3.00	9.0000	1.73205	5.47723	3.50	12.2500	1.87083	5.91608
3.01	9.0601	1.73494	5.48635	3.51	12.3201	1.87350	5.92453
3.02	9.1204	1.73781	5.49545	3.52	12.3904	1.87617	5.93296
3.03	9.1809	1.74069	5.50454	3.53	12.4609	1.87883	5.94138
3.04	9.2416	1.74356	5.51362	3.54	12.5316	1.88149	5.94979
3.05	9.3025	1.74642	5.52268	3.55	12.6025	1.88414	5.95819
3.06	9.3636	1.74929	5.53173	3.56	12.6736	1.88680	5.96657
3.07	9.4249	1.75214	5.54076	3.57	12.7449	1.88944	5.97495
3.08	9.4864	1.75499	5.54977	3.58	12.8164	1.89209	5.98331
3.09	9.5481	1.75784	5.55878	3.59	12.8881	1.89473	5.99166
3.10	9.6100	1.76068	5.56776	3.60	12.9600	1.89737	6.00000
3.11	9.6721	1.76352	5.57674	3.61	13.0321	1.90000	6.00833
3.12	9.7344	1.76635	5.58570	3.62	13.1044	1.90263	6.01664
3.13	9.7969	1.76918	5.59464	3.63	13.1769	1.90526	6.02495
3.14	9.8596	1.77200	5.60357	3.64	13.2496	1.90788	6.03324
3.15	9.9225	1.77482	5.61249	3.65	13.3225	1.91050	6.04152
3.16	9.9856	1.77764	5.62139	3.66	13.3956	1.91311	6.04979
3.17	10.0489	1.78045	5.63028	3.67	13.4689	1.91572	6.05805
3.18	10.1124	1.78326	5.63915	3.68	13.5424	1.91833	6.06630
3.19	10.1761	1.78606	5.64801	3.69	13.6161	1.92094	6.07454
3.20	10.2400	1.78885	5.65685	3.70	13.6900	1.92354	6.08276
3.21	10.3041	1.79165	5.66569	3.71	13.7641	1.92614	6.09098
3.22	10.3684	1.79444	5.67450	3.72	13.8384	1.92873	6.09918
3.23	10.4329	1.79722	5.68331	3.73	13.9129	1.93132	6.10737
3.24	10.4976	1.80000	5.69210	3.74	13.9876	1.93391	6.11555
3.25	10.5625	1.80278	5.70088	3.75	14.0625	1.93649	6.12372
3.26	10.6276	1.80555	5.70964	3.76	14.1376	1.93907	6.13188
3.27	10.6929	1.80831	5.71839	3.77	14.2129	1.94165	6.14003
3.28	10.7584	1.81108	5.72713	3.78	14.2884	1.94422	6.14817
3.29	10.8241	1.81384	5.73585	3.79	14.3641	1.94679	6.15630
3.30	10.8900	1.81659	5.74456	3.80	14.4400	1.94936	6.16441
3.31	10.9561	1.81934	5.75326	3.81	14.5161	1.95192	6.17252
3.32	11.0224	1.82209	5.76194	3.82	14.5924	1.95448	6.18061
3.33	11.0889	1.82483	5.77062	3.83	14.6689	1.95704	6.18870
3.34	11.1556	1.82757	5.77927	3.84	14.7456	1.95959	6.19677
3.35	11.2225	1.83030	5.78792	3.85	14.8225	1.96214	6.20484
3.36	11.2896	1.83303	5.79655	3.86	14.8996	1.96469	6.21289
3.37	11.3569	1.83576	5.80517	3.87	14.9769	1.96723	6.22093
3.38	11.4244	1.83848	5.81378	3.88	15.0544	1.96977	6.22896
3.39	11.4921	1.84120	5.82237	3.89	15.1321	1.97231	6.23699
3.40	11.5600	1.84391	5.83095	3.90	15.2100	1.97484	6.24500
3.41	11.6281	1.84662	5.83952	3.91	15.2881	1.97737	6.25300
3.42	11.6964	1.84932	5.84808	3.92	15.3664	1.97990	6.26099
3.43	11.7649	1 85203	5 85662	3.93	15.4449	1.98242	6.26897
3.44	11.8336	1.85472	5.86515	3.94	15.5236	1.98494	6.27694
3.45	11.9025	1.85742	5.87367	3.95	15.6025	1.98746	6.28490
3.46	11.9716	1.86011	5.88218	3.96	15.6816	1.98997	6.29285
3.47	12.0409	1.86279	5.89067	3.97	15.7609	1.99249	6.30079
3.48	12.1104	1.86548	5.89915	3.98	15.8408	1.99499	6.30872
3.49	12.1801	1.86815	5.90762	3.99	15.9201	1.99750	6.31664

n	n^2	$\sqrt{n}$	$\sqrt{10n}$	n	n^2	$\sqrt{n}$	$\sqrt{10n}$
4.00	16.0000	2.00000	6.32456	4.50	20.2500	2.12132	6.70820
4.01	16.0801	2.00250	6.33246	4.51	20.3401	2.12368	6.71565
4.02	16.1604	2.00499	6.34035	4.52	20.4304	2.12603	6.72309
4.03	16.2409	2.00749	6.34823	4.53	20.5209	2.12838	6.73053
4.04	16.3216	2.00998	6.35610	4.54	20.6116	2.13073	6.73795
4.05	16.4025	2.01246	6.36396	4.55	20.7025	2.13307	6.74537
4.06	16.4836	2.01494	6.37181	4.56	20.7936	2.13542	6.75278
4.07	16.5649	2.01742	6.37966	4.57	20.8849	2.13776	6.76018
4.08	16.6464	2.01990	6.38749	4.58	20.9764	2.14009	6.76757
4.09	16.7281	2.02237	6.39531	4.59	21.0681	2.14243	6.77495
4.10	16.8100	2.02485	6.40312	4.60	21.1600	2.14476	6.78233
4.11	16.8921	2.02731	6.41093	4.61	21.2521	2.14709	6.78970
4.12	16.9744	2.02978	6.41872	4.62	21.3444	2.14942	6.79706
4.13	17.0569	2.03224	6.42651	4.63	21.4369	2.15174	6.80441
4.14	17.1396	2.03470	6.43428	4.64	21.5296	2.15407	6.81175
4.15	17.2225	2.03715	6.44205	4.65	21.6225	2.15639	6.81909
4.16	17.3056	2.03961	6.44981	4.66	21.7156	2.15870	6.82642
4.17	17.3889	2.04206	6.45755	4.67	21.8089	2.16102	6.83374
4.18	17.4724	2.04450	6.46529	4.68	21.9024	2.16333	6.84105
4.19	17.5561	2.04695	6.47302	4.69	21.9961	2.16564	6.84836
4.20	17.6400	2.04939	6.48074	4.70	22.0900	2.16795	6.85565
4.21	17.7241	2.05183	6.48845	4.71	22.1841	2.17025	6.86294
4.22	17.8084	2.05426	6.49615	4.72	22.2784	2.17256	6.87023
4.23	17.8929	2.05670	6.50384	4.73	22.3729	2.17486	6.87750
4.24	17.9776	2.05913	6.51153	4.74	22.4676	2.17715	6.88477
4.25	18.0625	2.06155	6.51920	4.75	22.5625	2.17945	6.89202
4.26	18.1476	2.06398	6.52687	4.76	22.6576	2.18174	6.89928
4.27	18.2329	2.06640	6.53452	4.77	22.7529	2.18403	6.90652
4.28	18.3184	2.06882	6.54217	4.78	22.8484	2.18632	6.91375
4.29	18.4041	2.07123	6.54981	4.79	22.9441	2.18861	6.92098
4.30	18.4900	2.07364	6.55744	4.80	23.0400	2.19089	6.92820
4.31	18.5761	2.07605	6.56506	4.81	23.1361	2.19317	6.93542
4.32	18.6624	2.07846	6.57267	4.82	23.2324	2.19545	6.94262
4.33	18.7489	2.08087	6.58027	4.83	23.3289	2.19773	6.94982
4.34	18.8356	2.08327	6.58787	4.84	23.4256	2.20000	6.95701
4.35	18.9225	2.08567	6.59545	4.85	23.5225	2.20227	6.96419
4.36	19.0096	2.08806	6.60303	4.86	23.6196	2.20454	6.97137
4.37	19.0969	2.09045	6.61060	4.87	23.7169	2.20681	6.97854
4.38	19.1844	2.09284	6.61816	4.88	23.8144	2.20907	6.98570
4.39	19.2721	2.09523	6.62571	4.89	23.9121	2.21133	6.99285
4.40	19.3600	2.09762	6.63325	4.90	24.0100	2.21359	7.00000
4.41	19.4481	2.10000	6.64078	4.91	24.1081	2.21585	7.00714
4.42	19.5364	2.10238	6.64831	4.92	24.2064	2.21811	7.01427
4.43	19.6249	2.10476	6.65582	4.93	24.3049	2.22036	7.02140
4.44	19.7136	2.10713	6.66333	4.94	24.4036	2.22261	7.02851
4.45	19.8025	2.10950	6.67083	4.95	24.5025	2.22486	7.03562
4.46	19.8916	2.11187	6.67832	4.96	24.6016	2.22711	7.04273
4.47	19.9809	2.11424	6.68581	4.97	24.7009	2.22935	7.04982
4.48	20.0704	2.11660	6.69328	4.98	24.8004	2.23159	7.05691
4.49	20.1601	2.11896	6.70075	4.99	24.9001	2.23383	7.06399

n	n^2	$\sqrt{n}$	$\sqrt{10n}$	n	n^2	$\sqrt{n}$	$\sqrt{10n}$
5.00	25.0000	2.23607	7.07107	5.50	30.2500	2.34521	7.41620
5.01	25.1001	2.23830	7.07814	5.51	30.3601	2.34734	7.42294
5.02	25.2004	2.24054	7.08520	5.52	30.4704	2.34947	7.42967
5.03	25.3009	2.24277	7.09225	5.53	30.5809	2.35160	7.43640
5.04	25.4016	2.24499	7.09930	5.54	30.6916	2.35372	7.44312
5.05	25.5025	2.24722	7.10634	5.55	30.8025	2.35584	7.44983
5.06	25.6036	2.24944	7.11337	5.56	30.9136	2.35797	7.45654
5.07	25.7049	2.25167	7.12039	5.57	31.0249	2.36008	7.46324
5.08	25.8064	2.25389	7.12741	5.58	31.1364	2.36220	7.46994
5.09	25.9081	2.25610	7.13442	5.59	31.2481	2.36432	7.47663
5.10	26.0100	2.25832	7.14143	5.60	31.3600	2.36643	7.48331
5.11	26.1121	2.26053	7.14843	5.61	31.4721	2.36854	7.48999
5.12	26.2144	2.26274	7.15542	5.62	31.5844	2.37065	7.49667
5.13	26.3169	2.26495	7.16240	5.63	31.6969	2.37276	7.50333
5.14	26.4196	2.26716	7.16938	5.64	31.8096	2.37487	7.50999
5.15	26.5225	2.26936	7.17635	5.65	31.9225	2.37697	7.51665
5.16	26.6256	2.27156	7.18331	5.66	32.0356	2.37908	7.52330
5.17	26.7289	2.27376	7.19027	5.67	32.1489	2.38118	7.52994
5.18	26.8324	2.27596	7.19722	5.68	32.2624	2.38328	7.53658
5.19	26.9361	2.27816	7.20417	5.69	32.3761	2.38537	7.54321
5.20	27.0400	2.28035	7.21110	5.70	32.4900	2.38747	7.54983
5.21	27.1441	2.28254	7.21803	5.71	32.6041	2.38956	7.55645
5.22	27.2484	2.28473	7.22496	5.72	32.7184	2.39165	7.56307
5.23	27.3529	2.28692	7.23187	5.73	32.8329	2.39374	7.56968
5.24	27.4576	2.28910	7.23878	5.74	32.9476	2.39583	7.57628
5.25	27.5625	2.29129	7.24569	5.75	33.0625	2.39792	7.58288
5.26	27.6676	2.29347	7.25259	5.76	33.1776	2.40000	7.58947
5.27	27.7729	2.29565	7.25948	5.77	33.2929	2.40208	7.59605
5.28	27.8784	2.29783	7.26636	5.78	33.4084	2.40416	7.60263
5.29	27.9841	2.30000	7.27324	5.79	33.5241	2.40624	7.60920
5.30	28.0900	2.30217	7.28011	5.80	33.6400	2.40832	7.61577
5.31	28.1961	2.30434	7.28697	5.81	33.7561	2.41039	7.62234
5.32	28.3024	2.30651	7.29383	5.82	33.8724	2.41247	7.62889
5.33	28.4089	2.30868	7.30068	5.83	33.9889	2.41454	7.63544
5.34	28.5156	2.31084	7.30753	5.84	34.1056	2.41661	7.64199
5.35	28.6225	2.31301	7.31437	5.85	34.2225	2.41868	7.64853
5.36	28.7296	2.31517	7.32120	5.86	34.3396	2.42074	7.65506
5.37	28.8369	2.31733	7.32803	5.87	34.4569	2.42281	7.66159
5.38	28.9444	2.31948	7.33485	5.88	34.5744	2.42487	7.66812
5.39	29.0521	2.32164	7.34166	5.89	34.6921	2.42693	7.67463
5.40	29.1600	2.32379	7.34847	5.90	34.8100	2.42899	7.68115
5.41	29.2681	2.32594	7.35527	5.91	34.9281	2.43105	7.68765
5.42	29.3764	2.32809	7.36205	5.92	35.0464	2.43311	7.69415
5.43	29.4849	2.33024	7.36885	5.93	35.1649	2.43516	7.70065
5.44	29.5936	2.33238	7.37564	5.94	35.2836	2.43721	7.70714
5.45	29.7025	2.33452	7.38241	5.95	35.4025	2.43926	7.71362
5.46	29.8116	2.33666	7.38918	5.96	35.5216	2.44131	7.72010
5.47	29.9209	2.33880	7.39594	5.97	35.6409	2.44336	7.72658
5.48	30.0304	2.34094	7.40270	5.98	35.7604	2.44540	7.73305
5.49	30.1401	2.34307	7.40945	5.99	35.8801	2.44745	7.73951

n	n^2	$\sqrt{n}$	$\sqrt{10n}$	n	n^2	$\sqrt{n}$	$\sqrt{10n}$
6.00	36.0000	2.44949	7.74597	6.50	42.2500	2.54951	8.06226
6.01	36.1201	2.45153	7.75242	6.51	42.3801	2.55147	8.06846
6.02	36.2404	2.45357	7.75887	6.52	42.5104	2.55343	8.07465
6.03	36.3609	2.45561	7.76531	6.53	42.6409	2.55539	8.08084
6.04	36.4816	2.45764	7.77174	6.54	42.7716	2.55734	8.08703
6.05	36.6025	2.45967	7.77817	6.55	42.9025	2.55930	8.09321
6.06	36.7236	2.46171	7.78460	6.56	43.0336	2.56125	8.09938
6.07	36.8449	2.46374	7.79102	6.57	43.1649	2.56320	8.10555
6.08	36.9664	2.46577	7.79744	6.58	43.2964	2.56515	8.11172
6.09	37.0881	2.46779	7.80385	6.59	43.4281	2.56710	8.11788
6.10	37.2100	2.46982	7.81025	6.60	43.5600	2.56905	8.12404
6.11	37.3321	2.47184	7.81665	6.61	43.6921	2.57099	8.13019
6.12	37.4544	2.47386	7.82304	6.62	43.8244	2.57294	8.13634
6.13	37.5769	2.47588	7.82943	6.63	43.9569	2.57488	8.14248
6.14	37.6996	2.47790	7.83582	6.64	44.0896	2.57682	8.14862
6.15	37.8225	2.47992	7.84219	6.65	44.2225	2.57876	8.15475
6.16	37.9456	2.48193	7.84857	6.66	44.3556	2.58070	8.16088
6.17	38.0689	2.48395	7.85493	6.67	44.4889	2.58263	8.16701
6.18	38.1924	2.48596	7.86130	6.68	44.6224	2.58457	8.17313
6.19	38.3161	2.48797	7.86766	6.69	44.7561	2.58650	8.17924
6.20	38.4400	2.48998	7.87401	6.70	44.8900	2.58844	8.18535
6.21	38.5641	2.49199	7.88036	6.71	45.0241	2.59037	8.19146
6.22	38.6884	2.49399	7.88670	6.72	45.1584	2.59230	8.19756
6.23	38.8129	2.49600	7.89303	6.73	45.2929	2.59422	8.20366
6.24	38.9376	2.49800	7.89937	6.74	45.4276	2.59615	8.20975
6.25	39.0625	2.50000	7.90569	6.75	45.5625	2.59808	8.21584
6.26	39.1876	2.50200	7.91202	6.76	45.6976	2.60000	8.22192
6.27	39.3129	2.50400	7.91833	6.77	45.8329	2.60192	8.22800
6.28	39.4384	2.50599	7.92465	6.78	45.9684	2.60384	8.23408
6.29	39.5641	2.50799	7.93095	6.79	46.1041	2.60576	8.24015
6.30	39.6900	2.50998	7.93725	6.80	46.2400	2.60768	8.24621
6.31	39.8161	2.51197	7.94355	6.81	46.3761	2.60960	8.25227
6.32	39.9424	2.51396	7.94984	6.82	46.5124	2.61151	8.25833
6.33	40.0689	2.51595	7.95613	6.83	46.6489	2.61343	8.26438
6.34	40.1956	2.51794	7.96241	6.84	46.7856	2.61534	8.27043
6.35	40.3225	2.51992	7.96869	6.85	46.9225	2.61725	8.27647
6.36	40.4496	2.52190	7.97496	6.86	47.0596	2.61916	8.28251
6.37	40.5769	2.52389	7.98123	6.87	47.1969	2.62107	8.28855
6.38	40.7044	2.52587	7.98749	6.88	47.3344	2.62298	8.29458
6.39	40.8321	2.52784	7.99375	6.89	47.4721	2.62488	8.30060
6.40	40.9600	2.52982	8.00000	6.90	47.6100	2.62679	8.30662
6.41	41.0881	2.53180	8.00625	6.91	47.7481	2.62869	8.31264
6.42	41.2164	2.53377	8.01249	6.92	47.8864	2.63059	8.31865
6.43	41.3449	2.53574	8.01873	6.93	48.0249	2.63249	8.32466
6.44	41.4736	2.53772	8.02496	6.94	48.1636	2.63439	8.33067
6.45	41.6025	2.53969	8.03119	6.95	48.3025	2.63629	8.33667
6.46	41.7316	2.54165	8.03741	6.96	48.4416	2.63818	8.34266
6.47	41.8609	2.54362	8.04363	6.97	48.5809	2.64008	8.34865
6.48	41.9904	2.54558	8.04984	6.98	48.7204	2.64197	8.35464
6.49	42.1201	2.54755	8.05605	6.99	48.8601	2.64386	8.36062

351

n	n^2	$\sqrt{n}$	$\sqrt{10n}$	n	n^2	$\sqrt{n}$	$\sqrt{10n}$
7.00	49.0000	2.64575	8.36660	7.50	56.2500	2.73861	8.66025
7.01	49.1401	2.64764	8.37257	7.51	56.4001	2.74044	8.66603
7.02	49.2804	2.64953	8.37854	7.52	56.5504	2.74226	8.67179
7.03	49.4209	2.65141	8.38451	7.53	56.7009	2.74408	8.67756
7.04	49.5616	2.65330	8.39047	7.54	56.8516	2.74591	8.68332
7.05	49.7025	2.65518	8.39643	7.55	57.0025	2.74773	8.68907
7.06	49.8436	2.65707	8.40238	7.56	57.1536	2.74955	8.69483
7.07	49.9849	2.65895	8.40833	7.57	57.3049	2.75136	8.70057
7.08	50.1264	2.66083	8.41427	7.58	57.4564	2.75318	8.70632
7.09	50.2681	2.66271	8.42021	7.59	57.6081	2.75500	8.71206
7.10	50.4100	2.66458	8.42615	7.60	57.7600	2.75681	8.71780
7.11	50.5521	2.66646	8.43208	7.61	57.9121	2.75862	8.72353
7.12	50.6944	2.66833	8.43801	7.62	58.0644	2.76043	8.72926
7.13	50.8369	2.67021	8.44393	7.63	58.2169	2.76225	8.73499
7.14	50.9796	2.67208	8.44985	7.64	58.3696	2.76405	8.74071
7.15	51.1225	2.67395	8.45577	7.65	58.5225	2.76586	8.74643
7.16	51.2656	2.67582	8.46168	7.66	58.6756	2.76767	8.75214
7.17	51.4089	2.67769	8.46759	7.67	58.8289	2.76948	8.75785
7.18	51.5524	2.67955	8.47349	7.68	58.9824	2.77128	8.76356
7.19	51.6961	2.68142	8.47939	7.69	59.1361	2.77308	8.76926
7.20	51.8400	2.68328	8.48528	7.70	59.2900	2.77489	8.77496
7.21	51.9841	2.68514	8.49117	7.71	59.4441	2.77669	8.78066
7.22	52.1284	2.68701	8.49706	7.72	59.5984	2.77849	8.78635
7.23	52.2729	2.68887	8.50294	7.73	59.7529	2.78029	8.79204
7.24	52.4176	2.69072	8.50882	7.74	59.9076	2.78209	8.79773
7.25	52.5625	2.69258	8.51469	7.75	60.0625	2.78388	8.80341
7.26	52.7076	2.69444	8.52056	7.76	60.2176	2.78568	8.80909
7.27	52.8529	2.69629	8.52643	7.77	60.3729	2.78747	8.81476
7.28	52.9984	2.69815	8.53229	7.78	60.5284	2.78927	8.82043
7.29	53.1441	2.70000	8.53815	7.79	60.6841	2.79106	8.82610
7.30	53.2900	2.70185	8.54400	7.80	60.8400	2.79285	8.83176
7.31	53.4361	2.70370	8.54985	7.81	60.9961	2.79464	8.83742
7.32	53.5824	2.70555	8.55570	7.82	61.1524	2.79643	8.84308
7.33	53.7289	2.70740	8.56154	7.83	61.3089	2.79821	8.84873
7.34	53.8756	2.70924	8.56738	7.84	61.4656	2.80000	8.85438
7.35	54.0225	2.71109	8.57321	7.85	61.6225	2.80179	8.86002
7.36	54.1696	2.71293	8.57904	7.86	61.7796	2.80357	8.86566
7.37	54.3169	2.71477	8.58487	7.87	61.9369	2.80535	8.87130
7.38	54.4644	2.71662	8.59069	7.88	62.0944	2.80713	8.87694
7.39	54.6121	2.71846	8.59651	7.89	62.2521	2.80891	8.88257
7.40	54.7600	2.72029	8.60233	7.90	62.4100	2.81069	8.88819
7.41	54.9081	2.72213	8.60814	7.91	62.5681	2.81247	8.89382
7.42	55.0564	2.72397	8.61394	7.92	62.7264	2.81425	8.89944
7.43	55.2049	2.72580	8.61974	7.93	62.8849	2.81603	8.90505
7.44	55.3536	2.72764	8.62554	7.94	63.0436	2.81780	8.91067
7.45	55.5025	2.72947	8.63134	7.95	63.2025	2.81957	8.91628
7.46	55.6516	2.73130	8.63713	7.96	63.3616	2.82135	8.92188
7.47	55.8009	2.73313	8.64292	7.97	63.5209	2.82312	8.92749
7.48	55.9504	2.73496	8.64870	7.98	63.6804	2.82489	8.93308
7.49	56.1001	2.73679	8.65448	7.99	63.8401	2.82666	8.93868

n	n^2	$\sqrt{n}$	$\sqrt{10n}$	n	n^2	$\sqrt{n}$	$\sqrt{10n}$
8.00	64.0000	2.82843	8.94427	8.50	72.2500	2.91548	9.21954
8.01	64.1601	2.83019	8.94986	8.51	72.4201	2.91719	9.22497
8.02	64.3204	2.83196	8.95545	8.52	72.5904	2.91890	9.23038
8.03	64.4809	2.83373	8.96103	8.53	72.7609	2.92062	9.23580
8.04	64.6416	2.83549	8.96660	8.54	72.9316	2.92233	9.24121
8.05	64.8025	2.83725	8.97218	8.55	73.1025	2.92404	9.24662
8.06	64.9636	2.83901	8.97775	8.56	73.2736	2.92575	9.25203
8.07	65.1249	2.84077	8.98332	8.57	73.4449	2.92746	9.25743
8.08	65.2864	2.84253	8.98888	8.58	73.6164	2.92916	9.26283
8.09	65.4481	2.84429	8.99444	8.59	73.7881	2.93087	9.26823
8.10	65.6100	2.84605	9.00000	8.60	73.9600	2.93258	9.27362
8.11	65.7721	2.84781	9.00555	8.61	74.1321	2.93428	9.27901
8.12	65.9344	2.84956	9.01110	8.62	74.3044	2.93598	9.28440
8.13	66.0969	2.85132	9.01665	8.63	74.4769	2.93769	9.28978
8.14	66.2596	2.85307	9.02219	8.64	74.6496	2.93939	9.29516
8.15	66.4225	2.85482	9.02774	8.65	74.8225	2.94109	9.30054
8.16	66.5856	2.85657	9.03327	8.66	74.9956	2.94279	9.30591
8.17	66.7489	2.85832	9.03881	8.67	75.1689	2.94449	9.31128
8.18	66.9124	2.86007	9.04434	8.68	75.3424	2.94618	9.31665
8.19	67.0761	2.86182	9.04986	8.69	75.5161	2.94788	9.32202
8.20	67.2400	2.86356	9.05539	8.70	75.6900	2.94958	9.32738
8.21	67.4041	2.86531	9.06091	8.71	75.8641	2.95127	9.33274
8.22	67.5684	2.86705	9.06642	8.72	76.0384	2.95296	9.33809
8.23	67.7329	2.86880	9.07193	8.73	76.2129	2.95466	9.34345
8.24	67.8976	2.87054	9.07744	8.74	76.3876	2.95635	9.34880
8.25	68.0625	2.87228	9.08295	8.75	76.5625	2.95804	9.35414
8.26	68.2276	2.87402	9.08845	8.76	76.7376	2.95973	9.35949
8.27	68.3929	2.87576	9.09395	8.77	76.9129	2.96142	9.36483
8.28	68.5584	2.87750	9.09945	8.78	77.0884	2.96311	9.37017
8.29	68.7241	2.87924	9.10494	8.79	77.2641	2.96479	9.37550
8.30	68.8900	2.88097	9.11043	8.80	77.4400	2.96648	9.38083
8.31	69.0561	2.88271	9.11592	8.81	77.6161	2.96816	9.38616
8.32	69.2224	2.88444	9.12140	8.82	77.7924	2.96985	9.39149
8.33	69.3889	2.88617	9.12688	8.83	77.9689	2.97153	9.39681
8.34	69.5556	2.88791	9.13236	8.84	78.1456	2.97321	9.40213
8.35	69.7225	2.88964	9.13783	8.85	78.3225	2.97489	9.40744
8.36	69.8896	2.89137	9.14330	8.86	78.4996	2.97658	9.41276
8.37	70.0569	2.89310	9.14877	8.87	78.6769	2.97825	9.41807
8.38	70.2244	2.89482	9.15423	8.88	78.8544	2.97993	9.42338
8.39	70.3921	2.89655	9.15969	8.89	79.0321	2.98161	9.42868
8.40	70.5600	2.89828	9.16515	8.90	79.2100	2.98329	9.43398
8.41	70.7281	2.90000	9.17061	8.91	79.3881	2.98496	9.43928
8.42	70.8964	2.90172	9.17606	8.92	79.5664	2.98664	9.44458
8.43	71.0649	2.90345	9.18150	8.93	79.7449	2.98831	9.44987
8.44	71.2336	2.90517	9.18695	8.94	79.9236	2.98998	9.45516
8.45	71.4025	2.90689	9.19239	8.95	80.1025	2.99166	9.46044
8.46	71.5716	2.90861	9.19783	8.96	80.2816	2.99333	9.46573
8.47	71.7409	2.91033	9.20326	8.97	80.4609	2.99500	9.47101
8.48	71.9104	2.91204	9.20869	8.98	80.6404	2.99666	9.47629
8.49	72.0801	2.91376	9.21412	8.99	80.8201	2.99833	9.48156

n	n^2	$\sqrt{n}$	$\sqrt{10n}$	n	n^2	$\sqrt{n}$	$\sqrt{10n}$
9.00	81.0000	3.00000	9.48683	9.50	90.2500	3.08221	9.74679
9.01	81.1801	3.00167	9.49210	9.51	90.4401	3.08383	9.75192
9.02	81.3604	3.00333	9.49737	9.52	90.6304	3.08545	9.75705
9.03	81.5409	3.00500	9.50263	9.53	90.8209	3.08707	9.76217
9.04	81.7216	3.00666	9.50789	9.54	91.0116	3.08869	9.76729
9.05	81.9025	3.00832	9.51315	9.55	91.2025	3.09031	9.77241
9.06	82.0836	3.00998	9.51840	9.56	91.3936	3.09192	9.77753
9.07	82.2649	3.01164	9.52365	9.57	91.5849	3.09354	9.78264
9.08	82.4464	3.01330	9.52890	9.58	91.7764	3.09516	9.78775
9.09	82.6281	3.01496	9.53415	9.59	91.9681	3.09677	9.79285
9.10	82.8100	3.01662	9.53939	9.60	92.1600	3.09839	9.79796
9.11	82.9921	3.01828	9.54463	9.61	92.3521	3.10000	9.80306
9.12	83.1744	3.01993	9.54987	9.62	92.5444	3.10161	9.80816
9.13	83.3569	3.02159	9.55510	9.63	92.7369	3.10322	9.81326
9.14	83.5396	3.02324	9.56033	9.64	92.9296	3.10483	9.81835
9.15	83.7225	3.02490	9.56556	9.65	93.1225	3.10644	9.82344
9.16	83.9056	3.02655	9.57079	9.66	93.3156	3.10805	9.82853
9.17	84.0889	3.02820	9.57601	9.67	93.5089	3.10966	9.83362
9.18	84.2724	3.02985	9.58123	9.68	93.7024	3.11127	9.83870
9.19	84.4561	3.03150	9.58645	9.69	93.8961	3.11288	9.84378
9.20	84.6400	3.03315	9.59166	9.70	94.0900	3.11448	9.84886
9.21	84.8241	3.03480	9.59687	9.71	94.2841	3.11609	9.85393
9.22	85.0084	3.03645	9.60208	9.72	94.4784	3.11769	9.85901
9.23	85.1929	3.03809	9.60729	9.73	94.6729	3.11929	9.86408
9.24	85.3776	3.03974	9.61249	9.74	94.8676	3.12090	9.86914
9.25	85.5625	3.04138	9.61769	9.75	95.0625	3.12250	9.87421
9.26	85.7476	3.04302	9.62289	9.76	95.2576	3.12410	9.87927
9.27	85.9329	3.04467	9.62808	9.77	95.4529	3.12570	9.88433
9.28	86.1184	3.04631	9.63328	9.78	95.6484	3.12730	9.88939
9.29	86.3041	3.04795	9.63846	9.79	95.8441	3.12890	9.89444
9.30	86.4900	3.04959	9.64365	9.80	96.0400	3.13050	9.89949
9.31	86.6761	3.05123	9.64883	9.81	96.2361	3.13209	9.90454
9.32	86.8624	3.05287	9.65401	9.82	96.4324	3.13369	9.90959
9.33	87.0489	3.05450	9.65919	9.83	96.6289	3.13528	9.91464
9.34	87.2356	3.05614	9.66437	9.84	96.8256	3.13688	9.91968
9.35	87.4225	3.05778	9.66954	9.85	97.0225	3.13847	9.92472
9.36	87.6096	3.05941	9.67471	9.86	97.2196	3.14006	9.92975
9.37	87.7969	3.06105	9.67988	9.87	97.4169	3.14166	9.93479
9.38	87.9844	3.06268	9.68504	9.88	97.6144	3.14325	9.93982
9.39	88.1721	3.06431	9.69020	9.89	97.8121	3.14484	9.94485
9.40	88.3600	3.06594	9.69536	9.90	98.0100	3.14643	9.94987
9.41	88.5481	3.06757	9.70052	9.91	98.2081	3.14802	9.95490
9.42	88.7364	3.06920	9.70567	9.92	98.4064	3.14960	9.95992
9.43	88.9249	3.07083	9.71082	9.93	98.6049	3.15119	9.96494
9.44	89.1136	3.07246	9.71597	9.94	98.8036	3.15278	9.96995
9.45	89.3025	3.07409	9.72111	9.95	99.0025	3.15436	9.97497
9.46	89.4916	3.07571	9.72625	9.96	99.2016	3.15595	9.97998
9.47	89.6809	3.07734	9.73139	9.97	99.4009	3.15753	9.98499
9.48	89.8704	3.07896	9.73653	9.98	99.6004	3.15911	9.98999
9.49	90.0601	3.08058	9.74166	9.99	99.8001	3.16070	9.99500
				10.00	100.000	3.16228	10.0000

II Normal Curve z Scores and
Percentile Ranks

(The proportion of area given is that to the left of the given z score. It should be noted that all area values have been rounded to three decimal places.)

z	AREA	z	AREA	z	AREA
−4.0	.000	−1.3	.097	1.4	.919
−3.9	.000	−1.2	.115	1.5	.933
−3.8	.000	−1.1	.136	1.6	.945
−3.7	.000	−1.0	.159	1.7	.955
−3.6	.000	−0.9	.184	1.8	.964
−3.5	.000	−0.8	.212	1.9	.971
−3.4	.000	−0.7	.242	2.0	.977
−3.3	.001	−0.6	.274	2.1	982
−3.2	.001	−0.5	.308	2.2	.986
−3.1	.001	−0.4	.345	2.3	.989
−3.0	.001	−0.3	.382	2.4	.992
−2.9	.002	−0.2	.421	2.5	.994
−2.8	.003	−0.1	.460	2.6	.995
−2.7	.004	0.0	.500	2.7	.996
−2.6	.005	0.1	.540	2.8	.997
−2.5	.006	0.2	.579	2.9	.998
−2.4	.008	0.3	.618	3.0	.999
−2.3	.011	0.4	.655	3.1	.999
−2.2	.014	0.5	.692	3.2	.999
−2.1	.018	0.6	.726	3.3	.999
−2.0	.023	0.7	.758	3.4	1.000
−1.9	.029	0.8	.788	3.5	1.000
−1.8	.036	0.9	.816	3.6	1.000
−1.7	.045	1.0	.841	3.7	1.000
−1.6	.055	1.1	.864	3.8	1.000
−1.5	.067	1.2	.885	3.9	1.000
−1.4	.081	1.3	.903	4.0	1.000

z	.00	.01	.02	.03	.04	.05	.06	.07	.08	.09
0.0	.0000	.0040	.0080	.0120	.0160	.0199	.0239	.0279	.0319	.0359
0.1	.0398	.0438	.0478	.0517	.0557	.0596	.0636	.0675	.0714	.0753
0.2	.0793	.0832	.0871	.0910	.0948	.0987	.1026	.1064	.1103	.1141
0.3	.1179	.1217	.1255	.1293	.1331	.1368	.1406	.1443	.1480	.1517
0.4	.1554	.1591	.1628	.1664	.1700	.1736	.1772	.1808	.1844	.1879
0.5	.1915	.1950	.1985	.2019	.2054	.2088	.2123	.2157	.2190	.2224
0.6	.2257	.2291	.2324	.2357	.2389	.2422	.2454	.2486	.2517	.2549
0.7	.2580	.2611	.2642	.2673	.2704	.2734	.2764	.2794	.2823	.2852
0.8	.2881	.2910	.2939	.2967	.2995	.3023	.3051	.3078	.3106	.3133
0.9	.3159	.3186	.3212	.3238	.3264	.3289	.3315	.3340	.3365	.3389
1.0	.3413	.3438	.3461	.3485	.3508	.3531	.3554	.3577	.3599	.3621
1.1	.3643	.3665	.3686	.3708	.3729	.3749	.3770	.3790	.3810	.3830
1.2	.3849	.3869	.3888	.3907	.3925	.3944	.3962	.3980	.3997	.4015
1.3	.4032	.4049	.4066	.4082	.4099	.4115	.4131	.4147	.4162	.4177
1.4	.4192	.4207	.4222	.4236	.4251	.4265	.4279	.4292	.4306	.4319
1.5	.4332	.4345	.4357	.4370	.4382	.4394	.4406	.4418	.4429	.4441
1.6	.4452	.4463	.4474	.4484	.4495	.4505	.4515	.4525	.4535	.4545
1.7	.4554	.4564	.4573	.4582	.4591	.4599	.4608	.4616	.4625	.4633
1.8	.4641	.4649	.4656	.4664	.4671	.4678	.4686	.4693	.4699	.4706
1.9	.4713	.4719	.4726	.4732	.4738	.4744	.4750	.4756	.4761	.4767
2.0	.4772	.4778	.4783	.4788	.4793	.4798	.4803	.4808	.4812	.4817
2.1	.4821	.4826	.4830	.4834	.4838	.4842	.4846	.4850	.4854	.4857
2.2	.4861	.4864	.4868	.4871	.4875	.4878	.4881	.4884	.4887	.4890
2.3	.4893	.4896	.4898	.4901	.4904	.4906	.4909	.4911	.4913	.4916
2.4	.4918	.4920	.4922	.4925	.4927	.4929	.4931	.4932	.4934	.4936
2.5	.4938	.4940	.4941	.4943	.4945	.4946	.4948	.4949	.4951	.4952
2.6	.4953	.4955	.4956	.4957	.4959	.4960	.4961	.4962	.4963	.4964
2.7	.4965	.4966	.4967	.4968	.4969	.4970	.4971	.4972	.4973	.4974
2.8	.4974	.4975	.4976	.4977	.4977	.4978	.4979	.4979	.4980	.4981
2.9	.4981	.4982	.4982	.4983	.4984	.4984	.4985	.4985	.4986	.4986
3.0	.4987	.4987	.4987	.4988	.4988	.4989	.4989	.4989	.4990	.4990
3.1	.49903									
3.2	.49931									
3.3	.49952									
3.4	.49966									
3.5	.49977									
3.6	.49984									
3.7	.49989									
3.8	.49993									
3.9	.49995									
4.0	.50000									

(The proportion of area to the left of each *t* value is given at the top of the column.)

d. f.	.005	.010	.025	.050	.500	.950	.975	.990	.995
1	−63.66	−31.82	−12.71	−6.31	0.00	6.31	12.71	31.82	63.66
2	−9.92	−6.96	−4.30	−2.92	0.00	2.92	4.30	6.96	9.92
3	−5.84	−4.54	−3.18	−2.35	0.00	2.35	3.18	4.54	5.84
4	−4.60	−3.75	−2.78	−2.13	0.00	2.13	2.78	3.75	4.60
5	−4.03	−3.36	−2.57	−2.02	0.00	2.02	2.57	3.36	4.03
6	−3.71	−3.14	−2.45	−1.94	0.00	1.94	2.45	3.14	3.71
7	−3.50	−3.00	−2.36	−1.90	0.00	1.90	2.36	3.00	3.50
8	−3.36	−2.90	−2.31	−1.86	0.00	1.86	2.31	2.90	3.36
9	−3.25	−2.82	−2.26	−1.83	0.00	1.83	2.26	2.82	3.25
10	−3.17	−2.76	−2.23	−1.81	0.00	1.81	2.23	2.76	3.17
11	−3.11	−2.72	−2.20	−1.80	0.00	1.80	2.20	2.72	3.11
12	−3.06	−2.68	−2.18	−1.78	0.00	1.78	2.18	2.68	3.06
13	−3.01	−2.65	−2.16	−1.77	0.00	1.77	2.16	2.65	3.01
14	−2.98	−2.62	−2.14	−1.76	0.00	1.76	2.14	2.62	2.98
15	−2.95	−2.60	−2.13	−1.75	0.00	1.75	2.13	2.60	2.95
16	−2.92	−2.58	−2.12	−1.75	0.00	1.75	2.12	2.58	2.92
17	−2.90	−2.57	−2.11	−1.74	0.00	1.74	2.11	2.57	2.90
18	−2.88	−2.55	−2.10	−1.73	0.00	1.73	2.10	2.55	2.88
19	−2.86	−2.54	−2.09	−1.73	0.00	1.73	2.09	2.54	2.86
20	−2.84	−2.53	−2.09	−1.72	0.00	1.72	2.09	2.53	2.84
21	−2.83	−2.52	−2.08	−1.72	0.00	1.72	2.08	2.52	2.83
22	−2.82	−2.51	−2.07	−1.72	0.00	1.72	2.07	2.51	2.82
23	−2.81	−2.50	−2.07	−1.71	0.00	1.71	2.07	2.50	2.81
24	−2.80	−2.49	−2.06	−1.71	0.00	1.71	2.06	2.49	2.80
25	−2.79	−2.48	−2.06	−1.71	0.00	1.71	2.06	2.48	2.79
26	−2.78	−2.48	−2.06	−1.71	0.00	1.71	2.06	2.48	2.78
27	−2.77	−2.47	−2.05	−1.70	0.00	1.70	2.05	2.47	2.77
28	−2.76	−2.47	−2.05	−1.70	0.00	1.70	2.05	2.47	2.76
29	−2.76	−2.46	−2.04	−1.70	0.00	1.70	2.04	2.46	2.76
30	−2.75	−2.46	−2.04	−1.70	0.00	1.70	2.04	2.46	2.75
31	−2.75	−2.45	−2.04	−1.70	0.00	1.70	2.04	2.45	2.75
32	−2.74	−2.45	−2.03	−1.69	0.00	1.69	2.03	2.45	2.74
33	−2.74	−2.45	−2.03	−1.69	0.00	1.69	2.03	2.45	2.74

Appendix IV is abridged from Table III of Fisher & Yates, *Statistical Tables for Biological, Agricultural, and Medical Research*, published by Oliver & Boyd Ltd., Edinburgh, and by permission of the authors and publishers.

(The significance level for each value is given at the top of the column.)

d. f.	.01	.05	.10
1	6.64	3.84	2.71
2	9.21	5.99	4.60
3	11.34	7.82	6.25
4	13.28	9.49	7.78
5	15.09	11.07	9.24
6	16.81	12.59	10.64
7	18.48	14.07	12.02
8	20.09	15.51	13.36
9	21.67	16.92	14.68
10	23.21	18.31	15.99
11	24.72	19.68	17.28
12	26.22	21.03	18.55
13	27.69	22.36	19.81
14	29.14	23.68	21.06
15	30.58	25.00	22.31
16	32.00	26.97	23.54
17	33.41	27.59	24.77
18	34.80	28.87	25.99
19	36.19	30.14	27.20
20	37.57	31.41	28.41
21	38.93	32.67	29.62
22	40.29	33.92	30.81
23	41.64	35.17	32.01
24	42.98	36.42	33.20
25	44.31	37.65	34.38
26	45.64	38.88	35.56
27	46.96	40.11	36.74
28	48.28	41.34	37.92
29	49.59	42.56	39.09
30	50.89	43.77	40.26

Appendix V is abridged from Table IV of Fisher & Yates, *Statistical Tables for Biological, Agricultural, and Medical Research*, published by Oliver & Boyd Ltd., Edinburgh, and by permission of the authors and publishers.

(All values are at the .05 significance level.   The larger of n_1 and n_2 is to be read at the top and the smaller is to be read in the left margin.)

	5	6	7	8	9	10	11	12	13	14	15	16	17	18	19	20
2								2 6	2 6	2 6	2 6	2 6	2 6	2 6	2 6	2 6
3		2 8	2 8	2 8	2 8	2 8	2 8	2 8	2 8	3 8	3 8	3 8	3 8	3 8	3 8	3 8
4	2 9	2 9	2 10	3 10	3 10	3 10	3 10	3 10	3 10	3 10	3 10	4 10	4 10	4 10	4 10	4 10
5	2 10	3 10	3 11	3 11	3 12	3 12	4 12	4 12	4 12	4 12	4 12	4 12	4 12	5 12	5 12	5 12
6		3 11	3 12	3 12	4 13	4 13	4 13	4 13	5 14	5 14	5 14	5 14	5 14	5 14	6 14	6 14
7			3 13	4 13	4 14	5 14	5 14	5 14	5 15	5 15	6 15	6 16	6 16	6 16	6 16	6 16
8				4 14	5 14	5 15	5 15	6 16	6 16	6 16	6 16	6 17	7 17	7 17	7 17	7 17
9					5 15	5 16	6 16	6 16	6 17	7 17	7 18	7 18	7 18	8 18	8 18	8 18
10						6 16	6 17	7 17	7 18	7 18	7 18	8 19	8 19	8 19	8 20	9 20
11							7 17	7 18	7 19	8 19	8 19	8 20	9 20	9 20	9 21	9 21
12								7 19	8 19	8 20	8 20	9 21	9 21	9 21	10 22	10 22
13									8 20	9 20	9 21	9 21	10 22	10 22	10 23	10 23
14										9 21	9 22	10 22	10 23	10 23	11 23	11 24
15											10 22	10 23	11 23	11 24	11 24	12 25
16												11 23	11 24	11 25	12 25	12 25
17													11 25	12 25	12 26	13 26
18														12 26	13 26	13 27
19															13 27	13 27
20																14 28

Taken with permission from "Tables for Testing Randomness of Grouping in a Sequence of Alternatives" by C. Eisenhart and F. Smed, *Annals of Mathematical Statistics*, XIV (1943), 66.

VII Values of F

The 5% (Roman Type) and 1% (Boldface Type) Points for the Distribution of F

n_1 degrees of freedom (for greater estimate of variance)

n_2	1	2	3	4	5	6	7	8	9	10	11	12	14	16	20	24	30	40	50	75	100	200	500	∞
1	161 / 4,052	200 / 4,999	216 / 5,403	225 / 5,625	230 / 5,764	234 / 5,859	237 / 5,928	239 / 5,981	241 / 6,022	242 / 6,056	243 / 6,082	244 / 6,106	245 / 6,142	246 / 6,169	248 / 6,208	249 / 6,234	250 / 6,258	251 / 6,286	252 / 6,302	253 / 6,323	253 / 6,334	254 / 6,352	254 / 6,361	254 / 6,366
2	18.51 / 98.49	19.00 / 99.00	19.16 / 99.17	19.25 / 99.25	19.30 / 99.30	19.33 / 99.33	19.36 / 99.34	19.37 / 99.36	19.38 / 99.38	19.39 / 99.40	19.40 / 99.41	19.41 / 99.42	19.42 / 99.43	19.43 / 99.44	19.44 / 99.45	19.45 / 99.46	19.46 / 99.47	19.47 / 99.48	19.47 / 99.48	19.48 / 99.49	19.49 / 99.49	19.49 / 99.49	19.50 / 99.50	19.50 / 99.50
3	10.13 / 34.12	9.55 / 30.82	9.28 / 29.46	9.12 / 28.71	9.01 / 28.24	8.94 / 27.91	8.88 / 27.67	8.84 / 27.49	8.81 / 27.34	8.78 / 27.23	8.76 / 27.13	8.74 / 27.05	8.71 / 26.92	8.69 / 26.83	8.66 / 26.69	8.64 / 26.60	8.62 / 26.50	8.60 / 26.41	8.58 / 26.35	8.57 / 26.27	8.56 / 26.23	8.54 / 26.18	8.54 / 26.14	8.53 / 26.12
4	7.71 / 21.20	6.94 / 18.00	6.59 / 16.69	6.39 / 15.98	6.26 / 15.52	6.16 / 15.21	6.09 / 14.98	6.04 / 14.80	6.00 / 14.66	5.96 / 14.54	5.93 / 14.45	5.91 / 14.37	5.87 / 14.24	5.84 / 14.15	5.80 / 14.02	5.77 / 13.93	5.74 / 13.83	5.71 / 13.74	5.70 / 13.69	5.68 / 13.61	5.66 / 13.57	5.65 / 13.52	5.64 / 13.48	5.63 / 13.46
5	6.61 / 16.26	5.79 / 13.27	5.41 / 12.06	5.19 / 11.39	5.05 / 10.97	4.95 / 10.67	4.88 / 10.45	4.82 / 10.27	4.78 / 10.15	4.74 / 10.05	4.70 / 9.96	4.68 / 9.89	4.64 / 9.77	4.60 / 9.68	4.56 / 9.55	4.53 / 9.47	4.50 / 9.38	4.46 / 9.29	4.44 / 9.24	4.42 / 9.17	4.40 / 9.13	4.38 / 9.07	4.37 / 9.04	4.36 / 9.02
6	5.99 / 13.74	5.14 / 10.92	4.76 / 9.78	4.53 / 9.15	4.39 / 8.75	4.28 / 8.47	4.21 / 8.26	4.15 / 8.10	4.10 / 7.98	4.06 / 7.87	4.03 / 7.79	4.00 / 7.72	3.96 / 7.60	3.92 / 7.52	3.87 / 7.39	3.84 / 7.31	3.81 / 7.23	3.77 / 7.14	3.75 / 7.09	3.72 / 7.02	3.71 / 6.99	3.69 / 6.94	3.68 / 6.90	3.67 / 6.88
7	5.59 / 12.25	4.74 / 9.55	4.35 / 8.45	4.12 / 7.85	3.97 / 7.46	3.87 / 7.19	3.79 / 7.00	3.73 / 6.84	3.68 / 6.71	3.63 / 6.62	3.60 / 6.54	3.57 / 6.47	3.52 / 6.35	3.49 / 6.27	3.44 / 6.15	3.41 / 6.07	3.38 / 5.98	3.34 / 5.90	3.32 / 5.85	3.29 / 5.78	3.28 / 5.75	3.25 / 5.70	3.24 / 5.67	3.23 / 5.65
8	5.32 / 11.26	4.46 / 8.65	4.07 / 7.59	3.84 / 7.01	3.69 / 6.63	3.58 / 6.37	3.50 / 6.19	3.44 / 6.03	3.39 / 5.91	3.34 / 5.82	3.31 / 5.74	3.28 / 5.67	3.23 / 5.56	3.20 / 5.48	3.15 / 5.36	3.12 / 5.28	3.08 / 5.20	3.05 / 5.11	3.03 / 5.06	3.00 / 5.00	2.98 / 4.96	2.96 / 4.91	2.94 / 4.88	2.93 / 4.86
9	5.12 / 10.56	4.26 / 8.02	3.86 / 6.99	3.63 / 6.42	3.48 / 6.06	3.37 / 5.80	3.29 / 5.62	3.23 / 5.47	3.18 / 5.35	3.13 / 5.26	3.10 / 5.18	3.07 / 5.11	3.02 / 5.00	2.98 / 4.92	2.93 / 4.80	2.90 / 4.73	2.86 / 4.64	2.82 / 4.56	2.80 / 4.51	2.77 / 4.45	2.76 / 4.41	2.73 / 4.36	2.72 / 4.33	2.71 / 4.31
10	4.96 / 10.04	4.10 / 7.56	3.71 / 6.55	3.48 / 5.99	3.33 / 5.64	3.22 / 5.39	3.14 / 5.21	3.07 / 5.06	3.02 / 4.95	2.97 / 4.85	2.94 / 4.78	2.91 / 4.71	2.86 / 4.60	2.82 / 4.52	2.77 / 4.41	2.74 / 4.33	2.70 / 4.25	2.67 / 4.17	2.64 / 4.12	2.61 / 4.05	2.59 / 4.01	2.56 / 3.96	2.55 / 3.93	2.54 / 3.91
11	4.84 / 9.65	3.98 / 7.20	3.59 / 6.22	3.36 / 5.67	3.20 / 5.32	3.09 / 5.07	3.01 / 4.88	2.95 / 4.74	2.90 / 4.63	2.86 / 4.54	2.82 / 4.46	2.79 / 4.40	2.74 / 4.29	2.70 / 4.21	2.65 / 4.10	2.61 / 4.02	2.57 / 3.94	2.53 / 3.86	2.50 / 3.80	2.47 / 3.74	2.45 / 3.70	2.42 / 3.66	2.41 / 3.62	2.40 / 3.60
12	4.75 / 9.33	3.88 / 6.93	3.49 / 5.95	3.26 / 5.41	3.11 / 5.06	3.00 / 4.82	2.92 / 4.65	2.85 / 4.50	2.80 / 4.39	2.76 / 4.30	2.72 / 4.22	2.69 / 4.16	2.64 / 4.05	2.60 / 3.98	2.54 / 3.86	2.50 / 3.78	2.46 / 3.70	2.42 / 3.61	2.40 / 3.56	2.36 / 3.49	2.35 / 3.46	2.32 / 3.41	2.31 / 3.38	2.30 / 3.36
13	4.67 / 9.07	3.80 / 6.70	3.41 / 5.74	3.18 / 5.20	3.02 / 4.86	2.92 / 4.62	2.84 / 4.44	2.77 / 4.30	2.72 / 4.19	2.67 / 4.10	2.63 / 4.02	2.60 / 3.96	2.55 / 3.85	2.51 / 3.78	2.46 / 3.67	2.42 / 3.59	2.38 / 3.51	2.34 / 3.42	2.32 / 3.37	2.28 / 3.30	2.26 / 3.27	2.24 / 3.21	2.22 / 3.18	2.21 / 3.16

n1 degrees of freedom (for greater estimate of variance)

n2	1	2	3	4	5	6	7	8	9	10	11	12	14	16	20	24	30	40	50	75	100	200	500	∞
14	4.60/8.86	3.74/6.51	3.34/5.56	3.11/5.03	2.96/4.69	2.85/4.46	2.77/4.28	2.70/4.14	2.65/4.03	2.60/3.94	2.56/3.86	2.53/3.80	2.48/3.70	2.44/3.62	2.39/3.51	2.35/3.43	2.31/3.34	2.27/3.26	2.24/3.21	2.21/3.14	2.19/3.11	2.16/3.06	2.14/3.02	2.13/3.00
15	4.54/8.68	3.68/6.36	3.29/5.42	3.06/4.89	2.90/4.56	2.79/4.32	2.70/4.14	2.64/4.00	2.59/3.89	2.55/3.80	2.51/3.73	2.48/3.67	2.43/3.56	2.39/3.48	2.33/3.36	2.29/3.29	2.25/3.20	2.21/3.12	2.18/3.07	2.15/3.00	2.12/2.97	2.10/2.92	2.08/2.89	2.07/2.87
16	4.49/8.53	3.63/6.23	3.24/5.29	3.01/4.77	2.85/4.44	2.74/4.20	2.66/4.03	2.59/3.89	2.54/3.78	2.49/3.69	2.45/3.61	2.42/3.55	2.37/3.45	2.33/3.37	2.28/3.25	2.24/3.18	2.20/3.10	2.16/3.01	2.13/2.96	2.09/2.89	2.07/2.86	2.04/2.80	2.02/2.77	2.01/2.75
17	4.45/8.40	3.59/6.11	3.20/5.18	2.96/4.67	2.81/4.34	2.70/4.10	2.62/3.93	2.55/3.79	2.50/3.68	2.45/3.59	2.41/3.52	2.38/3.45	2.33/3.35	2.29/3.27	2.23/3.16	2.19/3.08	2.15/3.00	2.11/2.92	2.08/2.86	2.04/2.79	2.02/2.76	1.99/2.70	1.97/2.67	1.96/2.65
18	4.41/8.28	3.55/6.01	3.16/5.09	2.93/4.58	2.77/4.25	2.66/4.01	2.58/3.85	2.51/3.71	2.46/3.60	2.41/3.51	2.37/3.44	2.34/3.37	2.29/3.27	2.25/3.19	2.19/3.07	2.15/3.00	2.11/2.91	2.07/2.83	2.04/2.78	2.00/2.71	1.98/2.68	1.95/2.62	1.93/2.59	1.92/2.57
19	4.38/8.18	3.52/5.93	3.13/5.01	2.90/4.50	2.74/4.17	2.63/3.94	2.55/3.77	2.48/3.63	2.43/3.52	2.38/3.43	2.34/3.36	2.31/3.30	2.26/3.19	2.21/3.12	2.15/3.00	2.11/2.92	2.07/2.84	2.02/2.76	2.00/2.70	1.96/2.63	1.94/2.60	1.91/2.54	1.90/2.51	1.88/2.49
20	4.35/8.10	3.49/5.85	3.10/4.94	2.87/4.43	2.71/4.10	2.60/3.87	2.52/3.71	2.45/3.56	2.40/3.45	2.35/3.37	2.31/3.30	2.28/3.23	2.23/3.13	2.18/3.05	2.12/2.94	2.08/2.86	2.04/2.77	1.99/2.69	1.96/2.63	1.92/2.56	1.90/2.53	1.87/2.47	1.85/2.44	1.84/2.42
21	4.32/8.02	3.47/5.78	3.07/4.87	2.84/4.37	2.68/4.04	2.57/3.81	2.49/3.65	2.42/3.51	2.37/3.40	2.32/3.31	2.28/3.24	2.25/3.17	2.20/3.07	2.15/2.99	2.09/2.88	2.05/2.80	2.00/2.72	1.96/2.63	1.93/2.58	1.89/2.51	1.87/2.47	1.84/2.42	1.82/2.38	1.81/2.36
22	4.30/7.94	3.44/5.72	3.05/4.82	2.82/4.31	2.66/3.99	2.55/3.76	2.47/3.59	2.40/3.45	2.35/3.35	2.30/3.26	2.26/3.18	2.23/3.12	2.18/3.02	2.13/2.94	2.07/2.83	2.03/2.75	1.98/2.67	1.93/2.58	1.91/2.53	1.87/2.46	1.84/2.42	1.81/2.37	1.80/2.33	1.78/2.31
23	4.28/7.88	3.42/5.66	3.03/4.76	2.80/4.26	2.64/3.94	2.53/3.71	2.45/3.54	2.38/3.41	2.32/3.30	2.28/3.21	2.24/3.14	2.20/3.07	2.14/2.97	2.10/2.89	2.04/2.78	2.00/2.70	1.96/2.62	1.91/2.53	1.88/2.48	1.84/2.41	1.82/2.37	1.79/2.32	1.77/2.28	1.76/2.26
24	4.26/7.82	3.40/5.61	3.01/4.72	2.78/4.22	2.62/3.90	2.51/3.67	2.43/3.50	2.36/3.36	2.30/3.25	2.26/3.17	2.22/3.09	2.18/3.03	2.13/2.93	2.09/2.85	2.03/2.74	1.98/2.66	1.94/2.58	1.89/2.49	1.86/2.44	1.82/2.36	1.80/2.33	1.76/2.27	1.74/2.23	1.73/2.21
25	4.24/7.77	3.38/5.57	2.99/4.68	2.76/4.18	2.60/3.86	2.49/3.63	2.41/3.46	2.34/3.32	2.28/3.21	2.24/3.13	2.20/3.05	2.16/2.99	2.11/2.89	2.06/2.81	2.00/2.70	1.96/2.62	1.92/2.54	1.87/2.45	1.84/2.40	1.80/2.32	1.77/2.29	1.74/2.23	1.72/2.19	1.71/2.17
26	4.22/7.72	3.37/5.53	2.98/4.64	2.74/4.14	2.59/3.82	2.47/3.59	2.39/3.42	2.32/3.29	2.27/3.17	2.22/3.09	2.18/3.02	2.15/2.96	2.10/2.86	2.05/2.77	1.99/2.66	1.95/2.58	1.90/2.50	1.85/2.41	1.82/2.36	1.78/2.28	1.76/2.25	1.72/2.19	1.70/2.15	1.69/2.13

n_1 degrees of freedom (for greater estimate of variance)

n_2	1	2	3	4	5	6	7	8	9	10	11	12	14	16	20	24	30	40	50	75	100	200	500	∞
27	4.21 / 7.68	3.35 / 5.49	2.96 / 4.60	2.73 / 4.11	2.57 / 3.79	2.46 / 3.56	2.37 / 3.39	2.30 / 3.26	2.25 / 3.14	2.20 / 3.06	2.16 / 2.98	2.13 / 2.93	2.08 / 2.83	2.03 / 2.74	1.97 / 2.63	1.93 / 2.55	1.88 / 2.47	1.84 / 2.38	1.80 / 2.33	1.76 / 2.25	1.74 / 2.21	1.71 / 2.16	1.68 / 2.12	1.67 / 2.10
28	4.20 / 7.64	3.34 / 5.45	2.95 / 4.57	2.71 / 4.07	2.56 / 3.76	2.44 / 3.53	2.36 / 3.36	2.29 / 3.23	2.24 / 3.11	2.19 / 3.03	2.15 / 2.95	2.12 / 2.90	2.06 / 2.80	2.02 / 2.71	1.96 / 2.60	1.91 / 2.52	1.87 / 2.44	1.81 / 2.35	1.78 / 2.30	1.75 / 2.22	1.72 / 2.18	1.69 / 2.13	1.67 / 2.09	1.65 / 2.06
29	4.18 / 7.60	3.33 / 5.42	2.93 / 4.54	2.70 / 4.04	2.54 / 3.73	2.43 / 3.50	2.35 / 3.33	2.28 / 3.20	2.22 / 3.08	2.18 / 3.00	2.14 / 2.92	2.10 / 2.87	2.05 / 2.77	2.00 / 2.68	1.94 / 2.57	1.90 / 2.49	1.85 / 2.41	1.80 / 2.32	1.76 / 2.27	1.73 / 2.19	1.71 / 2.15	1.68 / 2.10	1.65 / 2.06	1.64 / 2.03
30	4.17 / 7.56	3.32 / 5.39	2.92 / 4.51	2.69 / 4.02	2.53 / 3.70	2.42 / 3.47	2.34 / 3.30	2.27 / 3.17	2.21 / 3.06	2.16 / 2.98	2.12 / 2.90	2.09 / 2.84	2.04 / 2.74	1.99 / 2.66	1.93 / 2.55	1.89 / 2.47	1.84 / 2.38	1.79 / 2.29	1.76 / 2.24	1.72 / 2.16	1.69 / 2.13	1.66 / 2.07	1.64 / 2.03	1.62 / 2.01
32	4.15 / 7.50	3.30 / 5.34	2.90 / 4.46	2.67 / 3.97	2.51 / 3.66	2.40 / 3.42	2.32 / 3.25	2.25 / 3.12	2.19 / 3.01	2.14 / 2.94	2.10 / 2.86	2.07 / 2.80	2.02 / 2.70	1.97 / 2.62	1.91 / 2.51	1.86 / 2.42	1.82 / 2.34	1.76 / 2.25	1.74 / 2.20	1.69 / 2.12	1.67 / 2.08	1.64 / 2.02	1.61 / 1.98	1.59 / 1.96
34	4.13 / 7.44	3.28 / 5.29	2.88 / 4.42	2.65 / 3.93	2.49 / 3.61	2.38 / 3.38	2.30 / 3.21	2.23 / 3.08	2.17 / 2.97	2.12 / 2.89	2.08 / 2.82	2.05 / 2.76	2.00 / 2.66	1.95 / 2.58	1.89 / 2.47	1.84 / 2.38	1.80 / 2.30	1.74 / 2.21	1.71 / 2.15	1.67 / 2.08	1.64 / 2.04	1.61 / 1.98	1.59 / 1.94	1.57 / 1.91
36	4.11 / 7.39	3.26 / 5.25	2.86 / 4.38	2.63 / 3.89	2.48 / 3.58	2.36 / 3.35	2.28 / 3.18	2.21 / 3.04	2.15 / 2.94	2.10 / 2.86	2.06 / 2.78	2.03 / 2.72	1.98 / 2.62	1.93 / 2.54	1.87 / 2.43	1.82 / 2.35	1.78 / 2.26	1.72 / 2.17	1.69 / 2.12	1.65 / 2.04	1.62 / 2.00	1.59 / 1.94	1.56 / 1.90	1.55 / 1.87
38	4.10 / 7.35	3.25 / 5.21	2.85 / 4.34	2.62 / 3.86	2.46 / 3.54	2.35 / 3.32	2.26 / 3.15	2.19 / 3.02	2.14 / 2.91	2.09 / 2.82	2.05 / 2.75	2.02 / 2.69	1.96 / 2.59	1.92 / 2.51	1.85 / 2.40	1.80 / 2.32	1.76 / 2.22	1.71 / 2.14	1.67 / 2.08	1.63 / 2.00	1.60 / 1.97	1.57 / 1.90	1.54 / 1.86	1.53 / 1.84
40	4.08 / 7.31	3.23 / 5.18	2.84 / 4.31	2.61 / 3.83	2.45 / 3.51	2.34 / 3.29	2.25 / 3.12	2.18 / 2.99	2.12 / 2.88	2.07 / 2.80	2.04 / 2.73	2.00 / 2.66	1.95 / 2.56	1.90 / 2.49	1.84 / 2.37	1.79 / 2.29	1.74 / 2.20	1.69 / 2.11	1.66 / 2.05	1.61 / 1.97	1.59 / 1.94	1.55 / 1.88	1.53 / 1.84	1.51 / 1.81
42	4.07 / 7.27	3.22 / 5.15	2.83 / 4.29	2.59 / 3.80	2.44 / 3.49	2.32 / 3.26	2.24 / 3.10	2.17 / 2.96	2.11 / 2.86	2.06 / 2.77	2.02 / 2.70	1.99 / 2.64	1.94 / 2.54	1.89 / 2.46	1.82 / 2.35	1.78 / 2.26	1.73 / 2.17	1.68 / 2.08	1.64 / 2.02	1.60 / 1.94	1.57 / 1.91	1.54 / 1.85	1.51 / 1.80	1.49 / 1.78
44	4.06 / 7.24	3.21 / 5.12	2.82 / 4.26	2.58 / 3.78	2.43 / 3.46	2.31 / 3.24	2.23 / 3.07	2.16 / 2.94	2.10 / 2.84	2.05 / 2.75	2.01 / 2.68	1.98 / 2.62	1.92 / 2.52	1.88 / 2.44	1.81 / 2.32	1.76 / 2.24	1.72 / 2.15	1.66 / 2.06	1.63 / 2.00	1.58 / 1.92	1.56 / 1.88	1.52 / 1.82	1.50 / 1.78	1.48 / 1.75
46	4.05 / 7.21	3.20 / 5.10	2.81 / 4.24	2.57 / 3.76	2.42 / 3.44	2.30 / 3.22	2.22 / 3.05	2.14 / 2.92	2.09 / 2.82	2.04 / 2.73	2.00 / 2.66	1.97 / 2.60	1.91 / 2.50	1.87 / 2.42	1.80 / 2.30	1.75 / 2.22	1.71 / 2.13	1.65 / 2.04	1.62 / 1.98	1.57 / 1.90	1.54 / 1.86	1.51 / 1.80	1.48 / 1.76	1.46 / 1.72
48	4.04 / 7.19	3.19 / 5.08	2.80 / 4.22	2.56 / 3.74	2.41 / 3.42	2.30 / 3.20	2.21 / 3.04	2.14 / 2.90	2.08 / 2.80	2.03 / 2.71	1.99 / 2.64	1.96 / 2.58	1.90 / 2.48	1.86 / 2.40	1.79 / 2.28	1.74 / 2.20	1.70 / 2.11	1.64 / 2.02	1.61 / 1.96	1.56 / 1.88	1.53 / 1.84	1.50 / 1.78	1.47 / 1.73	1.45 / 1.70

n_1 degrees of freedom (for greater estimate of variance)

n_2	1	2	3	4	5	6	7	8	9	10	11	12	14	16	20	24	30	40	50	75	100	200	500	∞	n_2
50	4.03 7.17	3.18 5.06	2.79 4.20	2.56 3.72	2.40 3.41	2.29 3.18	2.20 3.02	2.13 2.88	2.07 2.78	2.02 2.70	1.98 2.62	1.95 2.56	1.90 2.46	1.85 2.39	1.78 2.26	1.74 2.18	1.69 2.10	1.63 2.00	1.60 1.94	1.55 1.86	1.52 1.82	1.48 1.76	1.46 1.71	1.44 1.68	50
55	4.02 7.12	3.17 5.01	2.78 4.16	2.54 3.68	2.38 3.37	2.27 3.15	2.18 2.98	2.11 2.85	2.05 2.75	2.00 2.66	1.97 2.59	1.93 2.53	1.88 2.43	1.83 2.35	1.76 2.23	1.72 2.15	1.67 2.06	1.61 1.96	1.58 1.90	1.52 1.82	1.50 1.78	1.46 1.71	1.43 1.66	1.41 1.64	55
60	4.00 7.08	3.15 4.98	2.76 4.13	2.52 3.65	2.37 3.34	2.25 3.12	2.17 2.95	2.10 2.82	2.04 2.72	1.99 2.63	1.95 2.56	1.92 2.50	1.86 2.40	1.81 2.32	1.75 2.20	1.70 2.12	1.65 2.03	1.59 1.93	1.56 1.87	1.50 1.79	1.48 1.74	1.44 1.68	1.41 1.63	1.39 1.60	60
65	3.99 7.04	3.14 4.95	2.75 4.10	2.51 3.62	2.36 3.31	2.24 3.09	2.15 2.93	2.08 2.79	2.02 2.70	1.98 2.61	1.94 2.54	1.90 2.47	1.85 2.37	1.80 2.30	1.73 2.18	1.68 2.09	1.63 2.00	1.57 1.90	1.54 1.84	1.49 1.76	1.46 1.71	1.42 1.64	1.39 1.60	1.37 1.56	65
70	3.98 7.01	3.13 4.92	2.74 4.08	2.50 3.60	2.35 3.29	2.23 3.07	2.14 2.91	2.07 2.77	2.01 2.67	1.97 2.59	1.93 2.51	1.89 2.45	1.84 2.35	1.79 2.28	1.72 2.15	1.67 2.07	1.62 1.98	1.56 1.88	1.53 1.82	1.47 1.74	1.45 1.69	1.40 1.62	1.37 1.56	1.35 1.53	70
80	3.96 6.96	3.11 4.88	2.72 4.04	2.48 3.56	2.33 3.25	2.21 3.04	2.12 2.87	2.05 2.74	1.99 2.64	1.95 2.55	1.91 2.48	1.88 2.41	1.82 2.32	1.77 2.24	1.70 2.11	1.65 2.03	1.60 1.94	1.54 1.84	1.51 1.78	1.45 1.70	1.42 1.65	1.38 1.57	1.35 1.52	1.32 1.49	80
100	3.94 6.90	3.09 4.82	2.70 3.98	2.46 3.51	2.30 3.20	2.19 2.99	2.10 2.82	2.03 2.69	1.97 2.59	1.92 2.51	1.88 2.43	1.85 2.36	1.79 2.26	1.75 2.19	1.68 2.06	1.63 1.98	1.57 1.89	1.51 1.79	1.48 1.73	1.42 1.64	1.39 1.59	1.34 1.51	1.30 1.46	1.28 1.43	100
125	3.92 6.84	3.07 4.78	2.68 3.94	2.44 3.47	2.29 3.17	2.17 2.95	2.08 2.79	2.01 2.65	1.95 2.56	1.90 2.47	1.86 2.40	1.83 2.33	1.77 2.23	1.72 2.15	1.65 2.03	1.60 1.94	1.55 1.85	1.49 1.75	1.45 1.68	1.39 1.59	1.36 1.54	1.31 1.46	1.27 1.40	1.25 1.37	125
150	3.91 6.81	3.06 4.75	2.67 3.91	2.43 3.44	2.27 3.14	2.16 2.92	2.07 2.76	2.00 2.62	1.94 2.53	1.89 2.44	1.85 2.37	1.82 2.30	1.76 2.20	1.71 2.12	1.64 2.00	1.59 1.91	1.54 1.83	1.47 1.72	1.44 1.66	1.37 1.56	1.34 1.51	1.29 1.43	1.25 1.37	1.22 1.33	150
200	3.89 6.76	3.04 4.71	2.65 3.88	2.41 3.41	2.26 3.11	2.14 2.90	2.05 2.73	1.98 2.60	1.92 2.50	1.87 2.41	1.83 2.34	1.80 2.28	1.74 2.17	1.69 2.09	1.62 1.97	1.57 1.88	1.52 1.79	1.45 1.69	1.42 1.62	1.35 1.53	1.32 1.48	1.26 1.39	1.22 1.33	1.19 1.28	200
400	3.86 6.70	3.02 4.66	2.62 3.83	2.39 3.36	2.23 3.06	2.12 2.85	2.03 2.69	1.96 2.55	1.90 2.46	1.85 2.37	1.81 2.29	1.78 2.23	1.72 2.12	1.67 2.04	1.60 1.92	1.54 1.84	1.49 1.74	1.42 1.64	1.38 1.57	1.32 1.47	1.28 1.42	1.22 1.32	1.16 1.24	1.13 1.19	400
1000	3.85 6.66	3.00 4.62	2.61 3.80	2.38 3.34	2.22 3.04	2.10 2.82	2.02 2.66	1.95 2.53	1.89 2.43	1.84 2.34	1.80 2.26	1.76 2.20	1.70 2.09	1.65 2.01	1.58 1.89	1.53 1.81	1.47 1.71	1.41 1.61	1.36 1.54	1.30 1.44	1.26 1.38	1.19 1.28	1.13 1.19	1.08 1.11	1000
∞	3.84 6.64	2.99 4.60	2.60 3.78	2.37 3.32	2.21 3.02	2.09 2.80	2.01 2.64	1.94 2.51	1.88 2.41	1.83 2.32	1.79 2.24	1.75 2.18	1.69 2.07	1.64 1.99	1.57 1.87	1.52 1.79	1.46 1.69	1.40 1.59	1.35 1.52	1.28 1.41	1.24 1.36	1.17 1.25	1.11 1.15	1.00 1.00	∞

Reproduced by permission from *Statistical Methods, 6th Edition*, by George W. Snedecor and William G. Cochran. © 1967 by the Iowa State University Press.

The important formulas presented in the text are listed here for reference. Explanation of the symbols is given in the appropriate chapter as indicated by the number preceding each formula.

A. Computation of the Mean

    1. From ungrouped data.

(2.1)
$$\mu = \frac{\Sigma X}{N}$$

    2. From grouped data with original midpoints.

(6.2)
$$\mu = \frac{\Sigma FX}{\Sigma F}$$

    3. From grouped data using coded scores.

(6.3)
$$\mu = i\frac{\Sigma FU}{\Sigma F} + R$$

B. Computation of the Variance (the standard deviation is the positive square root of the variance).

    1. From ungrouped data.

(3.1)
$$\sigma^2 = \frac{\Sigma (X-\mu)^2}{N}$$

(3.2)
$$\sigma^2 = \frac{\Sigma X^2}{N} - \left(\frac{\Sigma X}{N}\right)^2 = \frac{\Sigma X^2}{N} - \frac{(\Sigma X)^2}{N^2} = \frac{\Sigma X^2}{N} - \mu^2$$

    2. From grouped data using original midpoints.

(6.5)
$$\sigma^2 = \frac{\Sigma FX^2}{N} - \left(\frac{\Sigma FX}{N}\right)^2$$

    3. From grouped data using coded scores.

(6.7)
$$\sigma^2 = i^2\left(\frac{\Sigma FU^2}{N} - \left(\frac{\Sigma FU}{N}\right)^2\right)$$

C. Standard Scores and z Scores

    1. The z score of the kth term.

(4.2)
$$z_k = \frac{X_k - \mu}{\sigma}$$

    2. The standard score of the kth term.

(4.3), (4.4)
$$T_k = 50 + 10z_k = 50 + 10\left(\frac{X_k - \mu}{\sigma}\right)$$

D. Computation of any percentile (including the median).

(6.1)
$$P_n = L + \frac{s}{f}(i)$$

E. Standard Deviation of Means of Samples of Size N.

(9.1)
$$\sigma_M = \frac{\sigma}{\sqrt{N}}$$

F. Testing Hypotheses Using Normal Curve Areas.

1. Value of z for dichotomous variable

(12.1)
$$z = \frac{O - NP}{\sqrt{NP(1 - P)}}$$

2. Value of z for sample mean.

(12.2)
$$z = \frac{\bar{X} - \mu}{\frac{\sigma}{\sqrt{N}}}$$

G. Estimation

1. Unbiased estimate of population variance.

(13.1)
$$s^2 = \frac{\Sigma (X - \bar{X})^2}{N - 1}$$

(13.2)
$$s^2 = \frac{\Sigma X^2 - \frac{(\Sigma X)^2}{N}}{N - 1} = \frac{N \Sigma X^2 - (\Sigma X)^2}{N(N - 1)}$$

2. 95 per cent confidence interval for the mean

(13.3)
$$\bar{X} - 1.96 \frac{\sigma}{\sqrt{N}} < \mu < \bar{X} + 1.96 \frac{\sigma}{\sqrt{N}}$$

3. 99 per cent confidence interval for the mean.

(13.4)
$$\bar{X} - 2.58 \frac{\sigma}{\sqrt{N}} < \mu < \bar{X} + 2.58 \frac{\sigma}{\sqrt{N}}$$

H. The t Distribution

1. Estimate of standard error of mean.

(14.2)
$$s_m = \frac{s}{\sqrt{N}}$$

2. Value of t.

(14.3)
$$t = \frac{\bar{X} - \mu}{s_m}$$

3. Estimate of standard deviation of distribution of differences of means of samples.

(14.5)
$$s_d = \sqrt{\frac{(N_1 - 1)s_1^2 + (N_2 - 1)s_2^2}{N_1 + N_2 - 2}} \cdot \sqrt{\frac{1}{N_1} + \frac{1}{N_2}}$$

I. Value of χ^2

(15.1)
$$\chi^2 = \sum \frac{(O-E)^2}{E}$$

J. Computation of the Correlation Coefficient.

    1. From z scores.

$$r = \frac{\sum z_x z_y}{N}$$

    2. From original data.

(17.1)
$$r = \frac{N \sum XY - (\sum X)(\sum Y)}{\sqrt{N \sum X^2 - (\sum X)^2} \sqrt{N \sum Y^2 - (\sum Y)^2}}$$

    3. Significance of correlation coefficient.

(18.5)
$$t = \frac{r\sqrt{N-2}}{\sqrt{1-r^2}}$$

K. Prediction

    1. Slope of raw score scattergram (Y vs. X) from r.

(18.1)
$$b = r\frac{\sigma_y}{\sigma_x}$$

    2. Slope of raw score scattergram from original data.

(18.2)
$$b = \frac{N \sum XY - (\sum X)(\sum Y)}{N \sum X^2 - (\sum X)^2}$$

    3. Predicting Y from X

(18.3)
$$\tilde{Y} = \mu_y + b(X - \mu_x)$$

L. Error Variance

(18.4)
$$\sigma_{y \cdot x}^2 = \sigma_y^2 (1 - r^2)$$

M. Coefficient of Rank Correlation

(19.1)
$$R = 1 - \frac{6 \sum (x-y)^2}{N(N^2-1)}$$

N. Value of F for the F distribution

(20.1)
$$F = \frac{S_1^2}{S_2^2}$$

O. Analysis of Variance

    1. Variance between groups.

(20.4)
$$S_B^2 = N S_M^2$$

    2. Variance within the groups.

366   (20.5)
$$S_W^2 = \frac{(N_1-1)S_1^2 + (N_2-1)S_2^2 + \cdots + (N_K-1)S_K^2}{N_T - K}$$

3. Computations for analysis of variance.

(20.6)
$$G = \frac{(\sum X_1)^2}{N_1} + \frac{(\sum X_2)^2}{N_2} + \cdots + \frac{(\sum X_K)^2}{N_K}$$

(20.7)
$$SS_B = G - \frac{(\sum X_T)^2}{N_T}$$

(20.8)
$$SS_W = \sum X_T{}^2 - G$$

(20.9)
$$S_B{}^2 = \frac{SS_B}{K-1}$$

(20.10)
$$S_W{}^2 = \frac{SS_W}{N_T - K}$$

(20.11)
$$SS_T = \sum X_T{}^2 - \frac{(\sum X_T)^2}{N_T}$$

P. Partitioning the Sums of Squares

(20.12)
$$\sum_{i,j} (X_{i,j} - \bar{X}_T)^2 = \sum_{i,j} (X_{i,j} - \bar{X}_i)^2 + \sum_{i,j} (\bar{X}_i - \bar{X}_T)^2$$

Chapter 2

2.1 93.68; 91; 88
2.2 74; 74; 75
2.3 41; 41; four modes: 36, 41, 46, 50
2.4 10.5; 11; 12 and 11
2.5 9.83
2.6 $16,680; $12,000; $12,000
2.7 46, 78, 54, 78, 118
2.8 49.3
2.9 49.3

Chapter 3

3.1 182.58; 13.51
3.2 173.0; 13.2
3.3 180.69; 13.4
3.4 1.01; 1.005
3.5 1433.21; 37.86
3.6 18.66; 4.3
3.7 5.75; 2.4
3.8 5.75; there is no difference in the variances
3.9 1.44; the new variance is one-fourth the size of the variance of the original scores
3.10 Yes; when the variance is equal to 1 or 0.

Chapter 4

4.1 (a) 2, (b) 40, (c) 66
4.2 (a) 22, (b) 50, (c) 74
4.3 (a) 34.3, (b) 45.4, (c) 55.7
4.4 (a) 40.2, (b) 50.6, (c) 58.0
4.5 (a) 23.5, (b) 67.6, (c) 85.3
4.6 (a) 47.1, (b) 76.5, (c) 94.1
4.7 (a) 41.9, (b) 54.9, (c) 63.0
4.8 (a) 50.0, (b) 58.1, (c) 64.6
4.9 (a) 14.7, (b) 26.5, (c) 44.1, (d) 79.4
4.10 (a) 20.6, (b) 32.4, (c) 61.8, (d) 88.2
4.11 (a) 38.5, (b) 44.8, (c) 51.0, (d) 59.4
4.12 (a) 40.6, (b) 46.9, (c) 53.1, (d) 61.5

Chapter 5

5.1 (a) 84.5–87.5; (b) 0.5–5.5; (c) 100.5–109.5; (d) 71.5–73.5

The answers to Problems 5.2 through 5.10 are only *examples* and are not unique. Your answers will depend on how you group the data.

5.2

Interval	Boundaries	Frequency
261–270	260.5–270.5	1
251–260	250.5–260.5	1
241–250	240.5–250.5	1
231–240	230.5–240.5	4
221–230	220.5–230.5	4

211–220	210.5–220.5	8
201–210	200.5–210.5	2
191–200	190.5–200.5	10
181–190	180.5–190.5	6
171–180	170.5–180.5	1
161–170	160.5–170.5	2

5.3

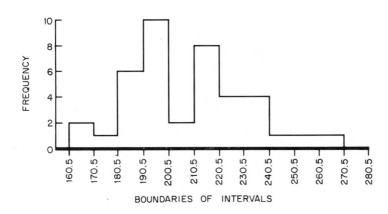

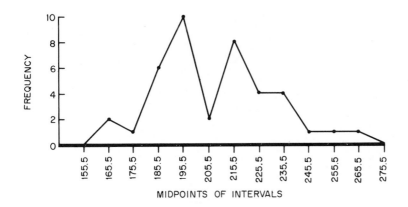

5.4

Interval	Boundaries	Frequency
251–270	250.5–270.5	2
231–250	230.5–250.5	5
211–230	210.5–230.5	12
191–210	190.5–210.5	12
171–190	170.5–190.5	7
151–170	150.5–170.5	2

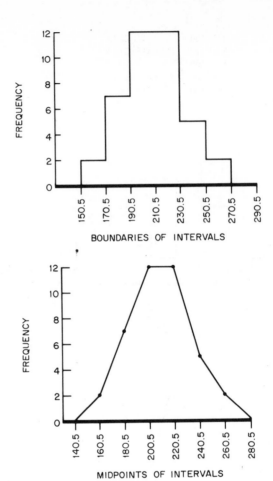

BOUNDARIES OF INTERVALS

MIDPOINTS OF INTERVALS

5.5

Interval	Boundaries	Frequency
266–270	265.5–270.5	1
261–265	260.5–265.5	0
256–260	255.5–260.5	0
251–255	250.5–255.5	1
246–250	245.5–250.5	0
241–245	240.5–245.5	1
236–240	235.5–240.5	3
231–235	230.5–235.5	1
226–230	225.5–230.5	3
221–225	220.5–225.5	1
216–220	215.5–220.5	5
211–215	210.5–215.5	3
206–210	205.5–210.5	1
201–205	200.5–205.5	1
196–200	195.5–200.5	6
191–195	190.5–195.5	4
186–190	185.5–190.5	2
181–185	180.5–185.5	4

176–180	175.5–180.5	1
171–175	170.5–175.5	0
166–170	165.5–170.5	2

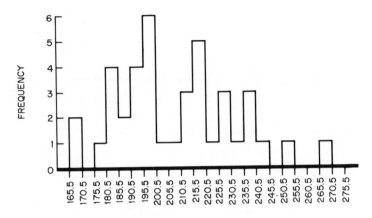

BOUNDARIES OF INTERVALS

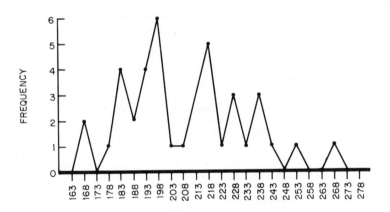

MIDPOINTS OF INTERVALS

5.6

Interval	Boundaries	Frequency
226–235	225.5–235.5	5
216–225	215.5–225.5	5
206–215	205.5–215.5	6
196–205	195.5–205.5	10
186–195	185.5–195.5	8
176–185	175.5–185.5	6
166–175	165.5–175.5	5
156–165	155.5–165.5	2
146–155	145.5–155.5	1

5.7

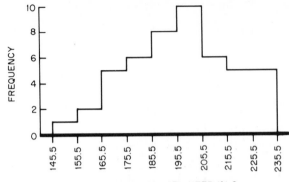

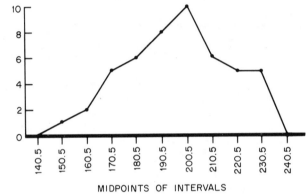

5.8

Interval	Boundaries	Frequency
216–235	215.5–235.5	10
196–215	195.5–215.5	16
176–195	175.5–195.5	14
156–175	155.5–175.5	7
136–155	135.5–155.5	1

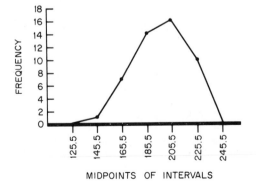

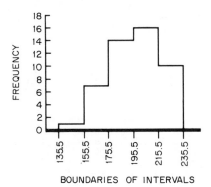

BOUNDARIES OF INTERVALS

5.9

Interval	Boundaries	Frequency
231–235	230.5–235.5	4
226–230	225.5–230.5	1
221–225	220.5–225.5	2
216–220	215.5–220.5	3
211–215	210.5–215.5	4
206–210	205.5–210.5	2
201–205	200.5–205.5	3
196–200	195.5–200.5	7
191–195	190.5–195.5	4
186–190	185.5–190.5	4
181–185	180.5–185.5	5
176–180	175.5–180.5	1
171–175	170.5–175.5	1
166–170	165.5–170.5	4
161–165	160.5–165.5	2
156–160	155.5–160.5	0
151–155	150.5–155.5	1

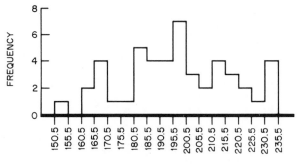

BOUNDARIES OF INTERVALS

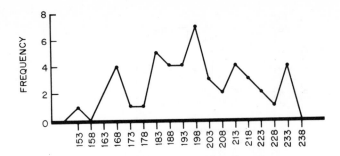

MIDPOINTS OF INTERVALS

5.10

Interval	Boundaries	Frequency
204–223	203.5–223.5	2
184–203	183.5–203.5	2
164–183	163.5–183.5	17
144–163	143.5–163.5	21
124–143	123.5–143.5	13
104–123	103.5–123.5	7
84–103	83.5–103.5	3

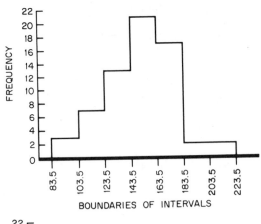

BOUNDARIES OF INTERVALS

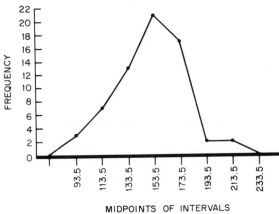

374

MIDPOINTS OF INTERVALS

Chapter 6

6.1   83; 80.2; 88.5; 81.2

6.2

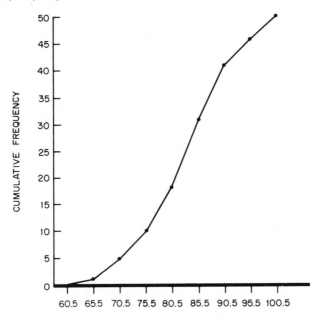

6.3   83; 82.6; 69.84
6.4   63.7; 61.0; 61.4; 64.4

6.5

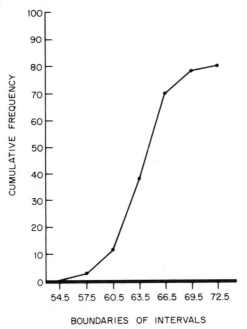

6.6 65; 63.5; 9.00
6.7 55.4; 42.8; 62.2; 67.7

6.8

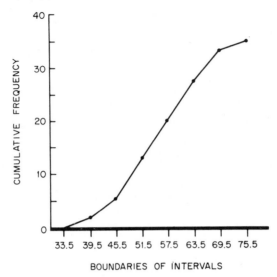

6.9 60.5; 55.01; 89.22
6.10 265; 222.5; 202.5; 325

6.11

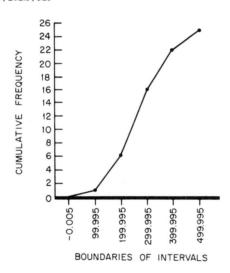

6.12 $250; $270; $10,399.92

Chapter 7

7.1

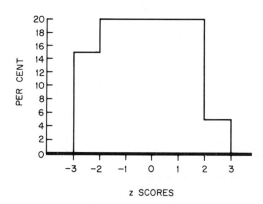

7.2

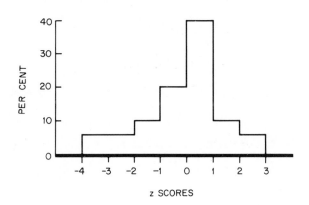

7.3

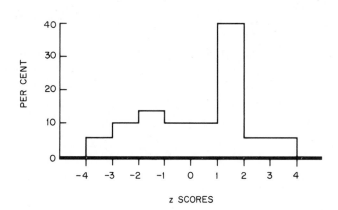

7.4

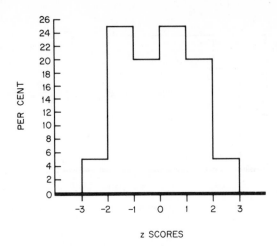

z SCORES

7.5 15; 45; 85
7.6 2.5; 37.5; 57.5; 92.5

7.7 z scores: −0.52; 0.52; 1.22; 1.57
        percentile ranks: 35;   65;   85;   95

7.8 (a) z scores: −0.37 −0.26 −0.19 2.98
           percentile ranks: 35   65   85   95
        (b) z scores: −.337 −.327 −.320 3.000
           percentile ranks: 35   65   85   95

7.9

	P_{10}	P_{30}	P_{50}	P_{80}
scores:	89.4	98.5	104.4	114.5
z scores:	−1.17	−0.51	−0.07	0.67

7.10

	P_{10}	P_{30}	P_{50}	P_{80}
scores:	87.3	95.6	102.4	113.2
z scores:	−1.43	−0.70	−0.11	0.84

7.11 (a) 101.9; 10.15

	P_5	P_{15}	P_{30}	P_{50}	P_{75}	P_{95}
IQ:	83.5	90.6	96.2	102.3	108.6	118.6
z scores:	−1.81	−1.11	−0.56	−0.04	0.66	1.65

7.12 (a) 85.24; 11.55

	P_5	P_{15}	P_{30}	P_{50}	P_{75}	P_{95}
IQ:	64.5	71.5	77.5	84.1	92.7	105.2
z scores:	−1.80	−1.19	−0.68	−0.10	0.65	1.73

Chapter 8

8.1 (a) 2, (b) 7, (c) 84, (d) 96
8.2 (a) 34, (b) 68, (c) 53, (d) 91
8.3 (a) 10, (b) 1, (c) 18, (d) 5
8.4 (a) −0.84, (b) −0.13, (c) +0.39, (d) +0.84
8.5 (a) −1.04, (b) −0.25, (c) +0.13, (d) +0.67

378

8.6   (a) 1, (b) 27, (c) 50, (d) 65, (e) 92
8.7   (a) 82, (b) 62, (c) 67–77
8.8   (a) 67, (b) 73, (c) 75, (d) 83, (e) 91
8.9   (a) 99, (b) 40, (c) 475, (d) 416–584, (e) 267
8.10 (a) 12, (b) 74, (c) 81, (d) 11, (e) 68, (f) 14, (g) 12
8.11 (a) 6.7, (b) 8.4
8.12 87
8.13 73
8.14 484

Chapter 9

9.1   (a) 2.28%, (b) 11.51%
9.2   (a) 1, (b) 8
9.3   (a) 5, (b) 16
9.4   (a) 90, 2.82, (b) 90, 1.42, (c) 90, 0.71
9.5   (a) 24, (b) 8, (c) 0
9.6   (a) 40, (b) 71, (c) 97
9.7   (a) 91.5, (b) 90.7, (c) 90.4
9.8   (a) 91.9–88.1, (b) 90.0–89.1, (c) 90.5–89.5
9.9   (a) 80, 0.9, (b) 0, (c) 89, (d) 80.8–79.2
9.10 (a) 0, (b) 100, (c) 179, (d) 181.5–178.5
9.11 98.5
9.12 100

Chapter 10

10.1   (a) 1/728, (b) 1/728, (c) 28/91, (d) 16/91, (e) 0, (f) 29/728, (g) 127/728, (h) 11/104
10.2   (a) .0630, (b) .9370, (c) .6178, (d) .3822, (e) .3830, (f) .0228
10.3   (a) 0.0143, (b) not random
10.4   (a) 0.8106, (b) 0.5021, (c) 0.6915, (d) 0.1894, (e) 0.4979
10.5   (a) 0.1112, (b) 0.8065, (c) 0.0119, (d) 0.6032
10.6   (a) 0.0013, (b) 0.6902, (c) 0.3085, (d) 0.9772
10.7   (a) 0.0475, (b) 0.0062, (c) 0.9463, (d) 0.4525
10.8   (a) 0, (b) 0.0016, (c) 0.1190, (d) 0.9232
10.9   (a) 0, (b) 0.9868, (c) 0.0005
10.10 (a) 0, (b) 0.9772, (c) 0.0013
10.11 (a) 0, (b) 0.9616, (c) 0.0038
10.12 (a) 0, (b) 0.9192, (c) 0.0026

Chapter 11

11.1   200, 10
11.2   12.5–19.5
11.3   .32
11.4   124–140
11.5   0.43, 0.57
11.6   160, 7.8
11.7   282.6–317.4
11.8   .69
11.9   130–155
11.10 0.04, 0.96

Chapter 12

12.1 Significant; reject hypothesis
12.2 Not significant; do not reject hypothesis
12.3 Not significant; do not reject hypothesis
12.4 Significant; reject hypothesis
12.5 Not significant; do not reject hypothesis
12.6 Significant; reject hypothesis
12.7 Significant; reject hypothesis
12.8 (a) Significant; reject; (b) not significant; do not reject; (c) significant; reject
12.9 (a) Significant; reject; (b) significant; reject; (c) not significant; do not reject
12.10 (a) Significant; reject hypothesis; (b) 984; (c) yes
12.11 .2542; hint: $p = q = 1/2$, $Np = 49 \times 1/2 = 24.5$, $\sigma = \sqrt{12.25} = 3.5$.
Should use 20.5 and 28.5 because of binomial probability, so
$z = (20.5 - 24.5)/3.5 = -1.14$
12.12 (a) 16–33; (b) 18–31; (c) $H_0:p = .5$, $H_1:p > .5$; reject H_0 if more than 30 white disks
are drawn, otherwise accept it; $[24.5 + (1.64)3.5 = 30.2]$

Chapter 13

13.1 (a) 101.2, (b) 24, (c) 190.2, (d) 13.8
13.2 (a) 181.9, (b) 9, (c) 301, (d) 17.35 lbs.
13.3 (a) 10.5, (b) 16, (c) 24.5, (d) 4.95
13.4 (a) 8.2, (b) 30, (c) 2.00, (d) 1.41
13.5 (a) 8.88, (b) 249, (c) 2.92, (d) 1.7
13.6 79.43–82.57
13.7 5'6.2"–5'7.8"
13.8 162.06–167.94
13.9 77.84–80.16
13.10 (a) 46.32"–47.68", (b) 46.10"–47.90"
13.11 $N = 96.$ $(\overline{X} \pm 20 = \overline{X} \pm 1.96\, \sigma/\sqrt{N})$
13.12 (a) 7.48–7.72, (b) 7.44–7.76

Chapter 14

14.1 $t = -2.67$; significant; reject hypothesis
14.2 $t = 2.15$; significant; reject hypothesis
14.3 $t = -2.14$; not significant; do not reject hypothesis
14.4 $t = 1.20$; not significant; do not reject hypothesis
14.5 $t = -2.84$; significant; reject hypothesis
14.6 $t = -2.67$; significant; reject hypothesis
14.7 $t = -4.33$; significant; reject hypothesis
14.8 $t = -1.47$; not significant; do not reject hypothesis
14.9 $t = 5.55$; significant; reject null hypothesis
14.10 $t = 1.55$; not significant; do not reject null hypothesis
14.11 $t = .227$; not significant; do not reject null hypothesis
14.12 $t = .36$; not significant; do not reject null hypothesis

Chapter 15

15.1 $\chi^2 = .38$, d.f. $= 1$; (a) not significant; do not reject null hypothesis;
(b) not significant; do not reject null hypothesis
15.2 $\chi^2 = 40.12$, df $= 3$; significant; reject null hypothesis

15.3 $\chi^2 = 9.46$, df = 1; significant; reject null hypothesis
15.4 $\chi^2 = 3.43$, d.f. = 4; not significant; do not reject null hypothesis
15.5 $\chi^2 = 6.10$, d.f. = 2; not significant; do not reject null hypothesis
15.6 $\chi^2 = .8$, d.f. = 2; not significant; do not reject null hypothesis

15.7

	B	C	D
A	4.24; not sig.	0.44; not sig.	0.00; not sig.
B		12.60; sig.	7.89; sig.
C			0.38; not sig.

15.8 $\chi^2 = 2.01$; not significant; do not reject null hypothesis
15.9 $\chi^2 = 9.58$; significant; reject null hypothesis
15.10 $\chi^2 = 17.49$; significant; reject null hypothesis

Chapter 16

16.1 (a) 6, (b) 24
16.2 (a) 10, (b) 3.5
16.3 6 years, 2.5; 7 years, 3.75; 8 years, 3.5; 9 years, 3.75; 10 years, 9.75

16.4

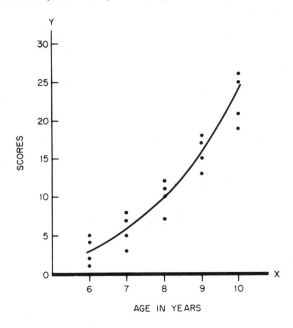

16.5 (a) 4.5, (b) 12.5, (c) 22
16.6 (a) 25, (b) 5
16.7 (a) 10, (b) 9.17
16.8 1 month, 33.25; 2 months, 17.25; 3 months, 13.5; 4 months, 3.75

16.9

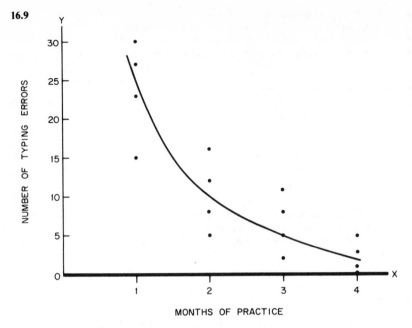

16.10 (a) 15, (b) 7, (c) 3

Chapter 17

17.1 (a)

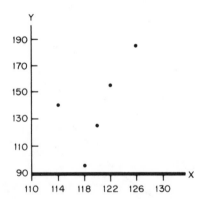

(b) r = .65, (c) r = .65

17.2 r = 1.00; perfect positive correlation
17.3 r = .00; no correlation
17.4 r = .88
17.5 r = .74
17.6 r = .91
17.7 r = .71
17.8 r = .25
17.9 r = .97
17.10 r = .87
17.11 r = .97
17.12 r = .57

Chapter 18

18.1 (a) $\tilde{Y} = 140 + 4.88 (x - 120)$; (b) 66.8, 125.36, 188.8; (c) 0.42
18.2 (a) $.795X + 12.3$; (b) 71.9, 77.4, 87.8; (c) 0.55
18.3 (a) $52.7X + 492$; (b) 640, 671, 708; (c) 0.14
18.4 (a) $0.539X + 60.2$; (b) 92.5, 109.7, 130.2; (c) 0.23
18.5 Significant; reject hypothesis
18.6 $r = .65$; not significant; do not reject null hypothesis
18.7 $r = .74$; significant; reject null hypothesis
18.8 $r = .37$; not significant; do not reject null hypothesis
18.9 $r = .48$; not significant; do not reject null hypothesis
18.10 $r = .25$; not significant; do not reject null hypothesis

Chapter 19

19.1 $R = .70$; not significant; do not reject null hypothesis
19.2 $R = .75$; not significant; do not reject null hypothesis
19.3 $R = .29$; not significant; do not reject null hypothesis
19.4 $R = .48$; not significant; do not reject null hypothesis
19.5 $z = .90$; not significant; do not reject null hypothesis
19.6 $z = 1.16$; not significant: do not reject null hypothesis
19.7 $u = 6$; significant; reject null hypothesis
19.8 $u = 27$; significant; reject null hypothesis
19.9 $u = 11$; not significant; do not reject null hypothesis
19.10 $u = 12$; not significant; do not reject null hypothesis

Chapter 20

20.1 $F = 2.5$; significant; reject null hypothesis
20.2 $F = 1.25$; not significant; do not reject null hypothesis
20.3 $F = 2.12$; not significant; do not reject null hypothesis
20.4 $F = 2.99$; significant; reject null hypothesis
20.5 $F = 3.33$; not significant; do not reject null hypothesis
20.6 $F = 18.61$; significant; reject null hypothesis
20.7 $F = 1.91$; not significant; do not reject null hypothesis
20.8 $F = 3.05$; significant; reject null hypothesis
20.9 $F = 5.64$; not significant; do not reject null hypothesis
20.10 $F = 2.06$; not significant; do not reject null hypothesis